STM and SFM
in Biology

STM and SFM in Biology

Edited by

Othmar Marti
Department of Physics
University of Konstanz
Konstanz, Germany

Matthias Amrein
Institute for Medical Physics and Biophysics
University of Münster
Münster, Germany

ACADEMIC PRESS, INC.
A Division of Harcourt Brace & Company
San Diego   New York   Boston   London   Sydney   Tokyo   Toronto

Cover photographs: (*Left*) 4'-*n*-Octyl-4-cyanobiphenyl (8CB) adsorbed on graphite. (Adapted from Figure 4.2; courtesy of D. P. E. Smith.) (*Top*) Surface of a diatom taken from a pond at the University of Konstanz, Germany. Image by SFM. (Courtesy of Achim Linder.) (*Right*) STM image in surface view representation of freeze-dried recA–DNA complexes coated with Pt-Ir-C. (Adapted from Figure 2.18; courtesy of Matthias Amrein.)

This book is printed on acid-free paper. ∞

Copyright © 1993 by ACADEMIC PRESS, INC.
All Rights Reserved.
No part of this publication may be reproduced or transmitted in any form or by any means, electronic or mechanical, including photocopy, recording, or any information storage and retrieval system, without permission in writing from the publisher.

Academic Press, Inc.
1250 Sixth Avenue, San Diego, California 92101-4311

United Kingdom Edition published by
Academic Press Limited
24–28 Oval Road, London NW1 7DX

Library of Congress Cataloging-in-Publication Data

STM and SFM in biology / edited by Othmar Marti, Matthias Amrein.
    p.   cm.
  Includes bibliographical references and index.
  ISBN 0-12-474500-8
  1. Scanning tunneling microscopy. 2. Scanning force microscopy.
I. Marti, Othmar. II. Amrein, Matthias.
QH212. S35S76   1993
578'.45--dc20
                                                                         92-43104
                                                                              CIP

PRINTED IN THE UNITED STATES OF AMERICA
93 94 95 96 97 98   BB   9 8 7 6 5 4 3 2 1

CONTENTS

Contributors ix
Preface xi

1 SXM: An Introduction 1
Othmar Marti

    1.1 Overview 2
    1.2 STM 2
    1.3 SFM 60
    1.4 SNOM 88
    1.5 Electrochemistry with SXM 92
    1.6 Local Experiments 94
    1.7 New Developments 97
    Appendix A: Background Plane Removal 99
    Appendix B: Correction of Linear Distortions in Two and Three Dimensions 101
    References 103

STM IN BIOLOGY

2 STM of Proteins and Membranes 127
M. Amrein, H. Gross, and R. Guckenberger

    2.1 Introduction 128
    2.2 STM for Biological Applications 129
    2.3 Tunneling Tips 130
    2.4 Sample Preparation 133
    2.5 STM of Uncoated Specimens 142
    2.6 STM of Metal-Coated Specimens 153
    2.7 Image Processing and Quantitative Analysis of STM Data 160
    2.8 Conclusions and Prospects 169
    References 172

3 Protein Assemblies and Single Molecules Imaged by STM — 177
M. J. Miles and T. J. McMaster

3.1 Introduction	178
3.2 Elongated Proteins and Polypeptides	179
3.3 Globular Proteins	193
3.4 Conclusions	202
References	202

4 Ordered Organic Monolayers Studied by Tunneling Microscopy — 205
D. P. E. Smith and J. E. Frommer

4.1 Introduction	206
4.2 Adsorbate Mobility	206
4.3 Chemisorption	207
4.4 Self-Assembly on van der Waals Surfaces	209
4.5 Langmuir–Blodgett Films	220
4.6 Conducting Solids	221
4.7 Conclusions	221
Appendix	222
References	223

5 Potentiostatic Deposition of Molecules for SXM — 229
S. M. Lindsay and N. J. Tao

5.1 Introduction	230
5.2 Theoretical Background	231
5.3 Experimental Procedures	235
5.4 The Au(111) Surface under Aqueous Electrolytes	238
5.5 Ligated DNA Oligomers on Au(111)	242
5.6 Discussion and Conclusions	253
References	256

6 STM of DNA and RNA — 259
Patricia G. Arscott and Victor A. Bloomfield

6.1 Introduction	260
6.2 Experimental Methods	261
6.3 Results	264
6.4 Discussion	266
References	270

SFM IN BIOLOGY

7 Investigation of the Na,K-ATPase by SFM   275
Hans-Jürgen Apell, Jaime Colchero, Achim Linder, and Othmar Marti

7.1 The Structure and Role of Biomembranes	276
7.2 Investigation of Biological Membranes by SFM	278
7.3 Na,K-ATPase	278
7.4 Sample Preparation for SFM	282
7.5 Imaging Techniques	286
7.6 Results	290
7.7 Outlook	301
Appendix: Model of the Tip-Sample Interaction	303
References	306

8 SFM and Living Cells   309
J. K. H. Hörber, W. Häberle, F. Ohnesorge, G. Binnig, H. G. Liebich, and C. P. Czerny

8.1 Introduction	310
8.2 Instrumentation	311
8.3 Preparation of Cells	313
8.4 Results and Discussion	314
8.5 Conclusions	321
References	322

Index   325

CONTRIBUTORS

Numbers in parentheses indicate the pages on which the authors' contributions begin.

M. Amrein (127), Institute for Medical Physics and Biophysics, University of Münster, 4400 Münster, Germany

Hans-Jürgen Apell (275), Department of Biology, University of Konstanz, D-7750 Konstanz, Germany

Patricia G. Arscott (259), Department of Biochemistry, University of Minnesota, St. Paul, Minnesota 55108

G. Binnig (309), IBM Research Division, IBM Physics Group Munich, 8000 Munich 40, Germany

Victor A. Bloomfield (259), Department of Biochemistry, University of Minnesota, St. Paul, Minnesota 55108

Jaime Colchero (275), Department of Physics, University of Konstanz, D-7750 Konstanz, Germany

C. P. Czerny (309), Tierärztliche Fakultät der Ludwig-Maximilians-Universität Munich, 8000 Munich 22, Germany

J. E. Frommer (205), IBM Almaden Research Center, San Jose, California 95120, and Institut für Physik der Universität, CH-4056 Basel, Switzerland

H. Gross (127), Institute for Cell Biology, Swiss Federal Institute of Technology, ETH-Hönggerberg, 8092 Zürich, Switzerland

R. Guckenberger (127), MPI for Biochemistry, 8033 Martinsried, Germany

W. Häberle (309), IBM Research Division, IBM Physics Group Munich, 8000 Munich 40, Germany

J. K. H. Hörber (309), EMBL, 6900 Heidelberg, Germany

H. G. Liebich (309), Tierärztliche Fakultät der Ludwig-Maximilians-Universität Munchen, 8000 Munchen 22, Germany

Achim Linder (275), Department of Biology, University of Konstanz, D-7750 Konstanz, Germany

S. M. Lindsay (229), Department of Physics, Arizona State University, Tempe, Arizona 85287

Othmar Marti (1, 275), Department of Physics, University of Konstanz, D-7750 Konstanz, Germany

T. J. McMaster (177), H. H. Wills Physics Laboratory, University of Bristol, Bristol BS8 1TL, United Kingdom

M. J. Miles (177), H. H. Wills Physics Laboratory, University of Bristol, Bristol BS8 1TL, United Kingdom

F. Ohnesorge (309), IBM Research Division, Physics Group Munich, 8000 Munich 40, Germany

D. P. E. Smith (205), IBM Physics Group Munich, 8000 Munich 40, Germany

N. J. Tao (229), Department of Physics, Florida International University, Miami, Florida 33199

PREFACE

The scanning tunneling microscope (STM) is the first of a new generation of microscopes in which various physical quantities of surfaces are locally probed by tiny sensors scanning in close proximity over them. The STM and its first offspring, the scanning force microscope (SFM), may resolve surface topography, as well as some electronic and mechanical properties at the atomic scale, by combining a highly precise scanning mechanism, a small probe size, and a localized site for probing surface properties. These microscopes attracted the immediate attention of the biological community, not least because they usually perform equally well in ultrahigh vacuum, in ambient atmosphere, and even in aqueous solution, the natural environment of most biological macromolecules.

The chapters in this book discuss in detail the application of STM and SFM in biology. The book is addressed to everyone who wants to become acquainted with the applications of scanning probe techniques in a wide field ranging from the most simple organic molecules, beautifully arranged in crystalline films, to dynamic events on the outer membranes of living cells in their natural environments.

In the Introduction to this book, the fundamentals of STM, SFM, and scanning probe microscopy (SXM), in general, are presented, and major aspects of the instrumental designs are discussed. Researchers who intend to design their own experimental setups will find this section especially valuable. The extended bibliographies guide the reader to the source publications and to further readings on every subject.

In the next parts of this book, biological applications of STM and SFM are presented. Each self-contained chapter is a review of a specific topic by scientists active in research on the subject of their contribution. This book covers most aspects of the field with respect to both the investigated specimens and the experimental techniques presented. The experiments are described such that they can be reproduced in an appropriate environment. Some views expressed in the book are controversial, and sometimes different authors express mutually incompatible interpretations. The reader should accept this as an inevitable side effect of any quickly evolving subject. While there have been a few books on SXM in general, this is the first to be dedicated fully to biological applications.

Othmar Marti
Matthias Amrein

CHAPTER 1
SXM: An Introduction

Othmar Marti
Department of Physics
University of Konstanz
Konstanz, Germany

1.1 Overview
1.2 STM
    1.2.1 Theory of the STM
    1.2.2 The STM
    1.2.3 Image Processing: An Introduction
    1.2.4 Selected Experiments
    1.2.5 Related Techniques
    1.2.6 Additional References
1.3 SFM
    1.3.1 Theory of Force Microscopy
    1.3.2 How to Measure Small Forces
    1.3.3 The Force Microscope
    1.3.4 Selected Experiments
1.4 SNOM
1.5 Electrochemistry with SXM
1.6 Local Experiments
    1.6.1 Light Mixing
    1.6.2 Additional References
1.7 New Developments
    1.7.1 Miniaturization of the STM
    1.7.2 Other SXM
Appendix A: Background Plane Removal
Appendix B: Correction of Linear Distortions in Two and Three Dimensions
References

1.1 OVERVIEW

All scanning probe microscopes (SXM) are based on similar principles. The aim of this chapter is to discuss their common aspects, to elucidate their differences, and to point to their possible applications in the field of biology. The scanning tunneling microscope (STM), invented by Binnig and Rohrer (1982), serves as a model system. A summary of the theory of the STM points out the different operating modes and techniques, deals with the problem of imaging, and gives resolution criteria.

A detailed introduction to the mechanical and electronic design of the STM is presented. Design rules are worked out to help the builders of an STM and to allow the users to judge their instruments. Important parts of any STM experiment are the data acquisition and the image processing. Critical points in the data acquisition systems and common image processing techniques are worked out. All the technical issues of the STM are equally valid for other SXM techniques.

The section on the STM concludes with the description of a few experiments. The application of the STM to the imaging of biological and organic matter is treated in depth in other chapters of this book.

The scanning force microscope (SFM) is the most successful offspring of the STM. The design principles worked out for the STM are equally valid for the SFM. The additional critical points of an SFM are treated. Special emphasis is given to the description of the various interaction forces and the force-sensing techniques, including the scanning force and friction microscope (SFFM). The section on the SFM is closed with the discussion of a few representative experiments. Again, Chapter 7 in this volume presents a review of the application of SFM to biology.

To conclude, a summary of other scanning probe techniques is given, together with an outlook. The interested reader might also want to consult review articles on scanning probe microscopy, such as those by Binnig and Rohrer (1986), Wickramasinghe (1989), Tersoff (1990b), and Rugar and Hansma (1990) and books such as that by Sarid (1991), or the series edited by Güntherodt and Wiesendanger (1992).

1.2 STM

The following sections give an overview of the theory of the STM, of design principles commonly used to build SXM, and of image processing techniques used in the field. Discussions of selected experiments and some related techniques complete the sections.

1.2.1 Theory of the STM

Some basic knowledge of the physics of STM is necessary to judge the relevance of experiments. The tunneling junction of an STM is a quantum mechanical system; hence a basic knowledge of quantum mechanics is required to understand the physics. An overview of methods and approximations used to model the tunneling process in STM is given by Baratoff (1984).

1.2.1.1 THE TUNNELING CURRENT—A SIMPLE THEORY

To get a first intuitive view about electron tunneling between the tip and the sample of an STM, I will consider the textbook case of quantum mechanical electron tunneling between infinite, parallel, plane metal electrodes. I only treat the simplest case with no time-dependent potentials. Excellent articles on this subject have been published by Simmons (1963), (1964), Gundlach (1966), Brinkman et al. (1969), Duke (1969), Hartman (1984), and Teague (1986). The axis perpendicular to the plane parallel electrodes is the z-axis, with its zero on the left side of the tunnel gap (see Fig. 1.1).

The electron motion is governed by the Schrödinger equation

$$i\frac{\partial}{\partial t}\Psi(\vec{z}, t) = H\Psi(\vec{z}, t), \tag{1.1}$$

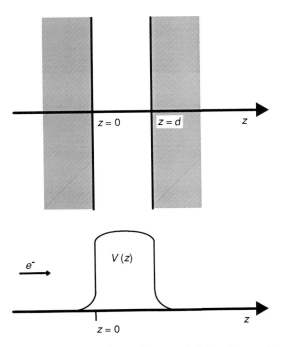

FIGURE 1.1 Coordinate system for calculating the transmissivity of a one-dimensional tunneling barrier. The electron plane wave is incident on the barrier from the left (negative z-axis). The two electrodes are separated by the distance d.

where H is the Hamiltonian of the system. The Hamiltonian for a simple tunnel junction consists of a kinetic energy part

$$-\frac{\hbar^2}{2m}\frac{\partial^2}{\partial z^2}\Psi(\vec{z}, t)$$

and a potential energy part

$$V(\vec{z})\,\Psi(\vec{z}, t).$$

The potential energy is equal to zero everywhere except in the barrier between the electrodes, from 0 to d, where d is the thickness of the barrier. The wave function $\Psi(\vec{z}, t)$ of the electrons is a solution of the equation

$$i\frac{\partial}{\partial t}\Psi(\vec{z}, t) = -\frac{\hbar^2}{2m}\frac{\partial^2}{\partial z^2}\Psi(\vec{z}, t) + V(\vec{z})\,\Psi(\vec{z}, t)$$

$$= \left(-\frac{\hbar^2}{2m}\frac{\partial^2}{\partial z^2} + V(\vec{z})\right)\Psi(\vec{z}, t). \tag{1.2}$$

The probability of finding a particle described by the wave function $\Psi(\vec{z}, t)$ at the position $\vec{z}$ at the time t is

$$P(\vec{z}, t) = \Psi(\vec{z}, t)\,\Psi^*(\vec{z}, t) = |\Psi(\vec{z}, t)|^2. \tag{1.3}$$

To simplify the calculation, consider the one-dimensional case of a tunneling barrier with a potential independent of time. The wave function $\Psi(z, t)$ is written as the product $\Psi_z(z)\,\Psi_t(t)$. Equation 1.2 can then be separated and written as

$$0 = \frac{\hbar^2}{2m}\frac{\partial^2}{\partial z^2}\Psi_z(z) + (E - V)\,\Psi_z(z). \tag{1.4}$$

We assume that the electrons are incident on the barrier from the left. There are three solutions to the Schrödinger equation: at the left of the barrier, in the barrier, and at the right. Our answer to this problem is

$$\Psi_z(z) = \begin{cases} Ae^{ipz/\hbar} + Be^{-ipz/\hbar}, & z < 0 \\ Ce^{-kz} + De^{kz}, & 0 \leq z \leq d \\ AS(E)e^{ip(z-d)/\hbar}, & z > d, \end{cases} \tag{1.5}$$

where $p = \sqrt{2mE}$, $\hbar k = \sqrt{2m(V - E)}$. (See, for instance, Baym [1969] for a detailed treatment of the problem.) At the boundaries of the three regions, these functions and their first derivative must be continuous. The function $S(E)$ is called the *tunneling matrix element*. It is a measure for the probability to tunnel from left to right for a particle being present at the left side of the junction.

Satisfying the boundary conditions in Eq. 1.5 leads to four simultaneous equations for the five parameters $A, B, C, D, S(E)$. We can chose an arbitrary value for the amplitude of the incoming electron wave; hence we set $A = 1$. The tunneling matrix element for $E < V$ is

$$S(E) = \frac{2i\hbar kp}{2i\hbar kp \cosh(kd) + (p^2 - \hbar^2 k^2) \sinh(kd)}. \tag{1.6}$$

The tunneling barrier has both a transmissivity and a reflectivity. In a measurement of the tunneling current, we can only detect the transmissivity R, which is given by

$$T(E) = |S(E)|^2 = \left[1 + \frac{\sinh^2(kd)}{4(E/V)(1 - E/V)}\right]^{-1}. \tag{1.7}$$

This equation can be simplified for electrons with a de Broglie wavelength much smaller than the barrier width d, or $kd \gg 1$. Equation 1.7 becomes

$$\begin{aligned}T(E) &\approx 16 \frac{E}{V}\left(1 - \frac{E}{V}\right) \exp(-2kd) \\ &= 16 \frac{E}{V}\left(1 - \frac{E}{V}\right) \exp\left[-\frac{2}{\hbar}\sqrt{2m(V-E)}\,d\right].\end{aligned} \tag{1.8}$$

To get a feeling for the magnitude of the transmission coefficient $T(E)$, we use values for V and E typical for a metal. The zero point of the energy scale is at the bottom of the conduction band for a metal, which is typically 12 eV below the Fermi energy. All electron states between the bottom of the conduction band and the Fermi energy are filled at zero temperature. The barrier height is, for a clean metal surface, about 4 eV above the Fermi energy; hence $V = 16$ eV and $E = 12$ eV. We further assume that the tunneling barrier width is $d = 1$ nm. Using these values we will get $T(E) \approx 10^{-9}$ for electrons at the Fermi energy. Figure 1.2 shows the transmission of the tunnel barrier as a function of the electron energy E for the above values of V and d.

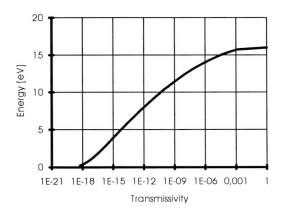

FIGURE 1.2 The transmission coefficient as a function of the electron energy. The zero energy corresponds to an electron at the bottom of the conduction band. The Fermi energy for this calculation is set to $E = 12$ eV, and the work function is 16 eV.

We can extend this simple picture by including the Fermi distribution to model the electron energy density. This allows us to calculate a tunneling current density for infinite, parallel, plane plates. We will assume that the temperature is 0 K. The calculation does not provide a description of all phenomena occurring in tunneling, but will elucidate some basic aspects. For the tunneling process only the velocities perpendicular to the sample surface matter. Assuming a tunneling barrier with a position-dependent barrier height $V(z)$, we obtain in the Wentzel, Kramers, Brillouin (WKB) approximation

$$T(E) = \exp\left(-\frac{2}{\hbar}\int_{s_1}^{s_2} \{2m[V(z) - E_z]\}^{1/2} dz\right). \quad (1.9)$$

Using the formalism of Simmons (1963), we can calculate the number of electrons tunneling from left to right:

$$N_{L \to R} = \frac{1}{m_e}\int_0^{E_M} n(v_z) T(E_z) dE_z, \quad (1.10)$$

where the z-coordinate is perpendicular to the tunneling junction, E_M is some maximum energy of the electrons, $n(v_z) dv_z$ is the number of electrons per unit volume with velocities between v_z and $v_z + dv_z$, and m_e is the electron mass. The transmission coefficient $T(E)$ is given by Eq. (1.9).

Next we assume that the electrons in the solids are distributed according to the Fermi statistics $f(E)$.

$$n(v) dv_x dv_y dv_z = \frac{m^3}{4\pi^3\hbar^3} f(E) dv_x dv_y dv_z. \quad (1.11)$$

We are only interested in the number of electrons in z-direction $n(v_z)dv_z$. We get this number by integrating over v_x and v_y.

$$n(v_z) = \frac{m_e^3}{4\pi^3\hbar^3} \int_{-\infty}^{\infty}\int_{-\infty}^{\infty} f(E) dv_x dv_y. \quad (1.12)$$

To facilitate the integration we change the integration from $dv_x dv_y$ to polar coordinates $v, dv, d\phi$ and then to energy using the relation $E_r = m_e v_r^2/2$.

$$n(v_z) = \frac{m_e^2}{2\pi^2\hbar^3}\int_0^{\infty} f(E) dE_r. \quad (1.13)$$

Combining Eqs. 1.10 and 1.13, we get for the number of electrons tunneling from left to right:

$$N_{L \to R} = \frac{m_e}{2\pi^2\hbar^3}\int_0^{E_M} T(E_z) dE_z \int_0^{\infty} f(E) dE_r. \quad (1.14)$$

The number of electrons tunneling from right to left, $N_{R \to L}$, is given by an analogous formula, except that the energy E in the Fermi distribution is replaced by $E + eV_t$, where V_t is the bias voltage for the tunneling junction.

The current density for the tunneling current is then given by

$$J = Ne$$
$$= (N_{L \to R} - N_{R \to L})e$$
$$= \frac{m_e e}{2\pi^2 \hbar^3} \int_0^{E_M} T(E_z) \, dE_z \times \int_0^\infty \{f(E) - f(E + eV_t)\} \, dE_r. \quad (1.15)$$

The average potential $V(z)$ in the barrier is expressed as a sum of the Fermi energy E_F and a potential $\Phi(z)$. For a single electrode, $\Phi(\infty)$ is commonly referred to as the *work function of the metal*. Simmons (1963) obtains analytical expressions for the tunneling current at a temperature of 0 K using an averaged barrier height $\overline{\Phi} = 1/(s_2 - s_1) \int_{s_1}^{s_2} \Phi(z) dz$.

$$J = \frac{e}{4\pi^2 \hbar (\beta \Delta s)^2} \left\{ \overline{\Phi} \exp\left[-\frac{2\beta \Delta s}{\hbar} (2m_e \overline{\Phi})^{1/2}\right] \right.$$
$$\left. - (\overline{\Phi} + eV) \exp\left[-\frac{2\beta \Delta s}{\hbar} (2m_e \overline{\Phi} + eV)^{1/2}\right] \right\}, \quad (1.16)$$

where $\Delta s = s_2 - s_1$ and β is a factor of order 1 to correct the substitution of the integral over the barrier by the average barrier height $\overline{\Phi}$. Equation 1.16 can be used to evaluate the current through a rectangular barrier (at zero bias voltage) of width s and height Φ_0:

$$J = \frac{(2m_e \Phi_0)^{1/2}}{4\pi^2 s} \left(\frac{e}{\hbar}\right)^2 V \exp\left[-\frac{2s}{\hbar} (2m_e \Phi_0)^{1/2}\right]$$
$$\text{for } V \simeq 0 \quad (1.17)$$

$$J = \left(\frac{e}{4\pi^2 \hbar s^2}\right)$$
$$\left[\left(\Phi_0 - \frac{eV}{2}\right) \exp\left\{-\frac{2s}{\hbar}\left[2m_e\left(\Phi_0 - \frac{eV}{2}\right)\right]^{1/2}\right\}\right.$$
$$\left. - \left(\Phi_0 + \frac{eV}{2}\right) \exp\left\{-\frac{2s}{\hbar}\left[2m_e\left(\Phi_0 + \frac{eV}{2}\right)\right]^{1/2}\right\}\right]$$
$$\text{for } V < \frac{\Phi_0}{e} \quad (1.18)$$

$$J = \left[\frac{e^3 (F/\beta)^2}{8\pi^2 \hbar \Phi_0}\right] \left\{ \exp\left(-\frac{2\beta}{\hbar eF} m_e^{1/2} \Phi_0^{3/2}\right) \right.$$
$$\left. - \left(1 + \frac{2eV}{\Phi_0}\right) \exp\left[-\frac{2\beta}{\hbar eF} m_e^{1/2} \left(1 + \frac{2eV}{\Phi_0}\right)^{3/2}\right] \right\}$$
$$\text{for } V > \frac{\Phi_0}{e}, \quad (1.19)$$

where for $V > \Phi_0/e$ the field strength is $F = V/s$ and the correction factor $\beta = 23/24$ (see Simmons, 1963). The tunneling current through the rectangular barrier around zero bias voltage follows Ohm's law, whereas the dependence on the barrier width is exponential. The current depends on the square of the voltage in the very high voltage range. At intermediate voltages, the tunneling resistance is highly nonlinear.

The theory outlined earlier does not account for the effect of the image potential. An electron in the barrier will create image charges in the two metal electrodes. These image charges will change the properties of the tunneling barrier. They will round off the barrier shape and lower its average value. The rounding of the tunneling barrier will increase the nonlinearity of the tunneling resistance. Simmons (1963) includes the image potential effects into his theory by calculating an effective tunneling distance that turns out to be smaller than the distance between the two electrodes that defined the surfaces of constant charge density.

The use of the WKB approximation further obscures the existence of resonances in the tunnel barrier. Gundlach (1966) calculates a better approximation than WKB and obtains an oscillatory behavior of the tunneling current. These resonances have been observed by the STM (see Binnig et al., 1985a; García et al., 1987). Gundlach (1966) finds that the energy and the periodicity of the resonances depend critically on the shape of the tunneling barrier.

The simple model outlined in this section does describe quite accurately the tunneling current between two metal surfaces. It can account for resonances due to image states and due to field states. It does not, however, provide any information on the spatial resolution of an STM. An estimate of the resolution based on the assumption of nearly plane parallel electrodes of the tunnel junction will be given in Section 1.2.1.3. The theory of Simmons also does not account for energy-dependent density of states, which is characteristic for semiconductors, superconductors, and semimetals. The transfer Hamiltonian method outlined in Section 1.2.1.2. will provide a more accurate approach.

1.2.1.2 TRANSFER HAMILTONIAN METHOD

The transfer Hamiltonian method takes into account the detailed electronic states of the electrodes. It also allows the calculation of effects resulting from the tip geometry. Tersoff and Hamann (1983) and Baratoff (1983) first applied this formalism to STM-related problems. The transfer Hamiltonian formalism originally was used by Bardeen (1961) to explain the first tunneling spectra obtained by Giaever (1960). Introductions to the formalism have been published by Kirtley (1978) and Wolf (1985). The short outline given here follows these two books.

The transfer Hamiltonian formalism is a perturbation theory. The electron waves in the two metal electrodes are considered to be independent. The coupling of the electron waves in the gap is treated as a perturbation, leading to a total Hamiltonian:

$$H = H_L + H_R + H_T, \tag{1.20}$$

where H_L and H_R are the unperturbed Hamiltonians of the two electrodes. The initial wave functions in WKB approximation are (outside and inside the barrier)

$$\psi_i^{out} \propto k_z^{-1/2} \exp[i(k_x x + k_y y)] \sin(k_z z + \gamma)$$

$$\psi_i^{in} \propto |k_z|^{-1/2} \exp[i(k_x x + k_y y)] \exp\left[-\int_0^d |k_z| dz\right], \tag{1.21}$$

where $k_z = (1/\hbar)\{2m[V(z) - E_z]\}^{1/2}$. d is the width of the barrier, and $V(z)$ is the potential in the barrier. To get the elastic tunneling current, we write a linear combination of an initial state and of a sum of final states:

$$\psi(t) = a(t)\psi_i \exp(-i\omega_0 t) + \sum_f b_f(t)\psi_f \exp(-i\omega_f t). \tag{1.22}$$

$\psi(t)$ is inserted into the time-dependent Schrödinger equation $H\psi = i\hbar(\partial/\partial t)\psi$ and solved to the first order in $b_f(t)$. One obtains for the transition rate per unit time

$$\omega_{if} = \frac{2\pi}{\hbar} |M_{if}|^2 \delta(\omega_0 - \omega_f). \tag{1.23}$$

The transfer matrix element M_{if} is given by

$$M_{if} = -\frac{\hbar^2}{2m} \int\int \left(\psi_i^* \frac{\partial}{\partial z} \psi_f - \psi_f \frac{\partial}{\partial z} \psi_i^*\right) dx \, dy \Big|_{z = \text{constant}}. \tag{1.24}$$

The integrals in Eq. 1.24 have to be evaluated on a plane anywhere inside the barrier. The total tunneling current is the sum over all initial and final states:

$$J = \frac{4\pi e}{\hbar} \sum_{k_i} \sum_{k_f} |M_{if}|^2 [f(E_i) - f(E_f - eV)] N_i(E_i) N_f(E_f + eV) \delta(E_i - E_f), \tag{1.25}$$

where $f(E)$ is the Fermi distribution. Equation 1.25 contains the densities of states $N(E)$, unlike the equations derived in the previous section. Since we take the unperturbed electron wave functions for the left and the right sides, we are calculating the tunneling current in a set of basis functions that are not orthonormal.

Tersoff and Hamann (1983) use surface electron wave functions with an exponential decay into the vacuum (z-direction). They expand these wave functions in the form

$$\psi_i = \Omega_s^{-1/2} \sum_G a_G \exp[-(k^2 + |\vec{k}_\parallel + \vec{G}|^2)^{1/2} z] \exp[i(\vec{k}_\parallel + \vec{G}) \cdot \vec{z}], \tag{1.26}$$

where $\vec{G}$ is the reciprocal-lattice vector at the surface, and $\vec{k}_\parallel$ is the surface Bloch wave vector. The factor $\Omega^{-1/2}$ ensures a proper normalization of the wave function. The inverse decay length k is given by $k = (2m\Phi)^{1/2}\hbar^{-1}$, where Φ is the work function as defined in the previous section. Tersoff and Hamann (1983) find that only the first few a_G are of order unity.

They further show that for many situations the tip can be approximated as an electron s-wave. The wave function modeling the tip is

$$\psi_f = \Omega_t^{-1} c_t kr \exp(kR) [k|\vec{r} - \vec{r}_0|]^{-1} \exp[-k|\vec{r} - \vec{r}_0|], \quad (1.27)$$

where Ω_t is the tip volume, $\vec{r}_0$ the center of curvature of the tip, and R the radius of curvature of the tip. The factor c_t depends on the exact tip geometry, its surface conditions, and the tip electronic structure, and is of the first order. This tip model is excellent for metal tips. The calculation of the tunneling current to or from semiconductor tips with nonvanishing higher order electron orbitals in the conduction band might have to include the dependence on angular and spin momentum.

The tip wave function is expanded in terms of the sample wave functions. The expanded wave functions are used to evaluate the transfer matrix element. Substituting this matrix element in the equation for the tunneling current yields

$$J = 32\pi^3 \hbar^{-1} e^2 V \Phi^2 D_t(E_F) R^2 k^{-4} \exp(2kR) \sum_\nu |\psi_\nu(\vec{r}_0)|^2 \delta(E_\nu - E_F), \quad (1.28)$$

where D_t is the density of states per unit volume of the tip at the center of its curvature $\vec{r}_0$. This result was derived for small bias voltages. Again it is found that the tunneling junction is ohmic at low voltages and that the tunneling current depends exponentially on the distance. The approximations used by Tersoff and Hamann (1983) make the model perform better for small tips (small R) than for large ones.

The local density of states can be accommodated by this model. Using this model Tersoff (1986) calculated the topography of graphite as imaged by STM.

The model of Tersoff and Hamann (1983) was extended by Lang (1985, 1986a–c) to model the tunneling from a chemisorbed atom on a uniform, featureless metal, called *jellium,* to another chemisorbed atom also on a jellium metal. Lang derives a formalism analogous to the transfer Hamiltonian formalism that not only yields the total tunneling current but also a spatial distribution of the tunneling current. The spatial distribution of the tunneling current is expected to be important in imaging materials with nonvanishing electron wave functions of p- or higher order. Lang (1986b) calculates the tunneling current for a sodium, a calcium, and a sulfur adatom on the jellium surface. The electron levels of sodium adatom are predominantly s-like. The diameter of the current distribution of the sodium adatom is narrower than the diameter of the current distribution of the calcium atom with its filled $4s$ shell. The p-resonance of the calcium lies above the Fermi level. The contribution of p-wave functions (or maybe, in other atoms, even higher order wave functions) tends to spread out the tunneling current, because these wave functions have a node on the adatom. Sulfur, on the other hand, has its p-resonance below the Fermi level. Its current distribution resembles closer that of sodium than that of calcium. Lang (1986b) states that most valence-p states away from the Fermi level are not visible in an STM experiment.

1.2.1.3 RESOLUTION OF AN STM

The resolution of an STM depends on the geometry of the tip and the sample and on their respective electronic structures. For large (μm) objects, the geometry of the tip on a micrometer length scale plays an important role in how the image is modified by the tip shape. On those length scales, it is usually possible to determine the tip shape and partially to correct the resulting image.

On an atomic scale, there are no general resolution criteria in STM like the well-known diffraction limit in optical microscopy. Even there the vertical resolution of phase-contrast microscopy has nothing to do with the diffraction limit except that the lateral extension of the object must be comparable with or larger than the wavelength of the light. In STM, a measure of the vertical resolution is the stability of the tunnel barrier width, since height variations of the sample surface smaller than the tip-sample vibration amplitudes are usually masked. The lateral resolution of an STM is governed by the width of the tunneling current filament, by the physical properties of the tip and the sample, and by the electronic states on the surface and on the tip. Various resolution criteria have been published in the literature:

1. A first criterion is based on the effective diameter, L_{eff}, of the tunnel current filament. It is applicable for theoretical approaches that provide a tunnel current density profile from which L_{eff} can be derived (García et al., 1983; Lang, 1986a).
2. For periodic structures equicurrent lines traced by the tip are calculated as a function of the tip radius, surface corrugation periodicity, amplitude and electronic structure, and tip-sample separation. The resolution limit is reached when the equicurrent lines become flatter than the STM system noise (Stoll, 1984; Stoll et al., 1984). This approach connects lateral and vertical resolution in a nontrivial way.
3. Consider an incident plane wave perpendicular to the average sample surface. If the tip is able to sample the directly transmitted wave $\vec{G} = 0$ but excludes the first-order waves, then the periodic structure cannot be resolved. The inverse of the largest detectable surface reciprocal lattice vector is taken as the resolution of the STM (Tersoff and Hamann, 1983, 1985; Baratoff, 1984).
4. The approach of Lang (1986a), who investigates two parallel metal planes with an adsorbed atom on each side, also yields a characteristic current variation when two atoms, each adsorbed on an infinite size electrode, are scanned past each other. Though not directly providing a resolution criterion, the resulting current variations give a good idea of the peculiar effects arising with adsorbates. It also gives indications of how to resolve them.

I will first derive a semiclassical argument (see Marti, 1986) on how the tip radius affects atomic resolution images. Then I will discuss the results of Tersoff and Hamann (1985) and of Lang (1986a). In a last paragraph, I will consider a micrometer-sized structure and investigate the effect of the tip shape.

In the following, I give a simple, instructive derivation of the resolution of the STM by calculating L_{eff} (criterion 1) from the Simmons (1963) model and compare it with the more rigorous treatments. For the calculation, I assume that electrons tunnel from the sample to the tip at an angle α from the surface normal and that they behave classically as far as the direction of their movement is concerned (see Fig. 1.3). Hence all tunneling paths aim at the center of curvature of the tip, and the tunneling probability of the electron depends on the energy of its motion perpendicular to the surface and on the projection, $d(x)$, of the tunneling path onto the surface normal.

The kinetic energy of the electron motion perpendicular to the surface is determined by the magnitude of the wave vector $\vec{k}$:

$$\vec{k} = \vec{K}_\parallel + \vec{k}_z. \tag{1.29}$$

Using

$$k^2 = K_\parallel^2 + k_z^2 \tag{1.30}$$

and

$$E_{\text{kin}} = \frac{\hbar^2 k^2}{2m_e} \tag{1.31}$$

for the relation between the electron energy and the magnitude of its $\vec{k}$, we can calculate the effective decay constant that determines the tunneling probability:

$$\kappa_{0,\text{eff}} = (\kappa_0^2 + K_\parallel^2)^{1/2}, \tag{1.32}$$

where κ_0 is the decay constant for electrons with $K_\parallel = 0$. Next we approximate $K_\parallel$ by

$$K_\parallel(x) = k\frac{x}{d+R}, \tag{1.33}$$

where d is the distance from the surface to the apex of the tip of radius R, and x

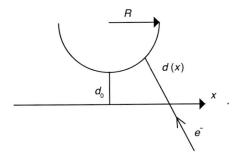

FIGURE 1.3 Semiclassic calculation of the resolution of the STM as a function of the tip radius and the tip–sample separation. The electrons are assumed to tunnel from the sample to the tip on straight lines crossing the center of curvature of the tip. x denotes the distance from the center of the current filament to the point where the electrons leave the sample.

is the lateral distance from the center of the tip. $\kappa_{0,\text{eff}}$ can then be expanded for small x, small bias voltages, and zero temperature:

$$\kappa_{0,\text{eff}} = \kappa_0 \left[1 + \frac{1}{2} \frac{k^2}{\kappa_0^2} \frac{x^2}{(d+R)^2} \right] = \kappa_0 \left[1 + \frac{1}{2} \frac{E_f}{\Phi_0} \frac{x^2}{(d+R)^2} \right], \quad (1.34)$$

where E_f is the Fermi energy measured from the bottom of the conduction band, and Φ_0 is the barrier height.

The tunneling distance as a function of the lateral position x is

$$d(x) = (d+R)\left\{ 1 - \frac{R}{[(d+R)^2 + x^2]^{1/2}} \right\}. \quad (1.35)$$

For small x, it can also be expanded. Neglecting higher order terms, we get

$$d(x) = d + \frac{R}{2} \frac{x^2}{(d+R)^2}. \quad (1.36)$$

The total current flowing in a filament of radius r is then

$$I(r) = A \int_0^r \exp\left\{ -2\kappa_0 \left[1 + \frac{1}{2} \frac{E_f}{\Phi_0} \frac{x^2}{(d+R)^2} \right] \left[d + \frac{R}{2} \frac{x^2}{(d+R)^2} \right] \right\} 2\pi x \, dx \quad (1.37)$$

All prefactors of the exponential are contained in A and set constant to ease the integration.

Keeping only terms up to order x^2 in the exponential slightly underestimates the tunneling current and the filament diameter. The total tunneling current after integration is then given by

$$I(r) = I_0 \left\{ 1 - \exp\left[-\kappa_0 \frac{r^2}{(d+R)^2} \left(d \frac{E_f}{\Phi_0} + R \right) \right] \right\}, \quad (1.38)$$

where I_0 is the total current. The effective diameter L_{eff} of the tunneling current filament carrying a fraction δ of the total current I_0 is

$$L_{\text{eff}} = 2 \left[-\frac{(d+R)^2}{\kappa_0 (d(E_f/\Phi_0) + R)} \ln(\varepsilon) \right]^{1/2}, \quad (1.39)$$

where $\varepsilon = 1 - \delta < 1$.

The Fermi energy E_f and the barrier height Φ_0 have approximately the same magnitude in metals. Hence Eq. 1.39 reduces to

$$L_{\text{eff}} = 2 \left[-\frac{d+R}{\kappa_0} \ln(\varepsilon) \right]^{1/2}. \quad (1.40)$$

The resolution of an STM on the atomic level is critically dependent on the radius of curvature of the apex of the tip. This suggests that very narrow tips with single atom apexes be used. These tips, however, are not stable during scanning. Landman and Luedtke (1991) calculate in a molecular dynamics calculation that the forces between the tip and the sample will tear apart single atom tips. Hence a high-resolution tip has to be a compromise between stability and high curvature. It is found that tip radii of a few nanometers to 10 nm give best results.

Tersoff (1989) calculates the resolution of the STM under the assumption of small corrugation. Under this assumption, the inherently nonlinear behavior of an STM can be treated like the linear imaging process of an optical microscope, where the measured image is a convolution of the ideal image and an instrument response function. The calculation by Tersoff (1989) yields a resolution function that is a Gaussian with a root-mean-square (RMS) width of

$$L_{\text{eff}} = \left(\frac{z_0}{2\kappa_0}\right)^{0.5}, \qquad (1.41)$$

where $z_0 = d + R$ is the distance between the sample and the center of curvature of the tip, and κ_0 is defined as above. Tersoff's solution for metals shows that our crude semiclassical calculation gives the correct functional behavior and will agree with Tersoff's result if we set $\varepsilon = 3.3 \times 10^{-4}$. Tersoff (1989) notes that for semiconductors or semimetals like graphite the resolution of an STM is higher than for metals imaged with the same tip.

The resolution of an STM for larger features with sizes of $\approx 1\ \mu$m or more is mainly determined by the tip shape (see also Stedman, 1988; Nishikawa et al., 1989). The variation of the tunnel barrier width is on the order of 1 nm or less and hence negligible. To demonstrate the influence of the tip shape, we are considering a conical tip with a tip angle of α. Figure 1.4 shows calculated traces for tip angles of 70° and 30°. The effect of a larger tip angle on features extending from the surface is to broaden the width of the feature. On ditches, however, a larger tip angle can prevent the imaging of the bottom of the ditch (see Fig. 1.4a).

If we assume that the tip opening angle is α, we can calculate the maximum depth $d_{\max}$ for a given width w of a rectangular ditch that can be imaged.

$$d_{\max} = \frac{w}{2\tan(\alpha/2)}. \qquad (1.42)$$

Using the same simple geometrical arguments, we can also calculate the apparent width W at half height for a rectangular feature of width w extending from the sample surface:

$$W = w + h * \tan(\alpha/2). \qquad (1.43)$$

Table 1.1 summarizes the results for a number of tip angles α. The first column gives the maximum ratio h/w with which the tip just reaches the bottom of the ditch. The remaining columns give, as a function of h/w, the broadening

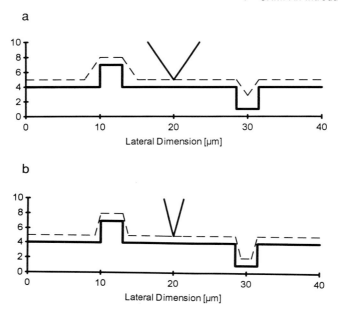

FIGURE 1.4 Tip shape dependency of the STM resolution for large features. **a:** A tip angle of 70°. **b:** A tip angle of 30°. The trajectory of the apex of the tip is shown.

TABLE 1.1 Large-Scale Resolution of an STM[a]

Tip angle (°)	Ditches (d_{max}/w)	Broadening (%)						
		0.1	0.2	0.5	1	2	5	10
120	0.29	17	35	87	173	346	866	1732
110	0.35	14	29	71	143	286	714	1428
100	0.42	12	24	60	119	238	596	1192
90	0.50	10	20	50	100	200	500	1000
80	0.60	8	17	42	84	168	420	839
70	0.71	7	14	35	70	140	350	700
60	0.87	6	12	29	58	115	289	577
50	1.07	5	9	23	47	93	233	466
40	1.37	4	7	18	36	73	182	364
30	1.87	3	5	13	27	54	134	268
20	2.84	2	4	9	18	35	88	176
10	5.72	1	2	4	9	17	44	87

[a] The tip angle is the angle between the surfaces in a cross section. The broadening is referenced to as the *full width at half maximum* and is given as a function of h/w, where w is the real width.

at half height of the structure due to the tip shape. Note that h and w are the real widths and heights.

1.2.1.4 LOCAL DENSITY OF STATES — TUNNELING SPECTROSCOPY

Tunneling spectroscopy on a local length scale is one of the strengths of STM. Images of semiconductor surfaces depend critically on the bias voltage at which the image is acquired. Approximations and recipes to analyze spectroscopic data exist, but a full theory of tunneling spectroscopy based on the electron wave functions in metals is under way (Tersoff, 1990a,b). Selloni et al. (1985) proposed a generalization to the transfer Hamiltonian method by noting that, for small voltages,

$$J(V) \approx \int_{E_F}^{E_F + V} \rho(E)\, T(E, V)\, dE \tag{1.44}$$

is a qualitative approximation. $\rho(E)$ is the local density of states near the sample surface. Tersoff (1990a) criticizes that this expression does not explain the relationship of dJ/dV with the spectrum of the electron density of states as determined, for example, by photoemission and inverse photoemission spectroscopy.

Stroscio et al. (1986) and Feenstra and Mårtensson (1988) proposed the use of $d\ln J/d\ln V = (dJ/dV)/(J/V)$ as a representation of the spectrum. The authors demonstrated a good agreement with known spectra of electron densities. The success of this representation of tunneling spectra depends on the cancellation of the exponential behavior of $T(E, V)$. Theoretical calculations by Lang (1986c) confirm the validity of this data representation and of Eq. 2.44.

1.2.1.5 ADDITIONAL REFERENCES

Tunneling theory is discussed by Feuchtwang et al. (1983), García and Flores (1984), Bono and Good (1985), García et al. (1986), Orosz and Balazs (1986), Sumetskii (1988), Tersoff (1990b), Noguera (1990), Doyen et al. (1990), Huang et al. (1990b), Sacks and Noguera (1991a,b), and Isshiki et al. (1991). Ezaki et al. (1990) discuss angle-resolved tunneling. Issues of the resolution of the STM have been discussed by Chung et al. (1987), Das and Mahanty (1987), Sacks et al. (1987), Stoll (1988), Laloyaux (1988), Aers and Leavens (1988), Hashizume et al. (1988), and Reiss et al. (1990a). The image potential and its influence on STM experiments are discussed by Huang et al. (1991). Leavens and Aers (1986) and Kubby et al. (1991) demonstrated the existence of transmission resonances. A resonant tunneling theory was put forward by Zheng and Tsong (1990). A multidimensional tunneling theory has been treated by Huang et al. (1989), Chen (1990, 1991a,b,) and Barniol et al. (1991). Louis and Flores (1987) calculated correlations between charge and current corrugations in STM. Çiraci and Batra (1987) discuss tunneling at small tip to surface distance. Louis et al. (1986) wrote an article on the current saturation through image surface states. Martin-Rodero *et al.* (1988), Guinea and García (1990), and García-García and García (1991) investigate quantum effects due to the saturation of the tunneling current and resulting possible current standards. Lang (1985, 1986a–c, 1987a,b) discusses extensively the tunneling process through single atoms. A microscopic view of STM is proposed by Çiraci and Tekman (1989), Ohnishi and Tsukuda (1989, 1990), and Chen (1991b). Self-consistent calculations were performed by Pollmann *et al.* (1987). Tunneling spectroscopy has been treated by García (1986), Baratoff *et al.* (1986), Noguera (1989), Fujita *et al.* (1990), Pitarke *et al.* (1990), and Davis *et al.* (1991). Binnig *et al.* (1985b), Leavens and Aers (1987), and Person and Baratoff (1987) investigated the role of inelastic tunneling processes. Pitarke *et al.* (1989) calculated the apparent barrier height for tunneling. Kuk and Silverman (1986), Tersoff (1990a), Tersoff and Lang (1990), Golubok and Tarasov (1990), and Tsukada *et al.* (1991) discuss the role of tip geometry on corrugation, work-function measurements, and related items. Kobayashi and Tsukada (1990), Tsukada *et al.* (1990), and Kobayashi *et al.* (1990) published a simulation of STM images. Issues of tunneling time are discussed by Büttiker and Landauer (1986) and by Cutler *et al.* (1987). Theoretical investigations of the effects of forces on STM images were presented by Çiraci *et al.* (1990a).

1.2.2 The STM

This section will focus on the critical points in the design of an STM system. I treat both the mechanics and the electronics of an STM. Most findings reported in this section are also applicable to other SXM.

1.2.2.1 MECHANICAL DESIGN

Scanning probe microscopes have to fulfill specifications that are contradictory. First, the scan range of an SXM should exceed the size of the largest features on the sample. This usually implies the use of large and thin-walled piezo electric translators. For many applications, it is desirable to have the ability to position the sample laterally versus the tip in coarse steps. This can be done by adding some hardware, thus increasing the size of the microscope. On the other hand, any SXM is, first of all, a system of mechanical resonators. This system has many resonances with varying quality factors. These resonances can be excited either by the surroundings or by the rapid movement of the tip or the sample. It is of paramount importance to optimize the design of the SXM for high-resonance frequencies. This usually means to decrease the size of the microscope. The article by Pohl (1986) analyzes design criteria for STM.

By using cube-like or ball-like structures for the microscope, one can considerably increase the lowest eigenfrequency. The eigenfrequency of any spring is given by

$$f = \frac{1}{2\pi}\sqrt{\frac{k}{m_{\text{eff}}}}, \qquad (1.45)$$

where k is the spring constant, and m_{eff} is the effective mass. The spring constant k of a cantilevered beam with uniform cross section is given by

$$k = \frac{3EI}{l^3}, \qquad (1.46)$$

where E is the Young's modulus of the material, l is the length of the beam, and I is the moment of inertia (see Thomson, 1988a). For a rectangular cross section with a width b (perpendicular to the deflection) and a height h, one obtains for I

$$I = \frac{bh^3}{12}. \qquad (1.47)$$

Combining Eqs. 1.45, 1.46, and 1.47, we get the final result for f:

$$f = \frac{1}{2\pi}\sqrt{\frac{3EI}{l^3 m_{\text{eff}}}} = \frac{1}{2\pi}\sqrt{\frac{Ebh^3}{4l^3 m_{\text{eff}}}}. \qquad (1.48)$$

The effective mass can be calculated using Raleigh's method. The general formula for the kinetic energy T using Raleigh's method is

$$T = \frac{1}{2} m_{\text{eff}} \left(\frac{\partial y}{\partial t}\right)^2 \qquad (1.49)$$

or, for a bar,

$$T = \int_0^l \frac{m}{l} \left(\frac{\partial y(x)}{\partial t}\right)^2 dx. \qquad (1.50)$$

For the case of a uniform beam with a constant cross section and length l, one obtains for the deflection $y(x) = y_{\text{max}}[1 - (3x)/(2l) + (x^3)/(2l^3)]$. Inserting y_{max} into Eq. 1.49 and solving the integral yields

$$T = \int_0^l \frac{m}{l} \left[\frac{\partial y_{\text{max}}(x)}{\partial t}\left(1 - \frac{3x}{2l} + \frac{x^3}{2l^3}\right)\right]^2 dx$$

$$= \frac{1}{2} m_{\text{eff}} \left(\frac{\partial y_{\text{max}}}{\partial t}\right)^2 \qquad (1.51)$$

$$m_{\text{eff}} = \frac{9}{20} m \qquad (1.52)$$

for the effective mass. Combining Eqs. 1.48 and 1.51 and noting that $m = \rho l b h$, where ρ is the density of mass, one obtains for the eigenfrequency

$$f = \left(\frac{1}{2\pi}\frac{\sqrt{5}}{3}\right)\sqrt{\frac{E}{\rho}}\frac{h}{l^2}. \qquad (1.53)$$

Further discussion on how to derive this equation was published by Thomson (1988a). It is evident from this equation that one way to increase the eigenfrequency is to choose a material with as high a ratio E/ρ as possible. Table

TABLE 1.2 Material Constants Used for Calculating Resonance Frequencies

Material	E (10^{10} N/m^2)	ρ (10^3 kg/m^3)	E/ρ ($10^7 m^2/s^2$)
Aluminum	7.0	2.7	2.6
Brass	9.0	8.5	1.1
Bronze	10.5	8.8	1.2
Copper	11.0	9.0	1.2
Crown glass	6.0	2.6	2.3
Nickel	21.0	8.88	2.4
Steel	20.0	7.8	2.6
Tungsten	39.0	19.3	2.0

(Modified from Anderson 1981, with permission.)

1.2 gives an overview over some materials and shows that aluminum and steel are the best construction materials in terms of eigenfrequencies.

Another way to increase the lowest eigenfrequency is also evident in Eq. 1.45. By optimizing the ratio h/l^2, one can increase the resonance frequency. However, it does not help to make the length of the structure smaller than the width or height. Their roles will just be exchanged. Hence the optimum structure is a cube. This leads to the design rule that long, thin structures like sheet metal should be avoided. If a given resonance frequency cannot be changed any more, its quality factor should be as low as possible. This means that an inelastic medium such as rubber should be present to convert kinetic energy into heat.

A typical SXM consists of many structures with coupled resonance frequencies. Hence, there might be low-frequency beats in the amplitude of an oscillation. Let us investigate the consequences of such beats for an SXM with linear response (such as certain force microscopes) and for an SXM with a nonlinear response such as an STM. Figure 1.5 shows the set-up for this calculation. The sample is supposed to be mounted on an oscillating block with an amplitude of

$$a(t) = a_0 + a_1 \sin(\omega_1 t) + a_2 \sin(\omega_2 t) \qquad (1.54)$$

opposite to a fixed probe. This probe is sensitive to variations in the distance between the sample and the tip. We assume that the frequencies ω_1 and ω_2 are larger than the frequency response of the probe, whereas the difference frequency $\omega_1 - \omega_2$ is well within. The probe will only respond to the average signal given by

$$s = \frac{1}{\Delta T} \int_{t-\Delta T}^{t} r[a(t')] \, dt', \qquad (1.55)$$

where $r(x)$ is the response function of the probe, and ΔT is the integration time. For a linear response function $r(x) = x$ and an integration time ΔT large compared with $1/\omega_1$ and $1/\omega_2$, the integral in Eq. 1.55 gives 0. No effect of the two vibration modes can be seen in the output signal for a linear system.

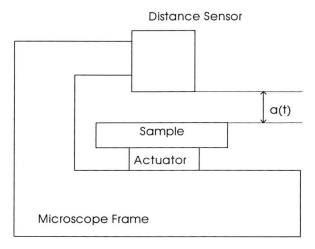

FIGURE 1.5 Influence of beat frequencies in SXM. Two resonating bodies act on the distance between the sample and the tip $a(t)$. For a system with a nonlinear response to $a(t)$, the response is also present at beat frequencies.

If we assume a nonlinear response function $r(x) = x + \beta x^2$, then $r[a(t)]$ becomes

$$r[a(t)] = a_1 \sin(\omega_1 t) + a_2 \sin(\omega_2 t) + \beta[a_1 \sin(\omega_1 t) + a_2 \sin(\omega_2 t)]^2$$
$$= a_1 \sin(\omega_1 t) + a_2 \sin(\omega_2 t) + \beta(a_1^2 + a_2^2)$$
$$- \frac{1}{2}\beta a_1^2 \cos(2\omega_1 t) - \frac{1}{2}\beta a_2^2 \cos(2\omega_2 t)$$
$$+ \beta a_0 a_1 \cos[(\omega_1 - \omega_2)t] - \beta a_0 a_1 \cos[(\omega_1 + \omega_2)t]. \qquad (1.56)$$

Inserting Eq. 1.56 into Eq. 1.55 and solving the integral gives nonzero results only for the term with constant amplitude $\beta(a_1^2 + a_2^2)$ and the term $\beta a_0 a_1 \cos[(\omega_1 - \omega_2)t]$ oscillating at $\omega_1 - \omega_2$. The first nonzero term represents a DC offset to probe response, whereas the second term is a low-frequency modulation that may well be within the control bandwidth of the feedback loop. Such an interference signal will appear as an apparent surface structure.

As a consequence, one has to be careful not to design two parts of an SXM with nearly the same frequencies. One case that is especially bad is when the tip has a resonance close to an eigenfrequency in some other part of the structure.

1.2.2.2 VIBRATION ISOLATION

I have discussed in Section 1.2.2.1 that an SXM viewed as a mechanical system has a variety of resonance frequencies, which may be even coupled. These resonant frequencies are excited both by the surroundings and by the piezoscanning within the SXM. The effect of the piezoscanning can be minimized through careful design, but it cannot be eliminated.

The microscope, however, can be isolated from the vibrations present in the surrounding environment. First, the microscope is affected by building vibrations and by the sounds people create when moving around. The frequency spectrum of these noises peaks between 10 and 100 Hz. The amplitudes of the ground movement are on the order of several micrometers. It is a wise practice to avoid any resonating body within the SXM with a resonance frequency ω_0 below a few 100 Hz to 1 kHz. Since the coupling to a mechanical harmonic oscillator scales like $1 - 2Q^2[(\omega - \omega_0)/\omega_0]^2$, a larger difference between the driving frequency and the resonance frequency decreases the coupling. This scaling law is correct for small deviations of ω from ω_0. The reference amplitude is the amplitude on resonance. The amplitude of the building vibrations reaching the microscope can be further decreased by isolating the microscope from these frequencies.

The first tunneling microscopes used a spring system with eddy current damping for vibration isolation. A description of this system is given by Binnig and Rohrer (1982). This vibration isolation system, though effective, has its disadvantages. The fundamental resonance frequency of the spring mass system has to be on the order of 1 Hz to be effective in suppressing the building vibrations. If this resonance frequency were aperiodically damped, then the increase in damping with frequency above the resonance frequency would be too slow. Furthermore, it is difficult to get aperiodic damping with an eddy current brake at room temperature. For an undercritical damping of the spring-mass system the vibration amplitudes at resonance are amplified! Therefore, using a vibration isolation system with a resonance frequency between 10 and 100 Hz, where the building vibrations peak, has to be avoided.

A second disadvantage of the spring-stage vibration isolation system is the lack of a well-defined position of the microscope with respect to the surrounding. This deficiency becomes a problem when using mechanical approach systems, *in situ* sample transfer, and tip exchange. All of these tasks demand spatially well-defined locations and the ability to exert forces on the microscope. One solution to this problem is to bridge the vibration isolation temporarily and to grip the microscope with a special tool. Another problem one might have is the very high Q of the springs on which the microscope is suspended. To reach the necessary low-resonance frequency, their spring constants have to be low, which also means that their resonance frequencies can be fairly low. Special problems might occur if the springs come into resonance with some other part of the microscope or the surroundings. Last, the spring-mass vibration isolation systems tend to be very bulky.

To decrease the size of the vibration isolation system, Gerber et al. (1986) proposed the use of stacks of metal plates separated by viton spacers. This set-up is still a mass-spring damping system, but it is most effective at high frequencies, where its multiple stages act like a higher order low-pass filter. Combined with an SXM of rigid design, such a viton–metal stack can provide superior performance. Such designs are widely used in ultrahigh vacuum (UHV)–STM systems and are known as pocket-sized STMs.

Outstanding results can be obtained using optical tables or at least the vibration isolation supports designed for these tables. Entire UHV chambers can be mounted on such vibration-isolated tables. An additional decrease of the vibration amplitudes can be achieved by placing the microscope in a room with low ground movements, such as a basement room or the separate basement of a microscopy room.

A cost-effective, though not very efficient, way to isolate the microscope from the building vibrations is to mount it on a platform suspended by bungee cords from the ceiling. Drake *et al.* (1986) describe such a system for STM and SFM. If the microscope is very rigidly designed, then it can be used with no further vibration isolation on these platforms or on optical tables. For an improved performance, the viton–metal stack can be combined with the external vibration isolation methods.

1.2.2.3 SOUND ISOLATION

Environmental noise not only consists of building vibrations, but also of sounds. Sources of sound may be the voices or actions of the experimenters, air conditioning, motors built into equipment like computers, or street noise, to name a few. SXM operating in vacuum are perfectly isolated against these noises. Microscopes working in air need added protection. A good way to isolate an ambient pressure microscope from sound is to build it into an airtight box with heavy walls, similar to the way sophisticated loudspeaker boxes are built. The closed room inside the box has its own resonances, which must be damped by using suitable materials. The foams used on the walls of sound recording studios or in soundproof rooms are good for this purpose.

When using metal as a coating on the sound isolation box or even metal as the construction material of the sound isolation box, it can also be used as an electrical shield, protecting the sensitive electronic circuits from electronic noise.

1.2.2.4 THERMAL DRIFT

Scanning probe microscopes operate in environments where the temperature is not constant. Any change in temperature causes a contraction or expansion of the various parts of the SXM. Since these parts are manufactured from different materials, their thermal expansion coefficients will be different. This causes the probe tip to be displaced both laterally and vertically from the old position with respect to the sample. To estimate the thermal drift, we will assume that an SXM is built of a steel body, with a piezotube holding the probe. We further assume that the size of the microscope, that is, the distance from the probe tip to the point where it is joined with the sample holder, is 5 cm. We further assume that the difference in the thermal expansion coefficients between the steel and the piezoceramic material is a very good 10^{-6} 1/K. The probe tip will move 50 nm for every K change in temperature. To perform experiments such as local tunneling spectroscopy, where the gap should be kept constant to better than 0.01 nm, a thermal environment with a temperature variation of less than 0.2 mK over the measurement time is required! For measurement times of a few seconds such a temperature stability is easily achieved. Problems arise for temperature-controlled samples. Only the best of these temperature controllers, together with sophisticated shielding techniques, will provide the necessary stability.

The sensitivity to temperature variations can be minimized by using a symmetric and balanced design (van Kempen and van de Walle, 1986; Albrektsen et al., 1989). As an example, consider an SXM with a cylindrical symmetry. We will use a piezotube as the tip translator. The tip is mounted in the center of the piezotube, which in turn is mounted in the center of the microscope body. Because of symmetry, the tip will remain at the center of the microscope body even if the temperature varies. It will move up or down with respect to the sample. The differential thermal expansion can be further minimized by employing the same materials for, in our example, the piezotube and the microscope body. This will give the best match for the thermal expansion coefficients and would for our example minimize the vertical movement of the tip with respect to the sample.

1.2.2.5 PIEZOSCANNERS

Almost all STM use piezotranslators to scan the tip or, seldomly, to scan the sample. Even the first STM by Binnig et al. (1982a,b) and some of the predecessor instruments (Young et al., 1971, 1972) used piezotranslators for scanning. Microscopes using magnetostrictive materials (Gerber and Marti, 1985) or electromagnetic drives (García-Cantu and Huerta-Garnica, 1987) have been proposed. I concentrate on piezoelectric materials.

An electric field applied across a piezoelectric material causes a change in the crystal structure, with expansion in some directions and contraction in others. Also, a net change in volume occurs. Detailed descriptions of the piezoelectric effect can be found in physics textbooks such as that of Ashcroft and Mermin (1976). The transverse piezoelectric effect is by far the most important for scanning probe microscopes. The expansion perpendicular to the applied electric field $\vec{E}$ for a long slab of material with the field applied across the small sides is

$$\Delta l = l|\vec{E}|d_{31} = l\frac{V}{t}d_{31}, \qquad (1.57)$$

where d_{31} is the piezoelectric constant, V the applied voltage, and t the thickness of the piezoslab or the distance between the electrodes where the voltage is applied. This allows the sensitivity of a piezoactuator to be chosen within the limits of its mechanical stability.

The first STM all used piezotripods for scanning (see, e.g., Binnig and Rohrer, 1982). The piezotripod (Fig. 1.6a) is an intuitive way to generate the three-dimensional movement of a tip attached to its center. However, to get a suitable stability and scanning range, the tripod needs to be fairly large (about 5 cm). Its size and its asymmetrical shape make it very susceptible to thermal drift. The design of van Kempen and van de Walle (1986) (Fig. 1.6b) tries to circumvent this problem by using a symmetrical design. Its thermal drift performance is much better than the simple tripod. However, a complicated assembly of many piezopieces is required. The tube scanner (Fig. 1.6c) is now widely used in scanning tunneling and SXM for its simplicity and its small size (Binnig and Smith 1986). The outer electrode is segmented in four equal sectors of 90°. Opposite sectors are driven by signals of the same magnitude, but opposite sign. This gives, through bending, a two-dimensional movement on, approximately, a sphere. The inner electrode is normally driven by the z-signal. It is possible, however, to use only the outer electrodes for scanning and for the z-movement. The main drawback of applying the z-signal to the outer electrodes is that the applied voltage is the sum of both the x- or y-movement plus the z-movement. Hence a larger scan size effectively reduces the available range for the z-control.

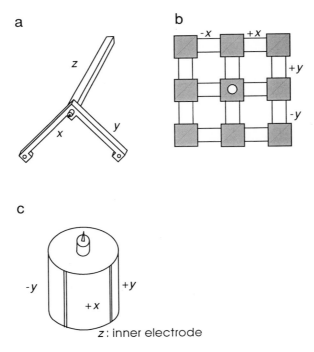

FIGURE 1.6 Types of piezoscanners. **a:** The tripod. **b:** The thermally compensated scanner. **c:** The piezotube.

Piezoscanners, -tubes, and -tripods, are made of piezoceramic material. Piezomaterials with a high conversion ratio (i.e., a large d_{31} or small distances between the electrodes, allowing large scan ranges with low driving voltages) do have substantial hysteresis resulting in a deviation from linearity by more than 10%. The sensitivity of the piezoceramic material (mechanical displacement divided by driving voltage) increases with reduced scanning range, whereas the hysteresis is reduced. Careful selections of the material for the piezoscanners, the design of the scanners, and the operating conditions are necessary to achieve optimum performance.

Several groups have published information on the calibration of piezoelectric materials (see, e.g., Vieira, 1986; Nishikawa *et al.*, 1987; Simpson and Wolfs, 1987; Bendre and Dharmadhikari, 1988; Riis *et al.*, 1989; Wetsel *et al.*, 1989; Poirier and White, 1990; van Leemput *et al.*, 1991). A finite element analysis of piezoscanners was published by Carr (1988). Electrostrictive materials are discussed by Nishikawa *et al.* (1987). Piezoelectric positioning devices are discussed by Okayama *et al.* (1985) and Blackford and Jericho (1990). Robinson (1990) addresses nonlinearities in the piezoelectric response.

1.2.2.6 CONTROL ELECTRONICS: BASICS

The control electronics and software play important roles in optimum performance of an SXM. Control electronics and software are now supplied with commercial SXM. Some manufacturers sell their control electronics and software without a microscope. Control electronics systems can use either analog or digital feedback. While digital feedback offers greater flexibility and the ease of configuration, analog feedback circuits might be better suited for ultralow-noise operation.

I describe here the basic electronics set-ups for STM and spectroscopy and then discuss the spectroscopy instrumentation. The concepts worked out in the following are also applicable to the control electronics of other SXM.

The main task for the control electronics of an STM is to maintain a distance of order 1 nm between the tip and the sample with an accuracy of < 0.01 nm. Figure 1.7 shows a block schematic of a typical STM feedback loop. The sample is connected to a bipolar, adjustable tunnel voltage source. The other side of the tunnel junction, the tip, is connected to a current to voltage converter (I/V converter). Alternatively, the tunnel voltage could have been connected to the tip and the I/V converter connected to the sample. In any case, the I/V converter should be connected to the side that is less affected by ambient noise and has less stray capacitance to earth. The tunnel voltage side is connected to a low-impedance voltage source and hence is not much affected by electrical interference. The I/V converter, on the other hand, may have a considerable input impedance and is connected to its voltage supply via the tunneling resistance and the feedback resistance (both on the order of MΩ to GΩ) and is therefore very much affected by stray electrical fields. The unavoid-

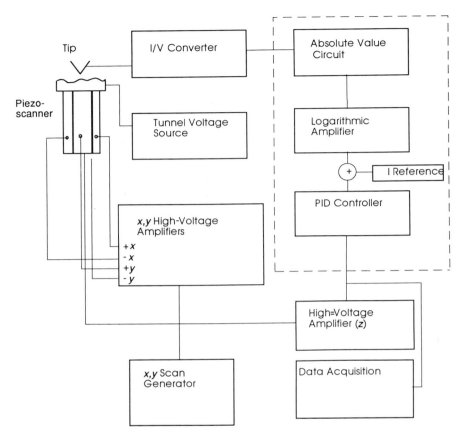

FIGURE 1.7 Block schematics of the electronics of STM. The tunneling current at the tip is converted to a voltage and processed by feedback electronics. The drive voltages for the x-, y-, and z-electrodes are provided by high-voltage amplifiers. The scan generator and the data acquisition system can be analog or digital.

able stray capacitance to earth will severely degrade the frequency response of the I/V converter and hence of the STM. Therefore the I/V converter should be located as close to the sample or to the tip as possible. In the case of additional mechanical or electrical equipment connected to the sample (heater, sample exchange mechanism, electrochemical cell, etc.), the I/V converter has to be connected to the tip.

The output voltage of the I/V converter is fed into an absolute value circuit. The absolute value circuit simplifies the use of both polarities of the bias voltage. The distance to current characteristics of the tunnel junction is exponential. Often it is advantageous to linearize the feedback loop after the rectifier. A set voltage (I Reference) is subtracted from the output of the linearizing unit or the rectifier. The resulting error voltage is then integrated and, optionally, amplified (PID controller). The gains of the integrator (high gain corresponds to short integration times) and of the proportional amplifier are set as high as possible without generating more than 1% overshoot. High gain minimizes the error margin of the current and forces the tip to follow the contours of constant densities of states as close as possible. This operating mode is known as the *constant current mode*. The outputs of the integrator and the amplifier are summed and amplified by a high-voltage amplifier. STM with piezotubes usually require ± 150 V at the output. Other designs might require amplifiers delivering more than ± 1000 V or as little as ± 15 V.

To scan the sample, additional voltages at high tension driving the piezo-scanners are required. For example, with a tube scanner, four scanning voltages are required, namely, $+V_x$, $-V_x$, $+V_y$, and $-V_y$. The x- and y-scanning voltages are generated in a scan generator (analog or computer controlled). Both voltages are input to the two respective power amplifiers. Two inverting amplifiers generate the input voltages for the other two power amplifiers.

The topography of the sample surface is determined by recording the input voltage to the z-high-voltage amplifier as a function of x and y (constant current mode). Recording devices such as chart recorders, analog storage oscilloscopes, videoframe grabbers, or computer data acquisition systems are used. The height variations on the sample surface can be determined quite accurately from the known piezocalibration.

Another operating mode is the *constant height mode*. It is feasible only on flat surfaces such as graphite. The gain in the feedback loop is lowered and the scanning speed increased such that the tip no longer follows the surface contours. It is held at constant height. Here the tunneling current is recorded as a function of x and y. This mode usually has a better signal to noise ratio than the constant current mode, mainly because the surface data appear in higher frequency bands in the tunneling current.

1.2.2.7 CONTROL ELECTRONICS: SPECTROSCOPY

The basic electronics circuits outlined in Section 1.2.2.6 can be expanded to allow spectroscopic imaging of the sample. Four modes of spectroscopic imaging are in common use: measuring $\partial I/\partial V$, $\partial I/\partial s$ spatially resolved, alternating the tunneling voltage between two values between scans, and measuring $I(V)$ curves at constant height for some or all points on the sample.

Figure 1.8 shows an electronics set-up for $\partial I/\partial V$ measurement. The output voltage of an oscillator (Reference Output) is added to the bias voltage. The sum is connected to the sample. The resulting tunneling current is converted to a voltage by the I/V converter, as usual. The low-frequency part is fed into the feedback electronics and controls the position of the tip. The modulation on the bias voltage, whose frequency should be well above the cut-off frequency of the feedback loop, is connected to the signal input of a lock-in amplifier. The

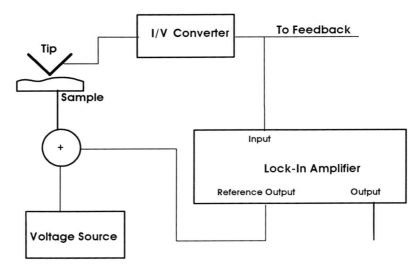

FIGURE 1.8 The block schematic for measuring $\partial I/\partial V$ with the STM. A small (≈ 1 mV) modulation voltage is added to the tunnel voltage. The resulting tunneling current is converted to a voltage. The feedback circuit operates on the DC component of the tunneling current, whereas the AC component is demodulated by the lock-in amplifier.

reference for demodulating the $\partial I/\partial V$ signal comes from the oscillator. The output of the lock-in amplifier can be recorded as a function of position alone or together with the z-signal.

Figure 1.9 shows the necessary connections to measure $\partial I/\partial s$. This time the output of the oscillator (Reference Osc.) is summed with the output of the feedback control electronics. The amplitude of the modulation voltage is chosen so as to move the tip rapidly with an amplitude of about 1–10 pm. The resulting modulation on the tunneling current is fed into the lock-in amplifier and detected. The lock-in amplifier output can be recorded in the data acquisition system, alone or together with the z-position. The frequency of the modulation should be smaller by a factor of 3 than the lowest resonance frequency of the piezotube. If the modulation frequency is too close to a resonance frequency, peaking will occur and the mechanical movement ∂s will no longer be given by the static sensitivity of the piezotube. The lower limit of the modulation frequency is given by the cut-off frequency of the feedback loop. The gain of the feedback loop will decrease the current modulation due to the mechanical modulation, and it will also introduce an additional gain-dependent phase shift. Since the gain depends on the local barrier height, it is hard to give a quantitative interpretation for too low of a modulation frequency. For microscopes with large piezoscans it is sometimes impossible to fulfill both requirements. An interesting implementation of this kind of spectroscopy using photothermal modulation of the tunnel gap has been published by Amer *et al.* (1986).

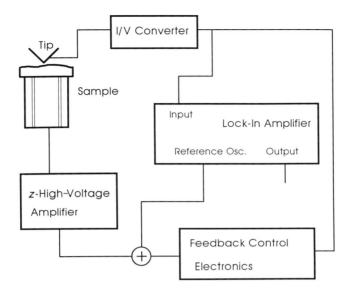

FIGURE 1.9 The block diagram for measuring $\partial I/\partial s$ with the STM. The spacing between the tip and the sample is modulated by adding a small AC voltage onto the output of the feedback amplifier. This results in a modulation of the tunneling current, which is detected by the lock-in amplifier.

Figure 1.10 shows the set-up for alternating two bias voltages, V_1 and V_2, on alternating scans. The basic feedback loop is the same as in previous experiments. This time the electronics generating the x-scanning signal must also be considered. The triangular output of the scan generator (function generator or computer output) is connected to the x and, through an inverting amplifier, to the $-x$ electrodes of the piezoscanner. Trigger pulses marking the begin or end of each x-scan are used to toggle a divider by 2 (see Tietze and Schenk, 1988). Its output in turn toggles the electronic switch connecting the sample to the voltage sources V_1 and V_2. Odd- and even-numbered scan lines are stored in two separate images, showing the topography at the two different voltages. Why does one alternate the voltages between the scan lines and not from frame to frame? The thermal drift always present would make it difficult to determine a registry between the two images. By interlacing the two voltages, the time from a scan line in one image to the corresponding scan line in the other is smaller by the number of scan lines. Hence, the drift is smaller by the same factor.

The set-up for the last mode of spectroscopy, current imaging tunneling spectroscopy (CITS), is depicted in Figure 1.11. To perform CITS, the basic feedback loop must be modified. The bias voltage source is a function generator that is triggered from a timing control. A trigger pulse starts the A/D conversion for the voltage sweep and also disconnects the integrator from the error signal. The integrator holds the output voltage and hence the tip stays at a fixed height

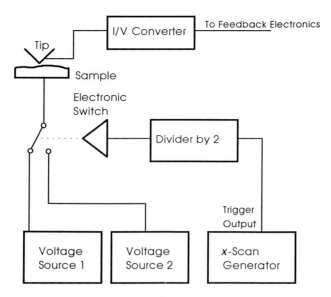

FIGURE 1.10 The block diagram for measuring with alternating bias voltages on alternating scan lines. The x-scan generator toggles the bias voltage between two preset values at the end of a scan line. The feedback electronics keeps the tunneling current constant. The resulting topographs are a function of the bias voltage. This set-up allows a direct comparison of two energies in the local density of states.

above the sample. The feedback loop is disabled until the end of the bias voltage sweep. The error signal V_E was zero at the begin of the voltage sweep. If there were no drift, it would be again zero after the voltage sweep. Because thermal drift is always present, it is necessary to optimize the microscope for low-thermal drift and to disable the feedback loop only for short times. Even if there is a nonvanishing error, the integrator will ramp smoothly to the new position. The proportional amplifier in the feedback loop should be omitted so that the output voltage does not jump to the new position. Information on the computer control systems have been published by Aguilar *et al.* (1986, 1987), Becker (1987), and Brown and Cline (1990).

1.2.2.8 TIP PREPARATION FOR STM

An important point in the successful operation of an STM is the tip preparation. The tips should have a minimal radius of curvature at the end, they should have a narrow diameter to penetrate into holes, but they should also be thick enough to keep the resonance frequency high. The tip material should be stable in high electric fields and, when operated in air or corrosive liquids, chemically stable. Tips have been made out of wires of tungsten, platinum–iridium, platinum, gold, and nickel.

A variety of tip preparation methods were used by STM researchers. In the initial use of STM, tips were prepared by grinding tungsten wire. However, this resulted in tips of minor quality. A simple preparation method is to cut a thin wire (0.1–0.3 mm) of the desired material by a cutter or scissors. Low-quality tools work best because they tear the material apart, forming sharp tips. Cutting is fast, but the success rate for working tips is not very high. Moreover, the shape of the tips is not well defined, making it harder to use them on strongly corrugated samples. The shape of the tip is less important when imaging atomically flat samples. The material suited best for this method seems to be platinum–iridium wires. Chapter 2 in this volume describes in detail the tip preparation methods. Fink (1986), van der Walle *et al.* (1986), Bryant *et al.*

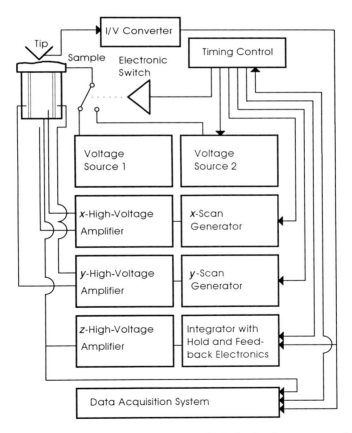

FIGURE 1.11 Current imaging tunneling spectroscopy (CITS). During the raster scan of the surface the tip is stopped at predetermined locations controlled by the timing device. At each such point the bias voltage is switched from voltage source 1 to voltage source 2. While the feedback loop is disabled, voltage source 2 ramps the voltage over a predefined range. The current variations are recorded as a function of the bias voltage. Then voltage source 1 is turned on again, and the feedback loop is enabled.

(1987), Colton *et al.* (1987), Schwoebel (1987), Gewirth *et al.* (1989), Binh (1988), Heben *et al.* (1988), Neddermeyer and Drechsler (1988), Biegelsen *et al.* (1989), Chen *et al.* (1989), Cricenti *et al.* (1989b), García-Cantú and Huerta-Garnica (1989), Yata *et al.* (1989), Nagahara *et al.* (1989), Akama *et al.* (1990), den Boef (1990), Lemke *et al.* (1990), Fiering and Ellis (1990), Garnaes *et al.* (1990), Ibe *et al.* (1990), Marcus *et al.* (1990), Möller *et al.* (1990b), Musselman and Russell (1990), Wengelnik and Neddermeyer (1990), Binh and García (1991), Miyamoto *et al.* (1991), Melmed (1991), and Vasile *et al.* (1991) have published reports on the production of STM tips and on their characterization.

1.2.2.9 COARSE SAMPLE POSITIONING

For many experiments it is important to position the tip over a well-defined area of the sample or to move the tip from one location to another. Biological samples, integrated circuits, and samples with surfaces showing localized phenomena demand a coarse positioning. On other samples, one has to compare images from a range of locations to ensure that the observed structures are indeed representative. This section discusses coarse positioning under optical control, whereas Section 1.2.2.10 gives a short introduction to positioning the tip under scanning electron microscope (SEM) control.

The coarse positioning devices in an STM have to fulfill contradicting requirements. First, they should be as small and rigid as possible to preserve the high-resonance frequencies of the original design. On systems having a vibration isolation, the drive for the coarse positioning device needs to be on the microscope, or, alternatively, its drive connection should be detachable.

The first STM (Binnig *et al.*, 1982a,b) employed a "louse" for coarse positioning and for the approach. It is a two-dimensional piezomotor, moving on a steel plate. In addition to linear movements, it also allows the sample to be turned. Detailed discussions of this device were published by Marti (1986) and by Gerber *et al.* (1986). The main disadvantage of a "louse" is that it has a rather large size. Newer designs use commercially available inchworms. The position of the tip with respect to the sample is observed by optical telescopes with a position resolution of about 10 μm.

Another type of coarse positioning is used in Besocke-type microscopes (Besocke, 1987). There the sample rests on three piezotubes. The tip is mounted onto a fourth piezotube in the center of the other three. By moving the outer three piezotubes rapidly in one direction and contracting them slightly during the movement, their contact points can be moved by a small amount. The three tubes then return slowly to their starting positions. By using a sample holder with a tilted border, movements in x, y, and z as well as a sample rotation can be accomplished. A high-power optical microscope can be used to position the tip, if the sample is transparent (see Guckenberger *et al.*, 1989).

Other coarse positioning devices employ micrometer screws and manual or micromotor drives. The manual drives require decoupling during measurement to avoid the transmission of vibrations to the microscopes. Motorized micrometers can be used on vibration isolation platforms without short-circuiting the isolation.

It is important to view the movement of the tip over the sample surface. A medium- to high-magnification optical microscope is usually incorporated into the design of the SXM. The SXM can be remotely operated if the optical microscope is connected to a videocamera. This allows the microscope to be mounted in a sound-isolated box, under inert gas, and on vibration isolation tables and still afford control.

1.2.2.10 INTEGRATING AN STM INTO AN SEM

The obtainable resolution of optical positioning is on the order 1 μm. One order of magnitude in resolution can be gained for conducting samples when using an SEM (Gerber et al., 1986; Edel'man et al., 1991) (see Fig. 1.12). There are two crucial points to be observed: First, the sample and the tip have to be arranged such that they are well visible from the electron gun of the SEM. The STM has to be connected rigidly to the SEM to ensure a good resolution of the SEM. The vibration isolation has to be located outside the SEM vacuum system. Second, the SEM has to have a UHV chamber. The rest gas in ordinary high-vacuum chambers contains hydrocarbons that are cracked by the electron beam at the sample surface. The resulting carbon film conducts poorly and can render an STM inoperable. (A hydrocarbon-rich rest gas together with an electron beam is used to write patterns on integrated circuits. The hydrocarbon film is known as *contamination resist*.)

The tunneling current in the STM is affected by the electron beam. If the electron beam current becomes comparable with the tunneling current, it might cause the feedback loop of the STM to become unstable. The SEM should always be operated at currents lower than the tunneling current unless the STM is not in the tunneling regime.

One advantage of the SEM is its extreme depth of view. The tip and the sample can be seen sharp simultaneously. This allows a very precise positioning of the tip, which must have a small opening angle so as not to obscure the sample. If the SEM is equipped with Auger electron analysis, it can determine the chemical composition of the sample surface on submicrometer scale.

1.2.2.11 APPROACHING THE TIP TO THE SAMPLE

One of the most important steps in operating an STM is the approach of the tip to the sample. A carefully prepared tip is of no use if it is damaged the moment it reaches the sample. The task is to bring the tip from a distance of 1 to about 3 mm down to a distance of about 1 nm and to establish a tunneling current of about 1 nA. (A scaled-up task would be to start a car 1 km away from the wall and to bring it to a halt at a distance of 1 mm from the wall. The additional

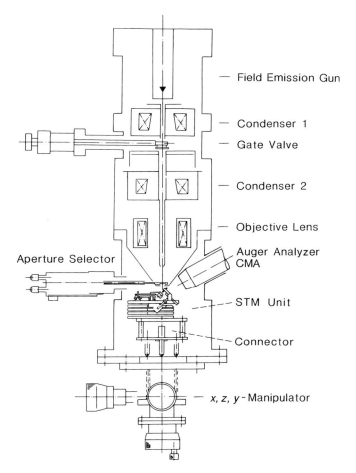

FIGURE 1.12 A schematic drawing showing the integration of an STM into a scanning electron microscope. (From Gerber *et al.*, 1986, with permission.)

difficulty would be that the driver does not see the wall until 2–3 mm away.) If we assume that the tip approaches the sample surface with 1 μm/s and that it is to stop within 1 nm, then the acceleration is 100 times the earth's acceleration for 1 ms.

If we were to use the speed of 1 μm/s for the whole distance of several millimeters, then the whole approach could easily take an hour or more. Most STM are equipped with a two-stage approach: first a coarse approach under optical control and then the fine approach under electronic control. The coarse approach is usually done with fine-pitched screws. A binocular with a long working distance is set up as depicted in Figure 1.13. On flat surfaces, one can see at the same time the tip and its mirror image. Depending on the magnification of the binocular, the tip can be approached to 2–4 μm of the surface.

One way to make the fine approach is to use similar fine-pitched screws but with a mechanical disadvantage between the screw displacement and the tip displacement (Schneir et al., 1986a). This fine approach screw can be driven by a stepper motor, by a synchronous AC motor, or by a DC motor. Stepper motors have the advantage that they can be stopped very rapidly, whereas synchronous motors and DC motors run more smoothly. The effect of the jumps from step to step can be minimized by selecting a stepper motor with gear reduction and by running it with a reduced driving voltage (e.g., 5 V instead of 12 V). In addition, stepper motors are easily interfaced with digital logic or with computers.

1.2.2.12 THE OPERATING MEDIUM

The first STM were operated in vacuum (Binnig and Rohrer, 1982). However, the STM was operated in other media, too. Park and Quate (1986) demonstrated the operation of the STM in air. Sonnenfeld and Hansma (1986) showed that atomic resolution was possible in water. Elrod et al. (1984a,b) operated their STM at cryogenic temperatures. Cryogenic STM may be operated either immersed in liquid helium or in small, cooled vacuum chambers.

STM operating in fluids are particularly appealing for the investigation of biological samples. The possibility of imaging molecules in their native environment is unparalleled at the resolutions the STM obtains. The chemical and electrochemical environments can be controlled accurately. Chemical reactions (see Drake et al., 1987, 1989) can be triggered while imaging, and their time evolution can be observed, provided it is not too fast. The operation of the STM in polar fluids like water requires the use of isolated tips to minimize the ionic

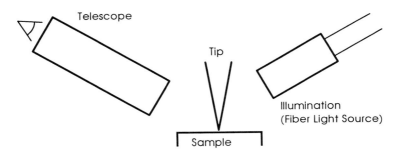

FIGURE 1.13 Typical set-up of a microscope. The tip and the sample are viewed through a telescope or a long-distance microscope as a stereomicroscope. The tip and its mirror image form an X-shaped pattern, which is separated at the center. The approach is finished when the tip and its mirror image seem to be touching. Typically the tip is then a few micrometers from the sample apart.

current between tip and sample. Such tips are readily available from commercial sources.

1.2.2.13 ADDITIONAL REFERENCES

Design examples of STM have been published by Gerritsen et al. (1985), Smith and Binnig (1986), Chiang and Wilson (1986), Demuth et al. (1986), Vieira et al. (1987), Hermsen et al. (1987), Blackford et al. (1987), Emch et al. (1988), Michel and Travaglini (1988), Lyding et al. (1988), Valdez et al. (1988), Xiao et al. (1988), Yao et al. (1988), Cox et al. (1989), Cricenti et al. (1989a), Khaikin (1989), Poirier and White (1989), Hosaka et al. (1990), Watanabe et al. (1990), Yasutake and Miyata (1990), Zeglinski et al. (1990), Kato and Tanaka (1990), Haase et al. (1990), Hipps et al. (1990), Shimizu et al. (1990), Griffith et al. (1990), García-Cantú and Huerta-Garnica (1990), Shang et al. (1991), Schmid and Kirschner (1991), and Kato et al. (1991). A remote tip exchange mechanism is described by Sugihara (1990) and Yokoyama et al. (1991). Micropositioning devices and systems for approaching the tunneling tip close to the sample surface are described by Corb et al. (1985), Mamin et al. (1985), Smith and Elrod (1985), Anders et al. (1987), Pohl (1987a,b), Jericho et al. (1987), Matey et al. (1987), Frohn et al. (1989), Stupian and Leung (1989), Meepagala et al. (1990), Shimizu et al. (1990), Okumura and Goshi (1990), Renner et al. (1990b), Yamagata et al. (1990), Chornik et al. (1991), Cortager et al. (1991), Jeon and Willis (1991a), Kato et al. (1991), Nishimura (1991), Park et al. (1991), McCord (1991), and Probst et al. (1991).

Information on the computer control systems have been published by Schroer and Becker (1986), Aguilar et al. (1986, 1987), Becker (1987), Laiho et al. (1987), Laegsgaard et al. (1988), DiLella et al. (1989), Piner and Reifenberger (1989), Fuchs et al. (1989), Hoeven et al. (1990), Grafström et al. (1990), Cutkosky (1990), Brown and Cline (1990), Schummers et al. (1991), Robinson et al. (1991a), and Maps (1991). Feedback loop electronics information has been published by Troyanovskii (1989); response functions are analyzed by Park and Quate (1987), Robinson et al. (1991b), and Jeon and Willis (1991b). Bryant et al. (1986) and Robinson (1988a,b) report on an ultrafast STM. Vibration isolation systems are discussed by Park and Quate (1987). A useful reference concerning noise sources is the book by Ott (1976). A technique to image weakly bonded surface deposits has been published by Jericho (1989). Ultrafast time resolution in scanning tunneling microscopy is discussed by Hamers and Cahill (1991). Tsukamoto et al. (1991), and Kawakatsu et al. (1991) describe a method to use a dual STM for precise metrological measurements. The sensitivity of the measurement can be enhanced with modulation techniques (see Abraham et al., 1988b; Stoll and Gimzewski, 1991). Procedures for imaging poorly conducting samples by scanning tunneling microscopy were proposed by Kochanski (1989) and Yuan et al. (1991). Leavens and Aers (1986) propose the use of tunneling resonances to calibrate the tip-to-sample distance. Kenny et al. (1991) built a micromachined tunnel sensor for the detection of motion.

1.2.3 Image Processing: An Introduction

The visualization and interpretation of images from SXM is intimately connected with the processing of these images. Reasons for using image processing algorithms in order to remove distortions (filtering in Fourier space and in real space) are discussed. Some methods to display data are then outlined.

1.2.3.1 WHY IMAGE PROCESSING?

An ideal SXM is a noise-free device that images a sample with perfect tips of known shape and has perfect linear piezoscanners. In reality, SXM are not ideal. The scanning device in SXM is affected by distortions. To do quantitative measurements like determining the unit cell size, these distortions have to be measured on test substances and have to be corrected for. The distortions are both linear and nonlinear. Linear distortions mainly result from imperfections in the machining of the piezotranslators causing cross talk from the z-piezo to the x- and y-piezos and vice versa. Among the linear distortions there are two kinds that are very important: first, piezoscanners invariably have different sensitivities along the different scan axes due to the variation of the piezomaterial and uneven sizes of the electrode areas. Second, the same reasons might cause the scanning axis not to be orthogonal. Furthermore, the plane in which the piezoscanner moves for constant z is hardly ever coincident with the sample plane. Hence a linear ramp is added to the sample data. This ramp is especially bothersome when the height z is displayed as an intensity map, also called *top view display*.

The nonlinear distortions are harder to deal with (see Libioulle *et al.*, 1991). They can affect SXM for a variety of reasons. First, piezoelectric ceramics do have a hysteresis loop, much like ferromagnetic materials. The deviations of piezoceramic materials from linearity increase with increasing amplitude of the driving voltage. The mechanical position for one voltage depends on the voltages previously applied to the piezo. Hence to get the best position accuracy one should approach a point on the sample always from the same direction.

Another type of nonlinear distortion of the images occurs when the scan frequency approaches the upper frequency limit of the x- and y-drive amplifiers or the upper frequency limit of the feedback loop (z-component). The distortion due to the feedback loop can only be minimized by reducing the scan frequency.

On the other hand, there is a simple way to reduce distortions due to the x- and y-piezodrive amplifiers. To keep the system as simple as possible, one normally uses a triangular waveform for driving the piezoscanners. However, triangular waves contain frequency components at multiples of the scan frequency. If the cut-off frequency of the x- and y-drive electronics or of the feedback loop is too close to the scanning frequency (two to three times the scanning frequency) the triangular drive voltage is rounded off as the turning points. This rounding error causes first a distortion of the scan linearity and second, through phase lags, the projection of part of the backward scan onto the forward scan. This type of distortion can be minimized by carefully selecting the scanning frequency and by using driving voltages for the x- and y-piezos with waveforms like trapezoidal waves, which are closer to a sine wave.

The values measured for x, y, or z are affected by noise. This noise can originate electronically or it might be due to disturbances through sound or to a property of the sample surface because of adsorbates. In addition to this incoherent noise, interference with mains and other equipment nearby might be present. Depending on the type of noise, one can filter it in real space or in Fourier space.

1.2.3.2 CORRECTING DISTORTED IMAGES

To improve the usefulness of the SXM data for measurements of distances and to enhance the visual appearance, the linear and nonlinear distortions have to be corrected. Normally, they will have well-defined physical origins and can be determined by independent methods. A common linear correction is the removal of a background plane by fitting a plane to the data. A mathematical formulation of this background subtraction is given in Section 1.9.

Another common distortion removal is the correction for nonorthogonal piezoscanners or for piezoscanners with an unequal sensitivity. An independent measurement of the distortion is usually required to obtain physically meaningful results. One excellent possibility is to correct the nonorthogonalities beforehand using an electronic matrix to mix the drive voltages of all three channels. Another method incorporates feedback control for all of the movements of the piezotube to reduce the nonlinearity (see Barrett and Quate, 1991b).

If one does not have access to such electronics it is possible to use a measurement with the same piezoscanners on a test substance such as graphite, a CD disk stamper, or an optical grating to determine the correction factors. With this technique, it is possible only to get a measure of the distortion in the xy-plane. The mathematics of the distortion removal in the sample plane and in all three dimensions is rather complicated and is treated in Section 1.10. Information on this subject has been published by Jericevic *et al.* (1988) and by Robinson (1990).

Nonlinear distortions pose far bigger problems to correct than the linear distortions. As said before, nonlinear distortions may stem from an insufficient analog bandwidth in the feedback system, from piezohysteresis, or from creep. Data hampered by insufficient bandwidth could be corrected by Fourier filtering and deconvolution in one dimension (Press *et al.*, 1989; Reiss *et al.*, 1990b). However, both the forward and the backward scan must be recorded at equal time intervals. The data gathered during the backward scan serve to determine the history of the feedback loop at the first few points of the forward scan.

The hysteretic behavior of the piezoceramic is most annoying at large scan ranges. A test measurement could determine a look-up table, with which the measured data could be translated into undistorted data. The look-up tables should be acquired in a few scan sizes. Intermediate scan sizes could then be interpolated. A very efficient method to linearize large scan ranges is based on the measurement of the actual piezodeflection (Barrett and Quate, 1991b).

1.2.3.3. FILTERING AND DATA ANALYSIS IN REAL SPACE

Real space filters are filters whose result for a point only depends on a few neighboring points. For large data sets they are more efficiently implemented than the corresponding filters in the spatial frequency domain (see Section 1.2.3.4). One of the most often occurring problems is to remove high-frequency noise from data. The origin of this noise can be electronical, come from adsorbates on the sample surface, or be digitizing noise. This high-frequency noise can be reduced by replacing each point by the weighed average of its neighboring points. If we only consider the nearest neighbors, that is, the points $z(x, y)$ with $x = [x_0 + 1 | x_0 | x_0 - 1]$ and $y = [y_0 + 1 | y_0 | y_0 - 1]$, then we can define the 3×3 convolution low pass filter by

$w_{-1,1}$	$w_{0,1}$	$w_{1,1}$
$w_{-1,0}$	$w_{0,0}$	$w_{1,0}$
$w_{-1,-1}$	$w_{0,-1}$	$w_{1,-1}$

The value at a point x_0, y_0 is the given by

$$z'(x, y) = \frac{\sum_{i=-1}^{1} \sum_{j=-1}^{1} w_{ij} z(x+i, y+j)}{\sum_{i=-1}^{1} \sum_{j=-1}^{1} w_{ij}}. \tag{1.58}$$

It should be noted that if the denominator in Eq. 1.58 is zero, then we should use the modified equation

$$z'(x, y) = \sum_{i=-1}^{1} \sum_{j=-1}^{1} w_{ij} z(x+i, y+j). \tag{1.59}$$

By setting all the w_{ij} to 1 we obtain the convolution averaging filter. Figure 1.14a shows the original, noisy data. The image in Figure 1.14b is filtered by the convolution averaging filter. We can construct directional averaging filters by setting selected w_{ij} to 1 and the others to 0. For instance, setting the $w_{ij=0}$ to 1 and all the other coefficients to 0, we obtain a filter along the x-axis. The coefficient need not necessarily be equal to 1 or 0. By increasing the center value one can emphasize the value of the point to be filtered compared with its neighborhood. The reader can easily verify that the filters in Table 1.3 do the respective tasks. Figure 1.14c shows the effect of the Laplacian filter and 1.14d the effect of a directional gradient filter.

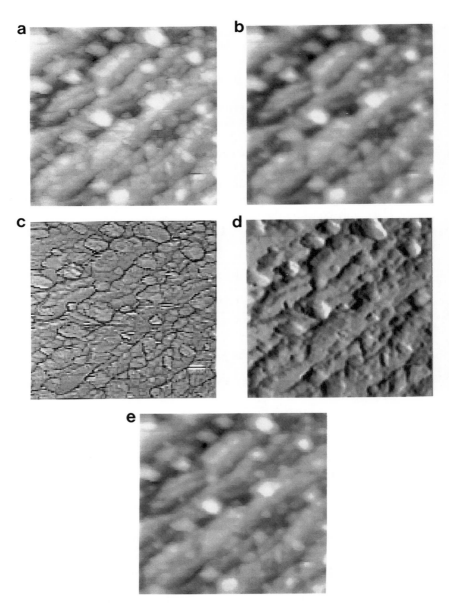

FIGURE 1.14 The effect of different real space filters. a: The original data. b: The convolution averaged data. c: Laplacian filter. d: Directional gradient filters. e: Filtered by the median filter.

TABLE 1.3 3 × 3 Convolution Kernels

Name	Filter		
Low pass filter	1	1	1
	1	1	1
	1	1	1
Low pass filter in horizontal direction	0	0	0
	1	1	1
	0	0	0
Low pass filter in vertical direction	0	1	0
	0	1	0
	0	1	0
Laplacian filter	1	1	1
	1	$-n$	1
	1	1	1
Edge detection (diagonal)	0	1	0
	1	-0	0
	0	0	-2

The convolution averaging filter is well suited for removing high-frequency, random noise. It has, however, the disadvantage that it also blurs steps and other well-defined variations from one pixel to another. A better filter to remove single-pixel impulses is the median filter. We do consider the same points in the 3 × 3 neighborhood as for the convolution averaging filter. But this time we do order the points and we take the middle value, the median. Figure 1.14e shows the noise removal by the median filter. Park and Quate (1987a) give a summary of digital procedures.

1.2.3.4 FILTERING AND DATA ANALYSIS IN THE SPATIAL FREQUENCY DOMAIN

Filtering in the spatial frequency domain (Fourier space) is a very powerful tool. All periodic surface data and many noise components such as mains interference have well-defined spatial frequencies, which show up as peaks in the Fourier transformed data:

$$F(k_x, k_y) = \int_{-\infty}^{\infty} \int_{-\infty}^{\infty} f(x, y) e^{i(xk_x + yk_y)} \, dx \, dy \tag{1.60}$$

is called the Fourier transform of $f(x, y)$. The inverse transform is given by

$$f(x, y) = \frac{1}{2\pi} \int_{-\infty}^{\infty} F(k_x, k_y) e^{-i(xk_x + yk_y)} \, dk_x \, dk_y. \tag{1.61}$$

These two transforms are the basis of Fourier filtering of images. An introduction to Fourier transforms on computers and on filters is provided by Press *et al.* (1989). On computers, one normally uses the fast Fourier Transform (FFT), an algorithm discovered by Danielson and Lanczos in 1942 and then rediscovered by Cooley and Tukey in 1965.

Fourier filtering and data analysis is especially powerful on periodic sample structures or on coherent noise. It is useful to display the Fourier-transformed data either as a power spectrum and a phase spectrum or as the real and imaginary parts of the spectrum. The FFT maps the data into k-space such that k_x and k_y both run from 0 to k_{max}. The spectrum for $k > k_{max}/2$ is the mirror image of the spectrum for $k < k_{max}/2$. Zero spatial frequency is therefore at all four corners of the Fourier-transformed data. Symmetries are not obviously in this representation. It is advantageous to move zero spatial frequency to the center of the display. This can be done by swapping the area in the way represented in Figure 1.15. The Fourier spectrum of surfaces in this representation is similar to the displays one gets from low-energy electron diffraction (LEED) (see Jona *et al.*, 1982).

Performing a filter, convolution, or deconvolution in the $\vec{k}$ space requires special attention: The FFT algorithm is based on periodic functions, with the maximum period being the size of the data or integral fractions thereof. Measured data will contain other frequency components that are of noninteger relation to the basic period of the FFT and that were truncated by the sampling process. Modifying the spectrum by a filter or convolution can introduce artifacts. Press *et al.* (1989) describe the use of data windowing or padding to minimize the unwanted content in the spectrum. If these procedures are not followed, meaningless data might be created.

An important filter in the $\vec{k}$ space is the Wiener filter. It is assumed that the scanning probe microscope has a response function $r(x, y)$ and a noise function $n(x, y)$. The real data $u(x, y)$ is first smeared out by $r(x, y)$ to

$$s(x, y) = \int_{-\infty}^{\infty} \int_{-\infty}^{\infty} r(\hat{x}, \hat{y}) u(x - \hat{x}, y - \hat{y}) \, d\hat{x} \, d\hat{y}. \tag{1.62}$$

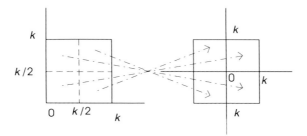

FIGURE 1.15 Fourier display: areas to be swapped to get a display like LEED displays. After the FFT the data points corresponding to low frequencies are located at the four edges of the data set. Fourier displays with 0 frequency at the center provide easy to get information on the symmetries of the surfaces.

The noise $n(x, y)$ is added to $s(x, y)$ to give

$$c(x, y) = s(x, y) + n(x, y). \qquad (1.63)$$

The Wiener filter $\Psi(k_x, k_y)$ tries to reconstruct the original data $u(x, y)$ by taking into account the effect of the noise. The reconstructed spectrum is

$$\tilde{U}(k_x, k_y) = \frac{C(k_x, k_y)\,\Psi(k_x, k_y)}{R(k_x, k_y)}, \qquad (1.64)$$

where $C(k_x, k_y)$ and $R(k_x, k_y)$ are the Fourier transforms of $c(x, y)$ and $r(x, y)$, respectively. The exact tip shape and the relevant interactions between the tip and the sample are not well known. Hence the assumption of a known response function $r(x, y)$ is usually not fulfilled in atomic resolution SXM. Therefore, one cannot hope to deconvolute such data using a Wiener filter. A noise reduction, however, is possible. For large scans (in the micrometer range), the tip shape can usually be determined by SEM or is known from the fabrication process (microfabricated cantilevers for scanning force microscopy) and the interaction details are of no concern on those length scales. In this situation, a successful noise reduction and deconvolution can be possible.

The filter function of the Wiener filter is

$$\Psi(k_x, k_y) = \frac{|S(k_x, k_y)|^2}{|S(k_x, k_y)|^2 + |N(k_x, k_y)|^2}. \qquad (1.65)$$

$\Psi(k_x, k_y)$ is determined by the power spectrum of the smeared data $s(x, y)$ and by the power spectrum of the noise function $n(x, y)$. The spectrum of the measured function $c(x, y)$ does not not enter into the calculation of the filter function. One way to get the additional information is to guess the noise spectrum from suitable plots of the spectrum of $c(x, y)$. Another way is to record images with the scanning motion disabled. This produces the true noise spectrum if there are no position-dependent noise components (see Stoll and Marti, 1986; Pancorbo et al., 1990, 1991).

Fourier-transform filters are very powerful for periodic data. The computational effort, however, increases sharply with the number of data points. The periodicity of the data can be used to define a unit cell, which is repeated all over the surface. Any point at a specific location x, y in this unit cell must have the same z-value as the corresponding points in the other unit cells. Figure 1.16 shows a sketch of a periodic surface. We can map all the points to one unit cell and average over them. The data in the one unit cell have a reduced noise background, since the coherent data, the structure of the unit cell, are passed unchanged, whereas the statistical noise, incoherent with the unit cell, is reduced by $\sqrt{n}$, where n is the number of averaged unit cells. The last step in this filter is the repetition of the filtered data over the original area. Correlation-averaging filters in conjunction with SXM have been described by Soethout *et al.* (1988).

Table 1.4 summarizes the various filter methods. It also provides indications of when to use which filter.

1.2.3.5 VIEWING THE DATA

The most important part of image processing is to visualize the measured data. Typical SXM data sets can consist of many thousands to over 1 million points per plane. There may be more than one image plane present. The STM data represent a topography. The output of the first STM was recorded on x,y-chart recorders. Usually, the z-value or the height of the tip was plotted against the tip position in the fast scan direction. Often the position in the slow scan direction

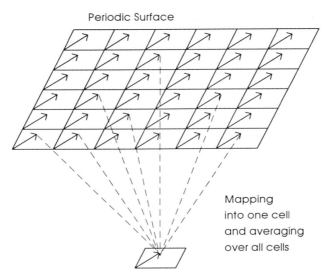

FIGURE 1.16 Unit cell filter. This figure depicts how an arbitrary point on a periodic surface is mapped to the unit cell.

TABLE 1.4 Use of Filters

Problem	Filter	Comments
Uneven height	Background subtraction	Fast, reversible
Random noise	Convolution low pass	Fast processing, limited action
Interference with a fixed frequency	Fourier filtering	Fast for small data sets; time consuming otherwise
Interference with a fixed frequency	Unit cell filter	Faster for large data sets; no complete suppression of the interference

was not recorded, but was assumed to be constant. A ramp added to the y-channel of the chart recorder helped to separate the scan lines. More sophisticated display systems added a fraction of the tip position in the slow scan direction to both the x- and y-channels of the chart recorder. This way, a pseudo three-dimensional display was achieved. Figure 1.17 displays a sample surface obtained with this technique. Chart recorders are slow devices; thus analog storage oscilloscopes, displaying the same line scan plots, were used.

FIGURE 1.17 A typical example of an SXM output using chart recorders. The data displayed are from a silicon surface with evaporated indium. The chart recorder is set up such that the horizontal axis display x and a fraction of y, the vertical axis z and a fraction of y.

A wire mesh display similar to the line scan display can be created on computer displays (Fig. 1.18). It is especially suitable for monochrome display systems with only two colors. The number of scan lines that can be displayed is usually well below 100, and the display resolution along the fast scanning axis x is much better than along the y.

If the computer display is capable of at least 64 shades of gray, then top-view images can be created (Fig. 1.19). In these images, the position on the screen corresponds to the position on the sample and the height is coded as a shade of gray. Usually the convention is that the brighter a point, the higher it is. The number of points that can be displayed is only limited by the number of pixels available. This view of the data is excellent for measuring distances between surface features. Periodic structures show up particularly well on such a top view. The human eye is not capable of distinguishing more than 64 shades of gray. If the average z-height of the tip varies from one side of the image to the other, then the interesting features usually have too little contrast. Hence, contrast equalization is needed. For data being affected by a large background slope, it is often possible still to detect some features in the line scan view. Some researchers prefer a simultaneous display of both line scan images and top-view images to get the most information in the shortest time. Top views require much

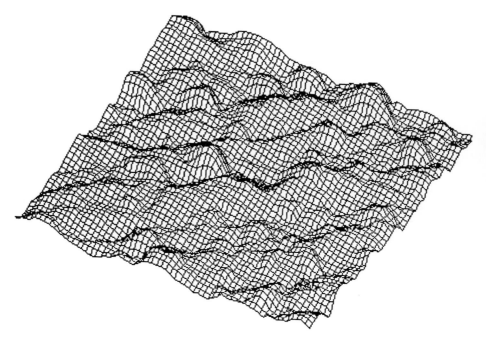

FIGURE 1.18 Wire mesh display of the data from Figure 1.17. The wire mesh display gives a quick look at the surface.

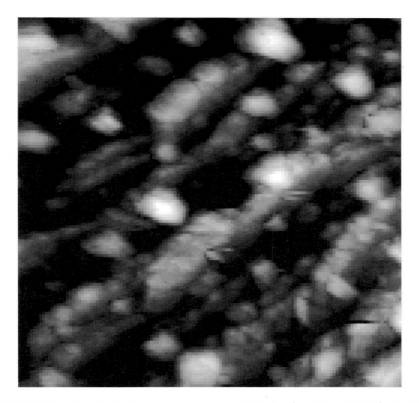

FIGURE 1.19 Top view display from a computer screen of the data from Figure 1.17. The top view display is the workhorse of the SXM data display methods. It allows a convenient judging of heights and sizes.

less calculation time than line scan images. Hence, computerized fast data acquisition systems usually display the data as a top view first.

The display can be made more illustrative by calculating the illuminated top view of the data, much like the way topographic maps are shaded. Figure 1.20a gives an example of a sample surface illuminated by a point light source at infinity. This technique is a powerful tool to enhance the appearance of a data set. But it can be abused! Changing the direction of the light source, as shown in Figure 1.20b, can obscure some undesired features. The effect of the illumination is similar to displaying the magnitude of the gradient of the sample surface along the direction to the light source. Features perpendicular to the illumination cannot be seen. Multiple light sources or extended light sources diminish this effect, but the illumination is much more complicated to calculate. If the illumination is not shown in conjunction with some other display method, then one is not able to judge the validity of such an image.

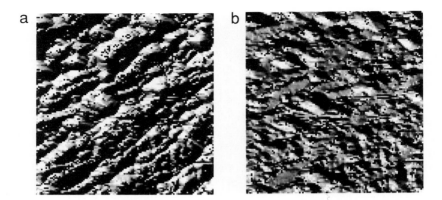

FIGURE 1.20 The same test surface as in Figures 1.17, 1.18, and 1.19 is illuminated from two different directions. a: Illuminated such that the structure seen in the previous figures is reproduced. b: By changing the illumination direction, information may be hidden. Data published with illumination, either flat or three-dimensionally rendered, should be viewed critically.

One can combine top views or illuminated top views and wire mesh scan displays to form solid surface models of the sample surface. Such images are usually only generated in the final processing stage before publication because they need quite a lot of computing time. Figure 1.21a shows a combination of the top-view display and the wire mesh display, a three-dimensional model in which the height is coded as a shade of gray. Figure 1.21b shows a combination of the illuminated top view and the wire mesh, a display much like a real landscape under the sun. Depending on the point of view, some features might be more prominent than others.

Additional information can be packed into an image by using color. Assume that an image has two planes of data. We can display the first plane with shades of green and the second one with shades of red on top of each other. Where the magnitude of both planes is high, one gets on orange color, where both are low, one gets black. But if the magnitude of one plane is larger than that of the other plane on one pixel, one gets red or green colors. This way, one can display the registry of two different quantities in the same image.

1.2.4 Selected Experiments

As an example, I describe very briefly experiments on graphite. STM experiments on a variety of biological samples are discussed in other chapters in this book.

1.2.4.1 GRAPHITE

Graphite was the first substance to be imaged in air (Park and Quate, 1986) and under liquids (Schneir *et al.*, 1986b) at atomic resolution. Investigations under

1 SXM: An Introduction 49

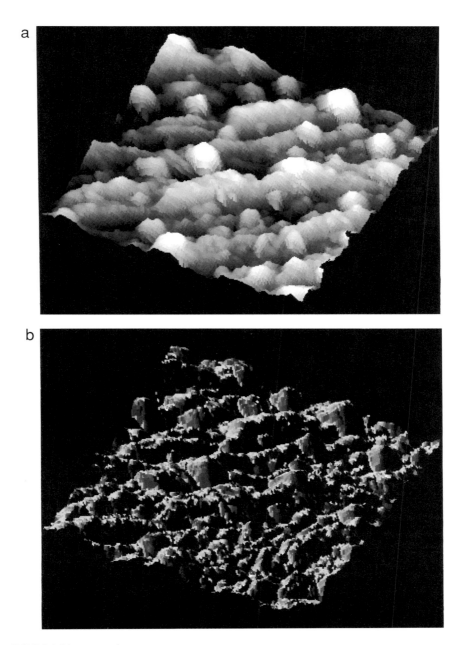

FIGURE 1.21 Two possible ways of a three-dimensional surface rendering of the surface shown in Figure 1.17. **a**: The shade of gray is determined by the height of the data, as in Figure 1.19. **b**: An illuminated three-dimensional rendering.

UHV conditions (Soler *et al.*, 1986) as well as low temperature experiments (Marti, 1986; Marti *et al.*, 1986) revealed the atomic scale structure of this surface. The relative ease of the imaging of the graphite surface under very different conditions has promoted its use as a calibration and test surface. Experiments going beyond the determination of the lateral scales revealed, however, that many unexpected effects play a role in the determination of the final appearance of the images.

Figure 1.22 shows a schematic view of the structure of the graphite crystal. The carbon atoms are organized in layers of hexagons, with only weak bonding between the layers. This structure permits an easy cleavage of the surface, for instance, by taping an adhesive tape to the surface and removing it carefully. The layered structure ensures, too, that there are large, atomically flat terraces. As an example, the topography of the graphite surface imaged at 6.8 K is shown in Figure 1.23. The image is similar to those obtained under other conditions. The hexagonal pattern of a graphite sheet is not resolved. Instead, a moundlike pattern with the repeat distance of the unit cell of the graphite surface 2.54 nm is observed. At other times, hexagonallike structures or other structures are observed, but all with the periodicity of the graphite surface. Moreover, the height of the observed corrugation is often much larger than the height determined by He-scattering (e.g., Carlos and Cole, 1980) or the calculated heights (Selloni *et al.*, 1985).

This behavior can be understood by noting that the STM images of the graphite surface are determined by the density of states of the surface. The Fermi surface of graphite (see Ashcroft and Mermin, 1976 is confined entirely to the edges of the hexagonal Brillouin zone (see Marti, 1986 and references therein). This means that, to a good approximation, the local density of states of the

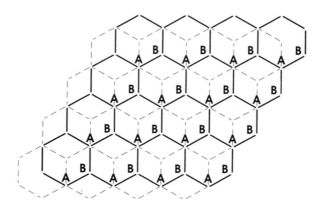

FIGURE 1.22 Structure of the graphite crystal. The carbon atoms in the hexagonal rings at the surface occupy two inequivalent sites: At the A site the carbon atom has a bond with the next lower layer, whereas at the B site there is no out-of-plane bond.

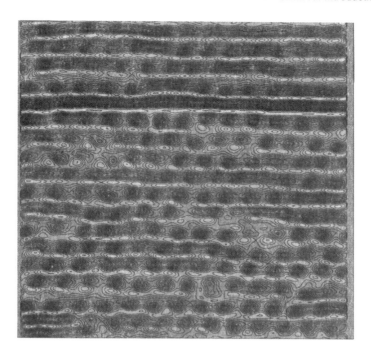

FIGURE 1.23 Low-temperature image of graphite. The sample is held at 6.8 K. The size of the image is 3.3 nm. The height varies by 0.54 nm, from black to white.

graphite surface at the Fermi energy can be represented by three standing waves (Mizes *et al.*, 1987). Tersoff (1986) showed that the particular form of the Fermi surface leads to a point with vanishing tunneling current in the unit cell. Since the STM traces curves of constant density of states, this would lead to large corrugations, limited by the actual tip shape and by the timing of the scanning and the feedback loop. Another theory put forward to explain the giant corrugations by Soler *et al.* (1986) first noted that the forces between the tip and the sample cannot be ignored. They argued that, because of the electronic structure of the graphite, the tip was so close to the surface that it would introduce a deformation of the surface. The different dependencies on the distance of the force and the tunneling current would explain the observed corrugations. Another explanation put forward by Mamin *et al.* (1986) and by Meepagala *et al.* (1991) involved contamination layers between the tip and the sample. These contamination layers in air consist partly of water and of other substances present in the air.

Mizes *et al.* (1987) also noted that the different possible images of the graphite surface can be explained by multiple tip effects. If two or more tips coherently sample the amplitudes of the three standing waves describing the graphite surface, a wide variety of images can result. It is also possible to imagine that entire leaflets of graphite are dragged across the graphite surface. Many tips in registry with the graphite surface sample the current. These images are, in principle, not distinguishable from single-tip images of a large ordered area. This mode of imaging is more likely to occur at higher tunneling currents, where the tip is closer to the graphite surface. Steps, however, can only be imaged with single or a few atom tips and not with a leaflet of graphite. Normally, steps are only seen at low tunneling currents and high-bias voltages, that is, at larger separations between tip and graphite.

In summary, graphite is an excellent material with which to check the operation of a microscope and to calibrate the x- and y-deflections. However, the details of the electronic structure and the intricacies of the surface condition of graphite make SXM images of this surface difficult to understand.

1.2.4.2 LOW-TEMPERATURE EXPERIMENTS

For structural investigations of proteins and biological membranes, it may be desirable to image the samples at low temperatures to fix the structures of the samples. STM have been used to image metal surfaces and superconductor surfaces at low temperatures from the start. Elrod *et al.* (1984a,b) reported the first measurements of a superconducting tunneling gap using an STM. They soon thereafter were able to image the spatial variation of the superconducting tunneling gap (Elrod *et al.*, 1986). The measurement of superconducting tunneling gaps by STM-like instruments has become an essential tool in characterizing high-TC superconductors (see Berthe *et al.* 1988; van Kempen, 1990). Figure 1.23, discussed in the previous section, shows an atomic resolution image of graphite obtained at 6.8 K by Marti (1986).

The design of a low-temperature STM is much more demanding than the design of a room temperature STM (see, e.g., Elrod *et al.*, 1984a,b; Marti, 1986; Fein *et al.* 1987; Lang *et al.*, 1989; Renner *et al.*, 1990a; Giessibl *et al.*, 1991). The microscope has to be shielded to keep the consumption of coolants such as liquid helium low. The encapsulation and the cool-down and warm-up times greatly increase the turnaround time. The mechanical systems have to be designed to work at 4.2 K without freezing. The sample has to be kept warmer than the rest of the microscope to prevent the condensation of the rest gas. The microscope normally is mounted in a tiny vacuum chamber, which is evacuated before the introduction to the cryostat. This pre-evacuation removes adsorbed water layers, which could mask the sample structure.

The control electronics must be modified, since normal electronic components do not work below 200 K. In addition, the temperature gradients will cause thermovoltages at all connections of wires. This thermovoltage is added to the tunneling voltage, which could move the energy position of features like the superconducting band gap.

1.2.4.3 OTHER EXPERIMENTS

On many samples the tip of an STM is close enough to the sample surface to induce considerable forces. References for this are Anders and Heiden (1988), Blackman et al. (1990), and Salmeron et al. (1991). The silicon surface has played an important role in the development of the STM. The outstanding experiments by Binnig et al. (1983) and Hamers et al. (1986) on the topography and spectroscopy of the silicon-(111)-7 $\times$ 7 surface gained acceptance of the STM as a new technique. Other experiments involving silicon or other semiconductor surfaces were published by Tromp et al., (1986) Park et al. (1988), van Loenen et al. (1988), Wang et al. (1988), Hamers and Köhler (1989), Nogami et al. (1989), Tomitori et al. (1990), Hamers and Markert (1990b), and Baski et al. (1991). Morita et al. (1986) measured the voltage dependence of metal surfaces. The effect of tip electronic states and of the tip apex on tunneling spectra was investigated by Park et al. (1987), Snyder et al. (1990), Klitsner et al. (1990), Kuk et al. (1990), Nishikawa et al. (1990), and Pelz (1991). The operation of an STM was monitored by Kuwabara et al. (1989) using reflection electron microscopy.

1.2.5 Related Techniques

There are a variety of techniques related to the STM. I discuss a few selected techniques that operate an STM in an unusual environment or an unusual way. I will begin with the oldest member, the scanning tunneling potentiometer.

1.2.5.1 SCANNING TUNNELING POTENTIOMETRY

On samples with a gradient of the electrical potential along the surface, it is important to know the potential as a function of position. The surface of an integrated chip with its diodes and transistors is a typical example. It is of the utmost importance for the designer of a chip that the potential gradients be not too steep. On exceeding a certain limit, avalanche breakdown can occur and destroy the respective junction. Local variations due to an imperfect processing of the chip might give high local gradients in certain junctions, even though the design would be adequate otherwise.

Figure 1.24 shows the experimental set-up used by Muralt and Pohl (1986). The voltage drop between the two lateral contacts is 5 V. The sample surface consists basically of two ohmic leads connected by a *pn*-junction, where most of the voltage drop occurs. How does one measure this voltage drop? Since the surface is not atomically flat and the junction covers an area of larger than 1000 nm^2, a constant height experiment, similar to those discussed above for graphite, is not possible. We have to measure two quantities simultaneously, the local potential and the surface topography. Muralt and Pohl (1986) solved the problem by measuring an AC voltage to control the z-position of the tip and the DC voltage to measure the local potential.

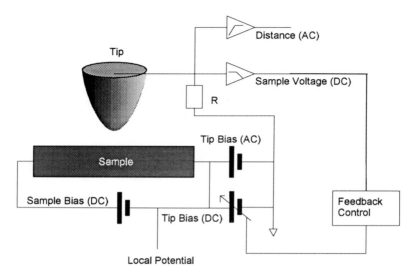

FIGURE 1.24 Experimental set-up for scanning tunneling potentiometry. A voltage is applied along the sample surface by the voltage source (sample bias). The distance between the tip and the sample is measured with a small AC voltage. The DC current is nulled by the feedback loop controlling the DC portion of the tip bias voltage. The voltage at the tip bias (DC) voltage source is equal to the local potential on the sample.

Since the resistance of the local potential is high and since it should not be loaded to give an accurate measurement, they used two feedback loops. The first feedback loop controls the z-position, as in any other STM. An AC voltage is applied between the sample and the tip. The AC current on the tip is high pass filtered and demodulated by a lock-in amplifier. The output of this amplifier is further processed by the normal STM feedback loop. The upper limit of this AC voltage is given by the DC resistance R_t of the tunnel junction and the stray capacitances C_s of the system tip sample, including the connecting wires. For a successful operation the operating frequency should be smaller than the cut-off frequency of the tunnel junction and the connecting wires, given by

$$f_{3dB} = \frac{1}{2\pi R_t C_s}. \tag{1.66}$$

Using typical values of 1 MΩ for the resistance of the tunnel junction R_t and of 5 pF for the stray capacitance C_s, we get a cut-off frequency of 32 kHz. We have to bear in mind that, at this frequency, the phase shift is 45° and very sensitive to the actual tunneling resistance (we can assume that the stray capacitance is a constant in this context). Hence a practical operating frequency might be 3 kHz. The lower limit of the operating frequency is given by the scanning speed and the bandwidth of the second feedback loop, which I will discuss in the next section. If the typical distance between surface features is a and the scanning speed v_s, then the operating frequency should be large compared with v_s/a. To maximize the scanning speed (and to minimize the waiting time for the operator!) the frequency of the AC voltage of the scanning tunneling potentiometer is set near the upper limit.

To measure the local surface potential with no loading, a potentiometer set-up is chosen. In this set-up, a second potential [tip bias (DC)] is varied so as to null the DC current. This potential is controlled by a feedback loop that keeps the DC component of the tunnel current zero. The time constants of the two feedback loops have to be sufficiently different (one being at least two orders of magnitude). Otherwise they would interfere with each other.

Figure 1.25 shows an example of a simultaneous measurement of the local potential and the surface topography. Theoretical considerations suggest that the ultimate resolution of the scanning tunneling potentiometer would be the same as that of an STM. The resolution of the potential map in Figure 1.25 is far worse. A minimal detectable potential change of 1 mV over 0.5 nm would give

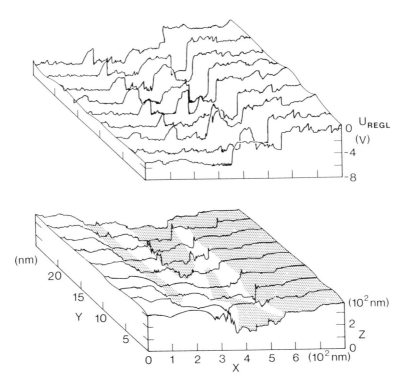

FIGURE 1.25 Scanning tunneling potentiometry. The top part shows the local potential on the sample, a metal insulator metal (MIM) structure. The bottom part of the figure is the corresponding topography. The image size is 800 × 25 nm. [From International Business Machines Corporation, (1986) with permission.]

an electrical field of 20,000 V/cm. Such fields only rarely exist in semiconductor devices. Besides the errors introduced by the dual feedback loop concept at least one other error source exists: thermoelectric potentials. These potentials are in series with the sought potential at the surface and are indistinguishable from it. High current densities in a *pn*-junction might cause considerable heating of the surface, which in turn could increase the error of the measured potential.

Other scanning tunneling potentiometry experiments and related issues are discussed by Le Due et al. (1986), Muralt et al. (1987), Chu and Sorbello (1990), Pelz and Koch (1990), and García-García and Sáenz (1991). Scanning potentiometry and the measurement of photovoltages by force microscopy has been performed by Anders et al. (1990), Weaver and Abraham (1991), and Weaver and Wickramasinghe).

1.2.5.2 SCANNING NOISE MICROSCOPY AND POTENTIOMETRY

I have shown in the previous sections that noise is present in any STM measurement. The origin of this noise is the tunneling current consisting of discrete charges, the electrons, and, additionally, the electronic components of the feedback system. The noise in the electronics can be minimized by a careful selection of the individual components and by the use of appropriate circuits. The noise in the tunneling current, on the other hand, is of far more fundamental origin and not yet fully understood. Möller et al. (1989, 1990a) have built and operated a scanning probe microscope based on noise measurements.

Figure 1.26 shows a measurement of the spectral noise density in the tunneling current between a tungsten tip and a GaAs surface as a function of the

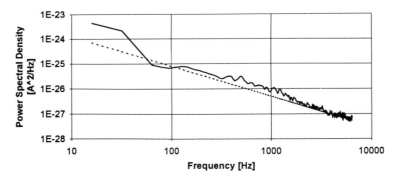

FIGURE 1.26 The spectral noise density in a tunneling junction between the tip and the sample in an STM. The noise curve has been measured on a GaAs sample by B. Koslowski. (Universität Konstanz, unpublished results.) Plotted is the spectral power density as a function of the frequency. The dotted line is a $1/f^{1.2}$ power law. (Used with permission, unpublished results.)

frequency. The spectral noise density and the frequency are plotted on a logarithmic scale. The electronic noise of the I/V converter had been carefully minimized and the bandwidth optimized for this measurement. At high frequencies, above a few kilohertz and outside the range shown in Figure 1.26, the spectral noise density is flat. The constant spectral noise density above a few kilohertz is characteristic of white noise. In all cases except the vanishing average tunnel current, the spectral noise density increases with decreasing frequency below a few kilohertz. The exact nature of this noise is not yet known. It is present in all electronic devices, diodes, and transistors, and it limits the performance of these devices. It has been suggested that the $1/f$ noise (named for its dependence on the frequency f) might be caused in the tunneling current by adsorbate atoms or molecules passing through the tunnel junction. These adsorbate atoms or molecules will diffuse through any tunnel junction whether there is a bias voltage or not. However, at zero bias voltage, there is no electric field to modify the trajectory of the particles under the tip. The high field gradients between the STM tip and the sample will induce dipoles in atoms and molecules diffusing around near it. The induced dipole will then be attracted by the high field region and be present for longer times under the tip than in the case of no field. This mechanism is one possibility to generate a $1/f$ dependence of the low frequency spectral noise density. The $1/f$ noise component distinguishes the tunnel junction from a common resistor.

Möller *et al.* (1991) have found that the zero bias white noise in the STM tunnel junction varies with the spacing of the tip and sample surface and with the tunneling resistance. Using low-noise electronics, they measured the noise density in a narrow frequency band and adjusted the relative spacing between tip and sample such that the total noise in this bandwidth was constant. The noise was also measured in a second narrow frequency band centered around a second frequency. If the noise is white, then the ratio of the amplitudes of these two bands is equal to 1. However, if the $1/f$ noise is present, then the amplitude of the lower frequency band is larger than that of the higher frequency band. With a feedback circuit, a small-bias voltage is applied to the sample to obtain white noise. The bias voltage is then exactly opposite the potential at the sample surface. This method, called *noise potentiometry,* is capable of measuring potentials down to the microvolt level, as exemplified in Figure 1.27. The upper part of Figure 1.27 shows the topography of a silver film consisting of a few grains and measured with noise microscopy. The lower part is a potential image of the same surface.

Scanning noise microscopy and potentiometry might be useful to image the activity of biological molecules. The transmission coefficient for electrons in all tunneling junctions critically depends on the exact arrangement of atoms and bonds present in the junction. A change in the conformation of a molecule hence changes the transmissivity of the tunnel junction. It is foreseen that displacements of a few picometers can show up in the noise spectrum.

FIGURE 1.27 Topography (upper part) and local potential (lower part) on a silver film. The scan size is 500 × 500 nm. The corrugation on the topography is 150 nm. The total potential variation is 56 μV. (Provided by R. Möller, with permission, unpublished results.)

1.2.5.3 BALLISTIC ELECTRON EMISSION MICROSCOPY

On most samples, an STM provides the experimenter with information on the surface properties of the electron states near the Fermi energy. The STM images are little affected by the underlying structure of the sample.

Figure 1.28 shows a sketch of the electron energy in a cross section of tip and sample. The sample is a Schottky diode. As discussed earlier, the theoretical section (see Fig. 1.2), mainly electrons near the Fermi energy of the negatively biased (higher energy) electrode tunnel through the tunnel barrier. Most of the electrons emerge from the barrier with the same energy they had before entering the tunnel region. Only a minor fraction (1 in 1000 or less) lose energy in the process of tunneling. These inelastic tunneling processes are not discussed in this chapter. The main fraction of the electrons entering the positively biased electrode have an excess energy equal to eV_t, the electron charge times the bias voltage. The electrons do not lose their excess energy instantly. They move ballistically through the sample with little interaction with the crystal structure. On the average electrons travel a distance called the *mean free path* in a crystal between two collisions. This mean free path is a characteristic number of the sample depending on its crystal structure, its chemical composition, the abundance of crystal defects, and the temperature. On the average, electrons collide after one mean free path with an ion core of the sample and lose their excess kinetic energy.

Now assume that the sample is very thin, smaller than the mean free path of the electrons, and that there is an energy analyzer at the back. We could then determine the number of electrons as a function of their energy loss in the sample. If we further had a point source of these hot electrons, we could characterize a material's properties to within a few nanometers diameter on the sample surface. Bell and Kaiser (1988) used the STM as their source of hot electrons. They used a Schottky diode with a distance of only a few nanometers between the Schottky barrier and the surface. By biasing the collector electrode negatively, one is able to measure a spectrum of the hot electrons arriving at this electrode. By comparing this current with the injected tunneling current, one gets a measure of the mean free path between the tip and the back electrode. By scanning the tip over the sample surface, this mean free path can be mapped on the surface. The presence of defects like dislocations, grain boundaries or a different species of atoms is likely to reduce the mean free path. This change manifests itself as a reduction of the current measured at the back electrode.

This technique promises to give additional information about technologically important devices on semiconductor surfaces in a non destructive way. Further information on the technique and its extension on holes have been published by Kaiser *et al.* (1989), Hecht *et al.* (1989, 1990), Bell *et al.* (1990), Hasegawa *et al.* (1991), Ludeke *et al.* (1991), and Schowalter and Lee (1991).

1.2.6 Additional References

Surface photovoltages using STM were investigated by Cahill and Hamers (1991), Weaver and Wickramasinghe (1991), and Hamers and Markert (1990a). Wiesendanger *et al.* (1990) observed the magnetic effects on the tunneling current.

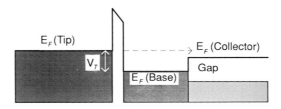

FIGURE 1.28 Ballistic electron emission microscopy. The electrons in the tip at the Fermi energy are injected into the base of a Schottky barrier. Some electrons move ballistically through the base and are collected by the collector. By varying the E_F of the collector with respect to the E_F of the tip transmission, properties of the Schottky barrier can be measured.

1.3 SFM

Scanning force microscopy by Binnig et al. (1986) was an early offspring of STM. The force between a tip and the sample was used to image the surface topography. The force between the tip and the sample, also called the *tracking force*, was lowered by several orders of magnitude compared with the profilometer (Whitehouse, 1974). The contact area between the tip and the sample also was reduced considerably. The force resolution was similar to that achieved by Israelachvili (1985). Soon thereafter atomic resolution in air was demonstrated by Binnig et al. (1987). Marti et al. (1987) demonstrated an SFM capable of atomic resolution under liquids. Kirk et al. (1988) operated an SFM successfully at 4.2 K, the temperature of liquid helium. The SFM measures either the contours of constant forces or force gradients or the variation of forces or force gradients with position, when the height of the sample is not adjusted by a feedback loop. These measurement modes are similar to the ones with the STM, with which contours of constant tunneling current or the variation of the tunneling current with position at fixed sample height are recorded.

The invention of the SFM demonstrated that forces can play an important role in other scanning probe techniques. Anders and Heiden (1988) and Blackman et al. (1990) published experimental evidence of forces in STM.

The type of force interaction between the tip and the sample surface is used to characterize SFM. The highest resolution is achieved when the tip is pressed against the sample surface, the so-called repulsive or contact mode. The forces in this mode basically stem from the Pauli exclusion principle, which prevents the spatial overlap of electrons. As with the STM, the force applied to the sample can be constant, the so-called constant force mode. If the sample z-position is not adjusted to the varying force, we speak of the constant z-mode. However, for weak cantilevers (0.01 N/m spring constant) and a static applied load of 10^{-8} N we get a static deflection of 10^{-6} m, which means that even structures of several nanometers in height will be subject to an almost constant force whether it is controlled or not. Hence for the contact mode with soft cantilevers, the distinction between constant force mode and constant z-mode is rather arbitrary. Additional information on the sample surface can be gained by measuring lateral forces (friction mode) or modulating the force to get dF/dz, which is nothing other than the stiffness of the surfaces. When using attractive forces one normally measures also dF/dz with a modulation technique. In the attractive mode the lateral resolution is at least one order of magnitude worse than that of the contact mode. The attractive mode is also referred to as the *noncontact mode*. Of widespread use also is the magnetic force microscope, another noncontact microscope.

I will introduce the theory of force microscopy and then discuss the techniques to measure small forces. I conclude this section with a discussion of selected experiments with inorganic samples. SFM experiments involving biological samples are discussed in Chapter 7.

1.3.1 Theory of Force Microscopy

There have been few theoretical articles published on the interaction between a tip and a sample surface. The interaction of the tip of an SFM with graphite and other samples has been treated by Abraham and Batra (1989), Gould *et al.* (1989), Tomanek *et al.* (1989), Zhong *et al.* (1991), and Overney *et al.* (1991). Girard (1991) published a theoretical study of SFM of stepped surfaces. Attractive mode SFM has been treated by Hartmann (1991a). Magnetic force microscopy was theoretically investigated by Wadas (1988).

1.3.1.1 THE INTERACTION OF A TIP AND THE SAMPLE

The forces in SFM in the absence of added magnetic or electrostatic potentials are governed by the interaction potentials between atoms. The interaction potential between two atoms usually has the form outlined in Figure 1.29. The interaction is attractive at large distances due to the van der Waals interaction. At short distances the repulsive forces have their origin in the quantum mechanical exclusion principle, which states that no two fermions can be in exactly the same state, that is, have the same spin, angular momentum, z-component of the angular momentum, and location.

The theoretical treatment of the force interaction between a sample and a tip is usually very complicated and requires large parallel computers. As an example, outlined is a simple model for the imaging of graphite by repulsive forces. Graphite is used because its potentials are well known and its structure is sufficiently simple. The theoretical treatment of the imaging of graphite using repulsive forces follows the treatment of Gould *et al.* (1989). The tip is assumed to consist of one to a few atoms. Its atoms are assumed to be rigid with respect to one another. The force between the tip atoms and the surface atoms can be repulsive or attractive, depending on the distance between two atoms. The total

FIGURE 1.29 Qualitative curve for the interaction potential between two atoms. This curve has been calculated using a Lennard–Jones potential.

potential is assumed to be the sum of two-body potentials. I first start with a single atom tip. The interaction is then given by

$$U(\vec{r}, \vec{r}_1, \ldots, \vec{r}_N) = \sum_i V(\vec{r} - \vec{r}_i) + \overline{V}(\vec{r}_1, \ldots, \vec{r}_N), \tag{1.67}$$

where $V(\vec{r} - \vec{r}_i)$ is the interaction potential between the tip atom and the ith atom of the surface, and $\overline{V}(\vec{r}_1, \ldots, \vec{r}_N)$ is the many-body potential in the absence of the tip. The vectors $\vec{r}_i$ denote the position of the ith atom in the sample. $\vec{r}$ is the position of the tip.

Gould et al. (1989) considered only interactions between an atom and its nearest four neighbors. They furthermore assumed that the distortion of the lattice under the influence of the tip was small so that the harmonic approximation was valid:

$$\overline{V}(\vec{r}_1, \ldots, \vec{r}_N) = \frac{1}{2} \sum_{i,j,\mu,\nu} u_\mu^{(i)} D_{\mu\nu}^{(i,j)} u_\nu^{(j)}, \tag{1.68}$$

where $u_\mu^{(i)}$ is the μth Cartesian component of the displacement from equilibrium. The $D_{\mu\nu}^{(i,j)}$ is the matrix of force constants in a solid. The interaction of the tip atom with the graphite is modeled with a Lennard-Jones potential

$$V(|\vec{r}_{tip} - \vec{r}_{atom}|) = V_0 \left[\frac{1}{2}(r_0/r)^{12} - (r_0/r)^6 \right]. \tag{1.69}$$

The interaction potential between atoms does not necessarily have the form of the Lennard-Jones potential. Other potentials could be used as well. Gould et al. (1989) chose this interaction potential because of the computational simplicity. They selected the constants $V_0 = 2.8 \times 10^{-21}$ J and $r_0 = 0.28$ nm to reproduce the measured corrugation and to obtain the correct tip-sample spacing, as calculated with the theories of Solet et al. (1986) and Batra and Çiraci (1988).

To calculate the measured topography, they positioned the tip over various locations in the unit cell and let the surface relax according to

$$\frac{\partial U}{\partial \vec{r}_i} = 0, \quad i = 1, \ldots, N, \tag{1.70}$$

$$-\frac{\partial U}{\partial z} = F_z. \tag{1.71}$$

Here z is the z-component of the tip position $\vec{r}$ and, F_z is the total force between tip and sample. The resulting values of z were then compared with the experiment. They found practically no height difference between the inequivalent A and B sites in graphite (see Fig. 1.22 for a sketch of the graphite crystal structure).

This theory is very simple to calculate. Gould et al. (1989) used a desktop computer to solve the relaxation equations. Theories of this kind do not give the full physical description of the processes between tip and sample, but they will give an idea of what to expect on graphite and, perhaps with some modifications, on other surfaces.

To include many-body interactions, Abraham and Batra (1989) used the effective potential that Stilinger and Weber (1985) developed for silicon. This potential gives a stable silicon lattice, whereas the Lennard-Jones potential used by Gould *et al.* (1989) does not yield a stable silicon lattice. The potential of Stilinger and Weber (1985) includes double and triple interactions. Abraham and Batra (1989) modified this potential to describe accurately the interaction of the carbon atoms in the graphite layer. For the weak bonding between layers, they still use the Lennard-Jones potential.

Abraham and Batra (1989) obtained a total corrugation of 10 pm or less, consistent with the calculation of the total charge density (Batra *et al.*, 1987). This corrugation is, however, smaller than the measured corrugations. Like Gould *et al.* (1989), Abraham and Batra did not find a significant height difference between the A and B sites.

1.3.1.2 FORCES

Experiments using a single atom at a time to interact with the sample surface are routinely done in atom-scattering experiments. As an example, I briefly outline the helium-scattering experiments. A well-collimated beam of helium atoms with a narrow distribution of velocities around a center velocity v_0 is aimed at the surface. The individual atoms hit the surface and in the case of helium bounce back elastically from it. The angle between the line of incidence and the line of emergence and the orientation of the plane defined by these two lines tells the experimenter about the structure and periodicity of the sample surface. These experiments a analogous to X-ray diffraction experiments. The scattering amplitudes and angles have to be transformed back to reveal the real surface structure. However, because it is a scattering experiment, little information can be gained regarding the defects of the periodic arrangement of the atoms. The area of interaction is on the order of 1 mm^2.

To get a local probe that need not be scattered off a sample surface, the probe atom is held on the apex of a tip. Figure 1.30 shows a once widely used example of such a tip, the cleavage planes of diamond. Diamond under stress is most likely to cleave along the (111) planes of its crystal structure.

For real SFM tips the assumption of a single interacting atom is not justified. Attractive forces such as van der Waals forces reach out for several nanometers. The attractive forces are compensated by the repulsion of the electrons when one atom tries to penetrate another. The decay length of the interaction and its magnitude depend critically on the type of atoms and on the crystal lattice in which they are bound. The shorter the decay length, the smaller the number of atoms that contribute a sizable amount to the total force. The decay length of

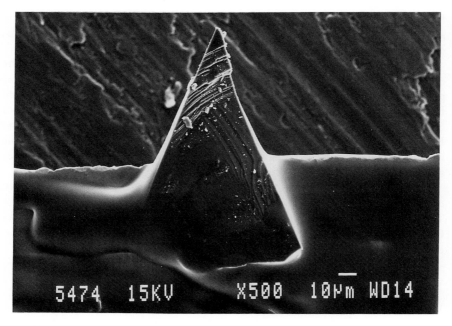

FIGURE 1.30 Cleaved diamond used as an SFM tip. (Courtesy of Charles Bracker, Purdue University.)

the potential, on the other hand, is directly related to the type of force. Repulsive forces between atoms at small distances are governed by an exponential law (like the tunneling current in the STM), by an inverse power law with large exponents, or by even more complicated forms. Hence, the highest resolution images are obtained using the repulsive forces between atoms in contact. The high inverse power exponent or even exponential decay of this distance dependence guarantees that the other atoms besides the apex atom do not significantly interact with the sample surface. Attractive van der Waals interactions, on the other hand, are reaching far out into space. Hence a larger number of tip atoms takes part in this interaction and hence the resolution cannot be as good. The same is true for magnetic potentials and for the electrostatic interaction between charged bodies.

A crude estimation of the forces between atoms can be obtained in the following way: assume that two atoms with mass m are bound in molecule. The potential at the equilibrium distance can be approximated by a harmonic potential or, equivalently, by a spring constant. The frequency of the vibration f of the atom around its equilibrium point is then a measure of the spring constant k:

$$k = (\omega)^2 \frac{m}{2}, \tag{1.72}$$

where we have to use the reduced atomic mass. The vibration frequency can be obtained from optical vibration spectra or from the vibration quanta $\hbar\omega$:

$$k = \left(\frac{(\hbar\omega)}{\hbar}\right)^2 \frac{m}{2}. \qquad (1.73)$$

As a model system we take the hydrogen molecule H_2. The mass of the hydrogen atom is $m = 1.673 \times 10^{-27}$ kg and its vibration quantum is $\hbar\omega = 8.75 \times 10^{-20}$ J. Hence the equivalent spring constant is $k = 560$ N/m. Typical forces for small deflections (1% of the bond length) from the equilibrium position are $\approx 5 \times 10^{-10}$ N. The force calculated in this way is an order of magnitude estimation of the forces between two atoms.

An atom in a crystal lattice on the surface is more rigidly attached since it is bound to more than one other atom. Hence the effective spring constant for small deflections is larger. The limiting force is reached when the bond length changes by 10% or more, which indicates that the forces used to image surfaces must be on the order of 10^{-8} N or less. The sustainable force before damage is dependent on the type of surfaces. Layered materials such as mica and graphite are more resistant to damage than are soft materials such as biological samples. Experiments have shown that, on selected inorganic surfaces such as mica, one can apply up to 10^{-7} N. On the other hand, some biological samples are destroyed by forces on the order of 10^{-9} N.

1.3.1.3 ADDITIONAL REFERENCES

Force issues in STM and SFM are discussed by Çiraci *et al.* (1990a). Girard *et al.* (1989), Maghezzi *et al.* (1991), Hartmann (1991b), Tomanek and Zhong (1991), and Chen (1991a) discuss attractive forces. The manifestation of zero-point fluctuations in SFM is treated by Hartmann (1990c). Goodman and García (1991) discuss the role of attractive and repulsive forces. Girard *et al.* (1990) published an article on the resolution of SFM. Experimental data on the contact forces in SFM are reported by Albrecht *et al.* (1989). Banerjea *et al.* (1990) discuss universal aspects of adhesion and SFM.

1.3.2 How to Measure Small Forces

The key to the successful operation of an SFM is the measurement of the interaction forces between a small probing structure, the tip, and the sample surface. The probing structure would ideally consist of only one atom, which is brought in the vicinity of the sample surface. The following discussion deals with how to approximate the single atom probing structure and how to detect the minute forces acting on this structure. Of the potential methods to detect minute distance changes, electron tunneling, interferometry, and the optical lever method are discussed. Capacitance measurements are not discussed.

1.3.2.1 CANTILEVER SPRINGS

The interaction forces between the SFM tip and the sample surface must be smaller than about 10^{-7} N for bulk materials and preferably well below 10^{-8} N for organic macromolecules. To obtain a measurable deflection larger than the inevitable thermal drifts and noise, the cantilever deflection for static measurements should be at least 10 nm. Hence the spring constants should be less than 10 N/m for bulk materials and less than 1 N/m for organic macromolecules. Experience shows that cantilevers with spring constants of about 0.01 N/m work best.

Building vibrations usually have frequencies in the range of 10–100 Hz. These vibrations are coupled to the cantilever. To get an estimate of the magnitude, we note that the resonance frequency of a structure in terms of its spring constant k and a lumped effective mass m_{eff} is

$$f_{res} = \frac{1}{2\pi} \sqrt{\frac{k}{m_{eff}}}. \tag{1.74}$$

Inserting 100 Hz for the resonance frequency and a spring constant of 0.1 N/m, we obtain an upper limit of the lumped effective mass m_{eff} of 0.25 mg. The quality factor of this resonance in air is typically between 10 and 100. To get a reasonable suppression of the excitation of cantilever oscillations, the cantilever's resonance frequency has to be at least a factor of 10 higher than the highest of the building vibration frequencies. This means that m_{eff} has under any circumstances to be no larger than 0.25 mg/100 = 2.5 µg. It would be preferable to limit the mass to 0.1 µg. A tungsten wire with 20 µm diameter must be shorter than 1.6 mm to have a mass of less then 0.1 µg. This lumped mass m_{eff}, however, is smaller than the real mass m by a factor that depends on the geometry of the cantilever. A good rule of thumb is that the effective mass m_{eff} is one-third of the real mass. Gluing tips or mirrors on cantilevers adds their masses to the effective mass. Since these additional gadgets are attached to the free end of the cantilever, they do not benefit from the factor one-third in calculating the effective mass.

Figure 1.31a gives approximate values for the spring constant and the resonance frequency of selected configurations. Of particular importance for understanding the performance of an SFM are configurations 5 (the free cantilever), 6 (the cantilever in repulsive contact with the sample), and 2 (lateral force measurement). Comparing 5 and 6, we see that a cantilever in repulsive contact with a sample has a resonance frequency 4.8 times that of the free cantilever. Figure 1.31b shows the moments of inertia for selected cross sections.

Micromachined cantilevers are commercially available and are used almost exclusively. The manufacturing process of cantilevers has been described by Pitsch *et al.* (1989), Akamine *et al.* (1990b), Grütter *et al.* (1990a), and Wolter *et al.* (1991).

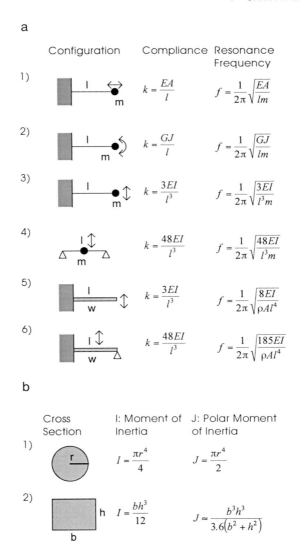

FIGURE 1.31 Selected configurations of levers and their resonance frequencies. a: Compliance of levers and their resonance frequencies. k is the compliance, f the resonance frequency, l the length of the lever, A the cross section, E the Young's modulus, G the shear modulus, ρ the density of the lever material, and m a concentrated mass at the end of the lever. The following configurations are shown: 1: a cantilevered mass-less beam in the compression mode with a weight concentrated at the end; 2: the torsion of a cantilevered mass-less beam with a concentrated weight; 3: the deflection of a cantilevered mass-less beam with a concentrated weight at the end; 4: a mass-less beam supported at the ends with a concentrated weight in the center; 5: a massive cantilevered beam; and 6: a massive cantilevered beam supported at the end. b: The moment of inertia I and the polar moment of inertia J are shown. r is the radius of a circular cross section and b and h the width and height of a rectangular cross section, respectively. (After Anderson, 1981, with permission.)

1.3.2.2 DETECTING THE SPRING DEFLECTION BY TUNNELING

The first SFM technique, published by Binnig et al., (1986), employed tunneling to detect the bending of the force-sensing cantilever. Figure 1.32 shows a sketch of the arrangement. The authors sandwiched a cantilever between the tip of an STM and the sample. The sensitivity of the tunneling detector in the SFM is the best of all possible detectors. On clean surfaces the change in tunneling current might approach 1 order of magnitude for every 0.1 nm change in deflection. In air there are a few critical points that might degrade the performance of the SFM.

An ideal deflection detector for an SFM should not have any sensitivity to the local surface structure of the force-sensing cantilever. The tunneling current between the back of the cantilever and the sensing electrode, however, is confined to a narrow region whose width is mainly determined by the local curvatures of the sensing electrode and the cantilever. If the sensing electrode is a sharp tip, as used for STM experiments, the SFM can be very susceptible to the lateral bending of the cantilever.

A solution of the resolution problem is to use as smooth a cantilever back as possible and to increase the area of the tunneling current. However, this aggravates a second problem that might occur in a tunneling SFM. Adsorbate layers are present on both the sensing electrode and the back of the cantilever. Typical distances between the two electrodes in a tunneling junction are on the order of 1 nm. If two monolayers are present both on the sample and on the cantilever tip, the distance between the sensing electrode and the back of the cantilever might be too small to allow a tunneling current to pass. Therefore the sensing electrode has to be pressed against the back of the cantilever, which will yield

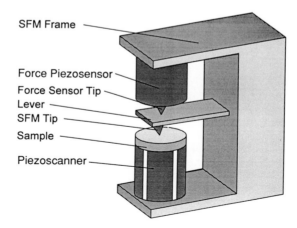

FIGURE 1.32 The principle of the SFM as proposed by Binnig et al. (1986). A lever with a spring constant of ≈ 1 N/m is pressed into a sample mounted on a piezotube. The deflection of the lever is measured by a tunnel junction. The tunnel gap is adjusted by a force piezosensor. Alternatively, the SFM tip on the lever can be operated in a noncontact mode via attractive forces.

due to its low spring constant. It is possible that no tunneling current can be established. Furthermore, the filled gap between the cantilever and the sensing electrode rigidizes the cantilever. The effective spring constant of the cantilever is then a function of the stiffness of the adsorbate layers. Whereas tunneling as a deflection detector has its deficiencies in air, it might become the method of choice in vacuum SFM. It is advantageous to use microfabricated cantilevers in an SFM. These cantilevers have a Q of about 100 in air and a $Q > 10^4$ in vacuum. Any sudden change in the surface topography starts a damped oscillation of the cantilever. The amplitude of the oscillation will decay to $1/e$ after Q oscillations. If the resonance frequency were 10 kHz, the time constant would be 1 second. This time constant would impose such a small scanning speed that the whole microscope would become impractical. Here the additional, highly non-linear force between the sensing electrode and the back of the cantilever could help to damp the oscillations of the cantilever.

1.3.2.3 HOMODYNE AND HETERODYNE INTERFEROMETRY

Soon after the first article on the SFM by Binnig *et al.* (1986), McClelland (1987) published details of an SFM obtained with interferometry. The sensitivity of the interferometer depends on the wavelength of the light employed in the apparatus. Figure 1.33 shows the principle of an interferometric design. The light incident from the left is focused by a lens on the cantilever. The reflected light is collimated by the same lens and interferes with the light reflected at the flat. To separate the reflected light from the incident light, a $\lambda/4$-plate converts the linearly polarized incident light into circular polarized light. The reflected light is again made linearly polarized by the $\lambda/4$-plate, but with a polarization ortho-

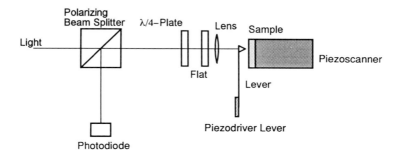

FIGURE 1.33 Principle of an interferometer SFM. The light of the laser light source on the left is polarized by the polarizing beam splitter and focused on the back of the force measuring cantilever on the right. The cantilever oscillates at or near its resonance frequency. The light passes twice through a $\lambda/4$-plate. The returning light is therefore polarized orthogonally to the incident light and reflected to the photodiode. The light reflected from the flat serves as the reference of the interferometer. The interference pattern is modulated at the oscillation frequency of the cantilever.

gonal to that of the incident light. The polarizing beam splitter then deflects the reflected light to the photodiode.

To improve the signal to noise ratio of the interferometer, the lever is driven by a piezo near its resonance frequency. The amplitude Δz of the lever is

$$\Delta z = \Delta z_0 \frac{1}{\sqrt{(\Omega^2 - \Omega_0^2) + \Omega^2/Q^2}}, \qquad (1.75)$$

where Δz_0 is the constant drive amplitude, Ω_0 the resonance frequency of the lever, Q the quality of the resonance, and Ω the drive frequency. The resonance frequency of the lever is given by the effective potential

$$\Omega_0 = \sqrt{(k + \frac{\partial^2}{\partial z^2} U)/m_{\text{eff}}}, \qquad (1.76)$$

where k is the spring constant of the free lever, U the interaction potential between the tip and the sample, and m_{eff} the effective mass of the cantilever. Equation 1.76 shows that an attractive potential decreases the resonance frequency Ω_0. The change in the resonance frequency Ω_0 in turn results in a change in the lever amplitude Δz (see Eq. 1.75).

The movement of the cantilever changes the path difference in the interferometer. The light reflected from the lever with the amplitude $A_{l,0}$ and the reference light with the amplitude $A_{r,0}$ interfere with the detector. The detected intensity $I(t) = [A_l(t) + A_r(t)]^2$ consists of two constant terms and a fluctuating term

$$\overline{2A_l(t)A_r(t)} = A_{l,0}A_{r,0} \sin\left[\omega t + \frac{4\pi\delta}{\lambda} + \frac{4\pi\Delta z}{\lambda} \sin(\Omega t)\right]. \qquad (1.77)$$

Here ω is the frequency of the light, and Δz is the instantaneous amplitude of the lever, given according to Eqs. 1.75 and 1.76 by the driving frequency Ω, the spring constant k, and the interaction potential U. The time average of Eq. 1.77 then becomes

$$\overline{2A_l(t)A_r(t)} \propto \cos\left[\frac{4\pi\delta}{\lambda} + \frac{4\pi\Delta z}{\lambda} \sin(\Omega t)\right]$$

$$\approx \cos\left(\frac{4\pi\delta}{\lambda}\right) - \sin\left[\frac{4\pi\Delta z}{\lambda} \sin(\Omega t)\right]$$

$$\approx \cos\left(\frac{4\pi\delta}{\lambda}\right) - \frac{4\pi\Delta z}{\lambda} \sin(\Omega t). \qquad (1.78)$$

Here all small quantities have been omitted, and functions with small arguments have been linearized. The amplitude of the lever oscillation Δz can be recovered with a lock-in technique. However, Eq. 1.78 shows that the measured amplitude is also a function of the path difference δ in the interferometer. Hence this path difference δ must be very stable. The best sensitivity is obtained when $\sin(4\pi\delta/\lambda) \approx 0$.

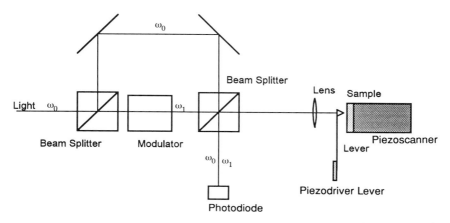

FIGURE 1.34 Heterodyne interferometer SFM. Light with a frequency of ω_0 is split into a reference path (upper light path) and a measurement path. The measurement light is shifted in frequency to ω_1 by a modulator. This light is reflected by the cantilever oscillating at or near its resonance frequency and interferes with the reference beam at ω_0 on the photodiode.

This influence is not present in the heterodyne detection scheme shown in Figure 1.34. Light incident from the left with a frequency ω is split in a reference path (upper path in Fig. 1.34) and a measurement path. Light in the measurement path is shifted in frequency to $\omega_1 = \omega + \Delta\omega$ and focused on the cantilever. The cantilever oscillates at the frequency Ω, as in the homodyne detection scheme. The reflected light $A_l(t)$ is collimated by the same lens and interferes on the photodiode with the reference light $A_r(t)$. The fluctuating term of the intensity is given by

$$2A_l(t)A_r(t) = A_{l,0}A_{r,0} \sin\left[(\omega + \Delta\omega)t + \frac{4\pi\delta}{\lambda} + \frac{4\Delta z}{\lambda}\sin(\Omega t)\right]\sin(\omega t), \quad (1.79)$$

where the variables are defined as in Eq. 1.77. Setting the path difference sin $(4\pi\delta/\lambda) \approx 0$ and taking the time average, omitting small quantities and linearizing functions with small arguments, we get

$$\overline{2A_l(t)A_r(t)} \propto \cos\left[\Delta\omega t + \frac{4\pi\delta}{\lambda} + \frac{4\pi\Delta z}{\lambda}\sin(\Omega t)\right]$$

$$= \cos\left(\Delta\omega t + \frac{4\pi\delta}{\lambda}\right)\cos\left[\frac{4\pi\Delta z}{\lambda}\sin(\Omega t)\right]$$

$$- \sin\left(\Delta\omega t + \frac{4\pi\delta}{\lambda}\right)\sin\left[\frac{4\pi\Delta z}{\lambda}\sin(\Omega t)\right]$$

$$\approx \cos\left(\Delta\omega t + \frac{4\pi\delta}{\lambda}\right)\left[1 - \frac{8\pi^2\Delta z^2}{\lambda^2}\sin(\Omega t)\right]$$

$$-\frac{4\pi\Delta z}{\lambda}\sin\left(\Delta\omega t+\frac{4\pi\delta}{\lambda}\right)\sin(\Omega t)$$

$$=\cos\left(\Delta\omega t+\frac{4\pi\delta}{\lambda}\right)-\frac{8\pi^2\Delta z^2}{\lambda^2}\cos\left(\Delta\omega t+\frac{4\pi\delta}{\lambda}\right)\sin(\Omega t)$$

$$-\frac{4\pi\Delta z}{\lambda}\sin\left(\Delta\omega t+\frac{4\pi\delta}{\lambda}\right)\sin(\Omega t)$$

$$=\cos\left(\Delta\omega t+\frac{4\pi\delta}{\lambda}\right)-\frac{4\pi^2\Delta z^2}{\lambda^2}\cos\left(\Delta\omega t+\frac{4\pi\delta}{\lambda}\right)$$

$$+\frac{4\pi^2\Delta z^2}{\lambda^2}\cos\left(\Delta\omega t+\frac{4\pi\delta}{\lambda}\right)\cos(2\Omega t)$$

$$-\frac{4\pi\Delta z}{\lambda}\sin\left(\Delta\omega t+\frac{4\pi\delta}{\lambda}\right)\sin(\Omega t)$$

$$=\cos\left(\Delta\omega t+\frac{4\pi\delta}{\lambda}\right)\left(1-\frac{4\pi^2\Delta z^2}{\lambda^2}\right)$$

$$+\frac{2\pi^2\Delta z^2}{\lambda^2}\left\{\cos\left[(\Delta\omega+2\Omega)t+\frac{4\pi\delta}{\lambda}\right]\right.$$

$$+\cos\left[(\Delta\omega-2\Omega)t+\frac{4\pi\delta}{\lambda}\right]\Big\}$$

$$+\frac{2\pi\Delta z}{\lambda}\left\{\cos\left[(\Delta\omega+\Omega)t+\frac{4\pi\delta}{\lambda}\right]\right.$$

$$+\cos\left[(\Delta\omega-\Omega)t+\frac{4\pi\delta}{\lambda}\right]\Big\}. \qquad (1.80)$$

Electronically multiplying the components oscillating at $\Delta\omega$ and at $\Delta\omega+\Omega$ and rejecting any product except the one oscillating at Ω, we obtain

$$A=\left(1-\frac{4\pi^2\Delta z^2}{\lambda^2}\right)2\Delta z\lambda\cos\left[(\Delta\omega+\Omega)t+\frac{4\pi\delta}{\lambda}\right]\cos\left(\Delta\omega t+\frac{4\pi\delta}{\lambda}\right)$$

$$=\left(1-\frac{4\pi^2\Delta z^2}{\lambda^2}\right)\Delta z\lambda\left\{\cos\left[(2\Delta\omega+\Omega)t+8\pi\delta\lambda\right]+\cos(\Omega t)\right\}$$

$$\approx\frac{\pi\Delta z}{\lambda}\cos(\Omega t). \qquad (1.81)$$

Equation 1.81 shows that the amplitude Δz of the cantilever motion can be recovered with a lock-in technique. Unlike in the homodyne detection scheme, the recovered signal is independent from the path difference δ of the interferometer. Furthermore, a lock-in amplifier with the reference set at $\sin(\Delta\omega t)$ can measure the path difference δ independent of the cantilever oscillation. If necessary, a feedback circuit can keep $\delta = 0$.

1.3.2.4 FIBEROPTIC INTERFEROMETER

The first solution proposed by Rugar *et al.* (1989) is to use fiberoptic interferometer. Its principle is sketched in Figure 1.35. The light of a laser is fed into an optical fiber. Laser diodes with integrated fiber pigtails are convenient light sources. The light is split in a fiberoptic beam splitter into two fibers. One fiber is terminated by index matching grease to avoid any reflections back into the fiber. The end of the other fiber is brought close to the cantilever in the SFM. The emerging light is partially reflected back into the fiber by the cantilever. Most of the light, however, is lost. This is not a complicated problem, since only 4% of the light is reflected at the end of the fiber, at the glass–air interface. The two reflected light waves interfere with each other. The product is guided back into the fiber coupler and again split into two parts. One-half is analyzed by the photodiode. The other half is fed back into the laser. Communications-grade laser diodes are sufficiently resistant against feedback to be operated in this environment. They have, however, a bad coherence length, which in this case does not matter since the optical path difference is no larger than 5 μm. Again, the end of the fiber has to be positioned on a piezodrive to set the distance between the fiber and the cantilever to $\lambda(n + 1/4)$.

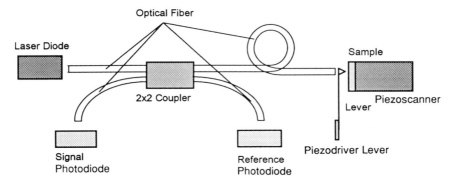

FIGURE 1.35 Principle of the fiberoptic interferometer SFM: The light of the laser diode is coupled into a fiber, and 50% of the light is transmitted to the back of the cantilever. The other half serves as an intensity reference (reference photodiode). Light reflected from the cantilever and light reflected from the end of the fiber interfere. The fiber coupler again splits the light traveling back evenly between the signal photodiode and the laser diode. The small path difference of ≈ 10 μm accounts for the excellent stability of the microscope.

1.3.2.5 NOMARSKI INTERFEROMETER

A third solution to minimize the optical path difference is based on the Nomarski principle and was presented by Schönenberger and Alvarado (1989). Figure 1.36 depicts a sketch of their microscope. The light of a laser is focused on the cantilever by a lens. A birefringent crystal between the cantilever and the lens with its optical axis 45° off the polarization direction of the light splits the light beam into two paths, offset by a distance given by the length of the birefringent crystal. Birefringent crystals have varying indexes of refraction. In calcite, one crystal axis has a lower index than the other two. This means that certain light rays will propagate at a different speed through the crystal than the others. By choosing a correct polarization, one can select the ordinary ray or the extraordinary ray, or one can get any distribution of the intensity among the two rays. A detailed description of birefringence can be found in textbooks (Shen, 1984). A calcite crystal deflects the extraordinary ray at an angle of 6° within the crystal. By choosing a suitable length of the calcite crystal, any separation can be selected.

The focus of one light ray is positioned near the free end of the cantilever, while the other is placed close to the clamped end. Both arms of the interferometer pass through the same space, except for the distance between the calcite crystal and the lever. The closer the calcite crystal is placed to the lever, the less influence disturbances (such as air currents) have.

1.3.2.6 DETECTING SPRING DEFLECTION BY THE OPTICAL LEVER METHOD

Still another spring detection system is the optical lever method as published by Meyer and Amer (1988) and Alexander *et al.* (1989). This method, depicted in

FIGURE 1.36 Principle of the Nomarski SFM. The circular polarized input beam is deflected to the left by a nonpolarizing beam splitter. The light is focused onto a cantilever. The calcite crystal between the lens and the cantilever splits the circular polarized light into two spatially separated beams with orthogonal polarizations. The two light beams reflected from the lever are superimposed by the calcite crystal and collected by the lens. The resulting beam is again circularly polarized. A Wollaston prism produces two interfering beams with a $\pi/2$ phase shift between them. The minimal path difference accounts for the excellent stability of this microscope. (Schönenberger and Alvarado, 1990b.)

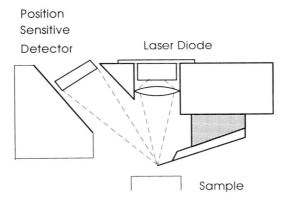

FIGURE 1.37 The principle of the lever SFM. Light from a laser diode is focused onto the back of a cantilever. The reflected light is deflected when the cantilever bends under an applied force. The deflection angle is measured by a position-sensitive detector.

Figure 1.37 employs the same technique as light beam deflection galvanometers used to have and still have. A fairly well-collimated light beam is reflected off a mirror and projected to a receiving target. Any change in the angular position of the mirror will change the location, where the light ray hits the target. Galvanometers use optical path lengths of several meters and scales projected to the target wall as a read-out help.

For the SFM using the optical lever method, a photodiode segmented into two closely spaced devices is used. Initially, the light ray is set to hit the photodiodes in the middle of the two subdiodes. Any deflection of the cantilever will cause an imbalance in the number of photons reaching the two halves. Hence the electrical currents in the photodiodes will be unbalanced too. The difference signal is further amplified and is the input signal to the feedback loop. Unlike the interferometric SFM, in which often a modulation technique is necessary to get a sufficient signal-to-noise ratio, most SFM employing the optical lever method are operated in a static mode. The domain of optical lever SFM are the measurements in the repulsive regime. It is the simplest method to construct an optical readout, and it can be confined in volumes smaller than 5 cm on the side. To evaluate the proper design parameters, let us calculate the sensitivity of the microscope. Figure 1.38 shows a cross section of the reflected light beam. For simplicity, we assume that the light beam is of uniform intensity with its cross section increasing proportional to the square of the distance between the cantilever and the quadrant detector. The movement of the center of the light beam is then given by

$$\Delta x = \Delta z \frac{d}{l}. \tag{1.82}$$

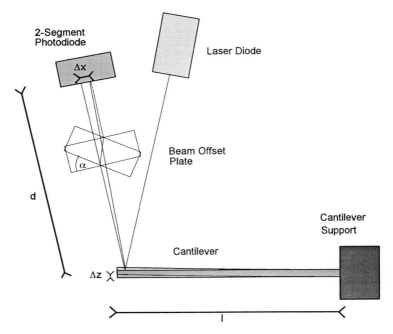

FIGURE 1.38 Sensitivity and calibration of a lever SFM. The deflection Δz of the cantilever translates into a displacement Δx on the two-segment photodiode. The photodiode is located at a distance d from the cantilever of length l. The deflection Δx can be calibrated by tilting the beam off-set plate by an angle α.

The photo current generated in a photodiode is proportional to the number of incoming photons hitting it. If the light beam contains a total number of N_0 photons, then the change in difference current becomes

$$\Delta(I_R - I_L) = \Delta I = \text{const } \Delta x \, dN_0. \tag{1.83}$$

Combining Eqs. 1.82 and 1.83, one sees that the difference current ΔI is independent of the separation of the quadrant detector and the cantilever. This relation is true if the light spot is smaller than the quadrant detector. If it is greater, the difference current ΔI becomes smaller with increasing distance. The light beam in reality has a Gaussian intensity profile. For small movements Δx (compared with the diameter of the light spot at the quadrant detector), Eq. 1.83 still holds. Larger movements Δx, however, will introduce a nonlinear response. If the SFM is operated in a constant force mode, only small movements Δx of the light spot will occur. The feedback loop will cancel out all other movements.

The scanning of a sample with an SFM can twist the microfabricated cantilevers because of lateral forces (see Mate *et al.*, 1987; Marti *et al.* 1990; Meyer and Amer, 1990) and affect the images (den Boef, 1991). When the tip is subjected to lateral forces, it will twist the lever, and the light beam reflected from the end of the lever will be deflected perpendicular to the ordinary deflection direction. For many investigations this influence of lateral forces is unwanted. The design of the triangular cantilevers stems from the desire to minimize the torsion effects. However, lateral forces open up a new dimension in force measurements. They allow, for instance, the distinction of two materials because of the different friction coefficient or the determination of adhesion energies. To measure lateral forces the original optical lever SFM has to be modified: Figure 1.39 shows a sketch of the instrument. The only modification compared with Figure 1.37 is the use of a quadrant detector photodiode instead of a two-segment photodiode and the necessary readout electronics. The electronics calculates the following signals:

$$U_{Force} = \alpha[(I_{Upper\ Left} + I_{Upper\ Right}) - (I_{Lower\ Left} + I_{Lower\ Right})]$$
$$U_{Friction} = \beta[I_{Upper\ Left} + I_{Lower\ Left}) - (I_{Upper\ Right} + I_{Lower\ Right})]. \quad (1.84)$$

The calculation of the lateral force as a function of the deflection angle does not have a simple solution for cross sections other than circles. Baumeister and Marks (1967) give an approximate formula for the angle of twist for rectangular beams:

$$\Theta = \frac{M_t l}{\beta G b^3 h}, \quad (1.85)$$

where $M_t = Fa$ is the external twisting moment due to friction, l is the length of the beam, b and h are the sides of the cross section, G is the shear modulus, and β

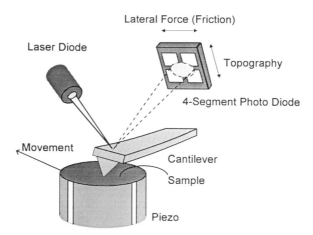

FIGURE 1.39 Scanning force and friction microscope (SFFM). The lateral forces exerted on the tip by the moving sample causes a torsion of the lever. The light reflected from the lever is deflected orthogonally to the deflection caused by normal forces.

is a constant determined by the value of h/b. For the equation to hold, h has to be larger than b.

Inserting the values for a typical microfabricated lever with integrated tips

$$b = 6 \times 10^{-7} \text{ m}$$
$$h = 10^{-5} \text{ m}$$
$$l = 10^{-4} \text{ m}$$
$$a = 3.3 \times 10^{-6} \text{ m}$$
$$G = 5 \times 10^{10} \text{ Pa}$$
$$\beta = 0.333$$

into Eq. 1.85, we obtain the relation

$$F = 1.1 \times 10^{-4} \, \Theta. \tag{1.86}$$

Typical lateral forces are on the order of 10^{-10} N.

1.3.2.7 ADDITIONAL REFERENCES

Details of SFM based on a tunneling deflection sensor were published by Bryant et al. (1988) and Probst et al. (1991). Sarid et al. (1989, 1990) describe the detection of the cantilever deflection by using the lever as the mirror in a laser diode cavity. An improved fiberoptic interferometer has been implemented by Mulhern et al. (1991). A capacitive detection scheme was incorporated into the SFM of Göddenhenrich et al. (1990a) and of Neubauer et al. (1990). A rocking balance was used by Miller et al. (1991) to measure small forces. Dürig et al. (1988) describe the detection of forces in an STM. Joyce and Houston (1991) built a force sensor with force feedback. Schmidt et al. (1990) describe force sensors based on quartz fibers. Umeda et al. (1991) used photothermal excitation to build an AC mode SFM.

1.3.3 The Force Microscope

This section will first focus on the design of a SFM and give some details on how to build such a device. Then, how to adjust an SFM will be discussed.

1.3.3.1 SPECIAL DESIGN CONSIDERATIONS

An SFM is very similar in design to an STM. A small, rigid design is even more important for an SFM than for an STM. The construction of the force-sensing unit imposes some changes to the arrangement of the piezoscanner and the sample location. The resonance frequency of the piezoscanner is decreased by loading it with an additional mass. In the case of the STM, this mass consists of the tip holder and the tip itself. It is smaller than the mass of the sample in most cases. In previous sections it was mentioned that the detection systems for the deflection of the force sensor can be quite bulky. Except for the tunneling detector, all are too large to be mounted on a piezotube. Even the tunneling detector requires additional distance adjustment, which would lower the piezoscanner's resonance frequency too much. As a consequence, most SFM have the sample mounted on the piezoscanner. The force-sensing unit is stationary, with the sample being scanned past the immobile tip. The structure of the force-sensing unit has to be as rigid as possible to minimize errors due to thermal drift. The tunneling detection method and the fiberoptic detection method are especially prone to this error. The problem is most severe in the tunneling deflection detector, since a rigid electrode, the sensing electrode, is about 1 nm from the cantilever.

To illustrate the problem of thermal drift, we calculate the requirements of the temperature stability for a microscope that works with repulsive forces and that does not employ heterodyne detection. If we assume a cantilever spring with a 1 N/m spring constant and if we set the force to 10^{-8} N, then the static deflection of the cantilever is 10 nm. The typical size of a tunneling force detector is 1 cm from the tunnel junction to the common attachment plane. A design with well-compensated thermal expansion coefficients will have a remaining thermal expansion coefficient of 10^{-6} m/km. This means that keeping the force within 10% requires a thermal stability of the microscope of 0.1 K. In a less well-compensated design, the allowable temperature fluctuations might be as low as 0.01 K.

If the temperature stability of the set-up is not sufficient, one can either use larger static deflections, which means larger forces, or a softer cantilever spring, which means degraded frequency response. For measurements with the smallest possible forces, a careful design of the force sensor with respect to thermal drift is a prerequisite.

The interferometric force sensors have the same drift problems. The classic Michelson or Mach-Zehnder interferometers are the worst, since their relevant distances for differential thermal expansions may be more than 10 cm long. The fiberoptic interferometer is comparable with the tunneling detector in its thermal performance, since the distances needed to position the end of the fiber are on the order of 1 cm. Much better is the Nomarski detector for the cantilever deflection. This detector is only sensitive to a thermally induced rotation of the cantilever spring. A crude estimate gives relevant distances for the thermal expansion of a few 10 μm. This increases the allowable temperature variations to more than 1 K.

Equally well suited is the optical lever method. This method is, to first order, only sensitive to the tilt of the reflecting mirror. For small angles between the incident and the reflected light beam, the change in distance between the plane defined by the quadrant detector and the light source is negligible. Any distance change between the light source and the quadrant detector affects directly the output signal. However, the deflection of the cantilever is amplified by a factor of up to 1000 due to the geometrical amplification. Hence the optical lever method is, to first order, insensitive to thermal drift.

1.3.3.2 HOW TO ADJUST A FORCE MICROSCOPE

Compared with an STM, an SFM needs some additional adjustments. I describe some procedures to facilitate the adjustment.

The adjustment of an SFM can be divided into the adjustment of the force sensor and the approach of the force sensor to the sample surface. The latter is similar to the approach of an STM tip to the sample and is discussed in the section on STM, earlier.

When mounting a new cantilever, first one has to position it correctly with respect to the deflection measurement sensor. The size of the cantilevers varies from 0.1 to 2 mm. The best sensitivity is obtained when the deflection sensor points to the end of the cantilever. This adjustment is best done under a microscope.

Tunneling Sensor

The sensing electrode in the tunneling sensor has to be brought to about 1 nm to the back of the cantilever. This requires an approach mechanism for the sensing electrode similar to that of an STM. One can use, for instance, the action of a differential spring system, as published by Marti *et al.* (1988b), or lever reduction systems. The surfaces of the sensing electrode and the cantilever should be as clean as possible to avoid a stiffening of the cantilever by the sandwiched adsorbates between the lever and the sensing electrodes. To keep the force accurate, the force sensor has to be constructed to minimize thermal drift. Alternatively, one could periodically readjust the force (Marti *et al.*, 1988a).

Interferometric Force Sensor

It is important to have the light reflected from the end of the cantilever. This adjustment can be done by using a microscope or by analyzing the diffraction patterns of the cantilever. *The experimenter should make sure that no visible or invisible laser radiation can reach his eye.* A good protective measure is to use a small TV camera mounted on the microscope. The TV monitor can be placed at any convenient location. In addition to the location one has also to adjust the phase of the reflected light to get the best sensitivity. This usually means moving the cantilever by fractions of 1 μm toward or away from the fiber or the interferometer flat or to shift the phase of one polarization in the Nomarski interferometer.

Optical Lever Sensor

To adjust the optical lever sensor, one has first to focus the laser diode light in the plane of the cantilever. Next, the laser diode is moved to bring the focal point on the end of the cantilever. Both the focus adjustment and the positioning of the focal point have to be done with a microscope. *Again, one must avoid eye damage.* It is best to use a small CCD-TV camera to transmit the image from the microscope to a monitor. The adjustment can be observed on the monitor. The last adjustment is the positioning of the quadrant detector diode. The correct position is found when the currents from all four segments are equal. This position guarantees also that the amplitude fluctuations do not influence the measurement (to first order) in the constant force mode. The force sensor should be treated with the utmost care after its adjustment. Strong accelerations should be avoided.

1.3.4 Selected Experiments

I now provide a few examples of experiments by SFM. I focus on the imaging of inorganic surfaces, since all items related to the imaging of biological samples are treated in Chapter 7 of this volume.

1.3.4.1 ATOMIC RESOLUTION IMAGING

The first surface to be imaged with atomic resolution by an SFM was the graphite surface (see Binnig *et al.*, 1987; Albrecht and Quate, 1987; Marti *et al.* 1987). The graphite surface is of great importance to SXM as a reference sample and a substrate with flat terraces of several hundred nanometer length. Calculations show that, for the SFM, the surface consists of hexagons of carbon atoms, each 0.146 nm apart. The centers of the rings are separated by 0.246 nm. There is a great variation in the appearance of the unit cell. The interpretation of the unit cell structure and the corrugation, however, is very complex. Abraham and Batra (1989) and Gould *et al.* (1989) explained the puzzling structures by multiple tips. These multiple tips create a superposition of several locations within the unit cell. They can produce an almost unlimited variation in the graphite unit cell appearance.

Another layered material important to the biologist is mica. Like graphite, it has long, flat terraces suitable for sample deposition. Since mica is an insulator, its binding properties with biological macromolecules are different. By comparing the appearance, the adhesion, and other properties of one sort of biological macromolecule bound to different substrates, one can learn about the molecule itself and its binding properties. Figure 1.40a shows an example of measurement by SFM.

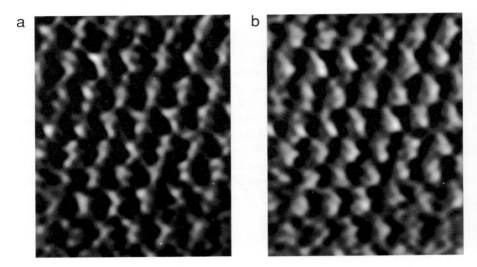

FIGURE 1.40 a: Force image of mica. b: Friction image of mica. The size of the image is 1.8 × 2.4 nm. The corrugation is 0.2 nm for the topography.

1.3.4.2 LATERAL FORCES AND FRICTION

The imaging of surfaces by the SFM in the repulsive mode is based on the dragging of a fine tip across the sample. There are lateral forces between the tip and the sample. At the beginning of a scan the tip sticks to the surface. Later it will move, but the lever will always feel a force parallel to the surface in addition to the normal force. If the lever is a simple wire, it will bend parallel to the surface. Mate et al. (1987) and Erlandsson et al. (1988) measured the sideways deflection by interferometry. They detected a variation in the lateral force with the periodicity of the graphite surface. Details of other lateral force microscopes have been published by Kaneko (1988).

Figure 1.40b shows a lateral force picture of the mica surface. These data were measured by an optical lever SFM, detecting the torsion of a microfabricated cantilever under the influence of friction. The mica periodicity is resolved with a lateral force modulation of 10^{-9} N. The band at the left side with no visible structure is due to the change in the scanning direction. The lateral force changes its sign; hence the width of the band is twice as large as the steady-state friction.

Systems studied by friction microscopy or combined force and friction microscopy include graphite by Mate et al. (1987), mica by Erlandson et al. (1988), and perfluorodiethylether by DeKoven and Meyers (1991). First theoretical works on friction in an SFM have been published by Tomanek et al. (1991).

1.3.4.3 SURFACE PROFILES

For many industrial applications one needs to know the exact profile of a surface. Several methods to obtain this information are possible. A scanning electron microscope will give the desired information, provided the structures under investigation are not too shallow and the sample is conducting. A second method is the use of a profilometer, a widely used, proven instrument in industrial applications. Its lateral resolution is limited to a few hundred nanometers, which may be insufficient. A third method is the STM. It has the desired sensitivity, but requires conducting surfaces. The most versatile tool is the SFM. Figure 1.41a shows a top view of an optical grating. Figure 1.41b shows the lateral force image measured concurrently. The profile of the topography and the lateral forces shown in Figure 1.41c.

1.3.4.4 ELECTROSTATIC FORCES

The SFM is sensitive not only to the interaction between uncharged bodies but also to any other force. Martin *et al.* (1988) demonstrated the use of an SFM to measure capacitances and local potentials. The ability to measure potentials parallels the scanning tunneling potentiometer of Muralt and Pohl (1986), but its theoretical resolution is inferior. The degraded resolution, however, is of no concern, since the smallest electronic devices available are still larger than the resolution limit.

The capacitance C between the tip and the sample is measured by applying a voltage between the tip and the sample. The stored charge on the capacitor plates causes an attractive force between the two electrodes, which is dependent on the dielectric constants of the materials within. Martin *et al.* (1988) measure the force by a heterodyne detection technique. The voltage they apply is a combination of a DC bias voltage and an AC modulation voltage. The modulation frequency is set to a few kilohertz. In addition, the cantilever is mechanically excited near its resonance frequency by a small piezoactuator. The two modulation frequencies are demodulated in two lock-in amplifiers. The signal near the resonance frequency of the cantilever is a measure of the distance between the cantilever and the surface. The amplitude of the voltage-modulation-induced signal, however, is determined by the capacitance between the tip and the sample and by the charges present in between. A change in the force gradient will change the amplitude of the oscillation of the cantilever.

The force gradient $f(z)$ is given by

$$f(z) = \frac{1}{2} V^2 \frac{\partial C}{\partial z}, \qquad (1.87)$$

where V is the applied voltage, C the capacitance, and z the separation between the tip and the sample.

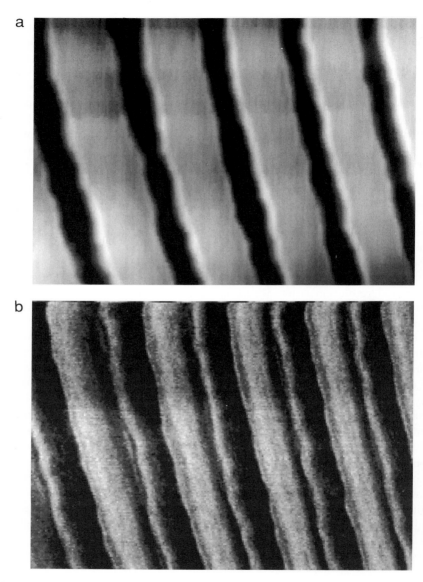

FIGURE 1.41 Image of an optical grating. **a**: The topograph of the image. **b**: A lateral force image. The size of the image is 1600 × 1200 nm. The corrugation of the topograph is ≈ 4 nm. **c**: A cross section through the topograph and the lateral force image.

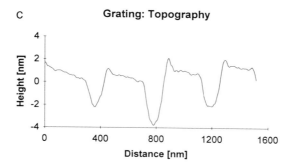

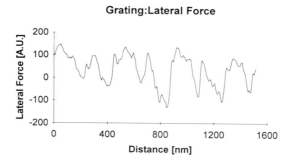

FIGURE 1.41 *(Continued)*

Martin *et al.* (1988) calculate that the minimum detectable capacitance is $C_m\text{in} = 8 \times 10^{-22}$ F in a 1 Hz bandwidth, measured with a silicon cantilever with a spring constant of 2.5 N/m having a resonance frequency of 33 MHz and a Q of 200.

An example of such a measurement is shown in Figure 1.42. Single charges were deposited triboelectrically by shooting small isolating spheres on the surface (Terris *et al.*, 1989). Further experiments involving charges or electrostatic forces were published by Terris *et al.* (1990), Saurenbach and Terris (1990), Schönenberger and Alvarado (1990a,c), and Hao *et al.* (1991). Anders *et al.* (1990), Weaver and Abraham (1991), and Weaver and Wickramasinghe (1991) used the SFM for spatially resolved potentiometric and photovoltage measurements.

1.3.4.5 Magnetic Forces

The first application of the SFM to forces other than the interatomic forces was the magnetic force microscope of Sáenz *et al.* (1987). The magnetic domain structure of the sample was measured by scanning it past a tip made of a ferromagnetic material. The interaction between the tip and the sample can be

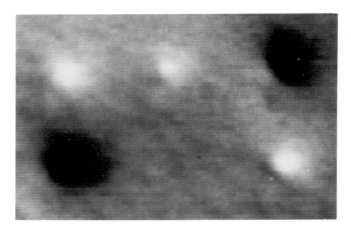

FIGURE 1.42 Positive and negative charge writing. Images of single electric charges by force microscopy. (From Terris *et al.*, 1989, with permission from the American Physical Society.)

selected such that the magnetic moments of the tip and the sample dominate the force. Sáenz et al. (1987) have shown that the force between the tip and the sample can be modeled as the magnetic dipole force. Both the tip and the sample are assumed to consist of microscopic magnetic domains with random orientation. The domains farther away from the nearest point between tip and sample tend to cancel their respective forces. Hence Sáenz et al. (1987) were calculating the force $F(z)$ between the tip and the sample using the force $f_z(\vec{r})$ between two dipoles

$$F(z) = \int_{tip} d\vec{r}_1 \int_{sample} d\vec{r}_2 f_z(\vec{r}_1 - \vec{r}_2) \qquad (1.88)$$

$$f_z(\vec{r}) = \left(\frac{\mu_0}{4\pi}\right) \frac{\partial}{\partial_z} \left(\frac{3(\vec{r}\vec{\mu}_1)(\vec{r}\vec{\mu}_2)}{r^5} - \frac{(\vec{\mu}_1\vec{\mu}_2)}{r^3}\right). \qquad (1.89)$$

μ_0 is the permeability of the vacuum; and $\vec{\mu}_1$ and $\vec{\mu}_2$ are the tip and sample magnetic dipoles, respectively. The two dipoles are assumed to be separated by the distance $\vec{r} = \vec{r}_1 - \vec{r}_2$.

The first consequence of Eq. 1.88 is that a sample with a uniform magnetization will not exert a force on the tip. Only domain walls will be visible in the magnetic force microscope.

By assuming a spherical tip, Sáenz et al (1987) showed that the force between the tip and the sample as a function of the distance x between the tip and the domain wall was given by

$$F(x) = F_0 \frac{ax_r + b}{x_r^2 + 1}, \tag{1.90}$$

where $x_r = x/L$ is the relative separation between the tip and the domain wall; $L \gg z$ is the radius of curvature of the tip; and $F_0 = 8/\pi\mu_0\mu_1\mu_2 L^2 \cdot a$ and b are two constants depending on the spin states of the sample and the tip. The magnitude of the force is calculated to be on the order of 10^{-11} to 10^{-10} N. This is 1 to 2 orders of magnitude smaller than the forces used in repulsive imaging.

A detailed account of the problems of magnetic force imaging can be found in the article by Schönenberger and Alvarado (1990b). Figure 1.43 shows an example of a magnetic force image.

1.3.4.6 Additional References

Details of combined SFM and STM have been published by Sugawara *et al.* (1990a,b). Review articles on SFM were published by Rugar and Hansma (1990) and by Sarid and Elings (1991). High-resolution imaging results in the contact regime have been published by Soethout *et al.* (1988), Yang *et al.* (1988), Chalmers *et al.* (1989), Bryant *et al.* (1988, 1990), and Barrett and Quate (1990, 1991a). Meyer *et al.* (1990) demonstrated a different response to charge density by SFM and STM. The force measurement and imaging using an AC-SFM is discussed by Ducker *et al.* (1990). Horie and Miyazaki (1990) measured the surface topography of graphite by van der Waals forces. Gleyzes *et al.* (1991)

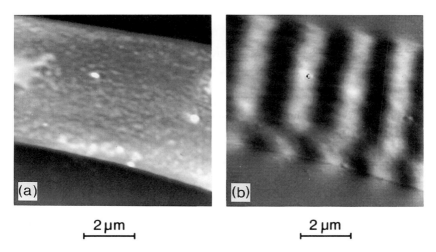

FIGURE 1.43 Magnetic force microscopy of tracks written on a magnetic storage medium. Part a shows the topography of a data track, whereas part b shows the simultaneously recorded magnetic bit pattern. The image is from Schönenberger and Alvarado, 1990b, with permission.

investigate the bistable behavior of a vibrating tip near the surface. Combined STM and SFM have been built by Bryant et al. (1988) and by Sugawara et al. (1990a,b). Burnham and Colton (1989) investigated the micromechanical properties of surfaces. Marti et al. (1988b) and Burnham et al. (1990) have investigated the surfaces (and their properties) of monolayer films. Interpretation issues of SFM are discussed by Burnham et al. (1991). The inspection of SFM tips by SFM is discussed by Hellemans et al. (1991). Rugar and Grütter (1991) designed a mechanical parametric amplifier to squeeze thermomechanical noise.

Review articles on magnetic force microscopy have been published by Rugar et al. (1990). Details of magnetic force microscopy have been published by Martin and Wickramasinghe (1987), Abraham et al. (1988a), Hartmann and Heiden (1988), Wadas (1988, 1989), Wadas and Grütter (1989), Mamin et al. (1989), Mansuripur (1989), Hobbs et al. (1989), Hartmann (1989a–c, 1990a–c), Grütter et al. (1989, 1990b,c, Moreland and Rice (1990), Wadas and Güntherodt (1990b), Abraham and McDonald (1990), and Hartmann et al. (1990). Block walls have been investigated by Göddenhenrich et al. (1988, 1990b). Göddenhenrich et al. (1990c) and Gibson and Schultz (1991) published works on the calibration of the magnetic field sensitivity. Probes for magnetic force microscopy have been discussed by Grütter et al. (1990a), den Boef (1990), and Sueoka et al. (1991). Scheinfein et al. (1989), Schönenberger et al. (1990), and Wadas and Güntherodt (1990a) investigated the effects of topography on magnetic force images. An alternative approach to the measurement of magnetic forces involving tunneling microscopy is outlined by Wandass et al. (1989). Manassen et al. (1989) observed the precession of individual spins by STM. Wiesendanger et al. (1990) observed magnetic effects on the tunneling current.

1.4 SNOM

Scanning tunneling microscopy STM and SFM are near field techniques. This means that the tip and the sample are spaced at a distance of the same order or less as the characteristic length of the interaction. In the case of the STM and the repulsive SFM, the characteristic length is the decay length of the electron density in the vacuum. The lateral resolution of all techniques operating in the near field is determined by the size of the probe and not by the characteristic wavelength of the interaction. Electromagnetic waves do have a near field also. Radio transmitters quite often work in the near field region. This region is characterized by $\lambda \gg l$, where l is a characteristic dimension of the antenna for radio waves or the tip size for optical near field microscopes, and λ is the wavelength.

Light waves are electromagnetic waves; hence there is also a near field here. The wavelength of visible light is between 400 and 700 nm. Therefore the emitter or the detector of an optical near field microscope must be smaller than this dimension. I treat two implementations of near field optical microscopy.

These techniques might become valuable for very localized optical spectroscopy of bulk material.

The scanning near field optical microscope (SNOM; also called the near field optical scanning microscope, the scanning tunneling optical microscope, and the scanning optical tunneling microscope) was the first SXM to be designed after the construction of the STM (see Pohl *et al.*, 1985; Dürig et al. 1986a,b). It is an extension of concepts first described by Ash and Nichols (1972) for small apertures for microwave radiation to light.

Dürig *et al.* (1986a,b) used a tiny aperture with a diameter much smaller than the wavelength λ of the light. The aperture, constructed on top of a sharpened glass rod, is illuminated with the light. A small amount of the light leaks out of the aperture with an exponentially decaying amplitude. This hemispherical evanescent wave is transformed back to a propagating wave by the sample surface, which is located a few nanometers from the aperture. Since the illuminated area of the sample surface is only slightly larger than the aperture, the resolution of the microscope is determined by the aperture and was demonstrated by Pohl *et al.* (1985) to be on the order of $\lambda/20$. The amount of light transmitted through the sample and detected by a photomultiplier is used as a control signal in a feedback loop that adjusts the distance between the emitter and the sample such that the transmitted intensity remains constant.

The aperture is scanned across the surface in the same way as in other scanning probe microscopes. The position of the glass rod then tracks the surface topography. Figure 1.44 shows one of the early results with such a microscope. Dürig *et al.* (1986b) have imaged a 30 nm tantalum film with 100 nm holes. In Figure 1.44 the near field optical micrographs are compared with standard SEM images. The aperture was formed by coating a sharpened glass rod by metal, except at the very apex of the tip. Since the sensitivity to distance changes was not sufficient, the authors used electric tunneling between the metal shield on the glass rod and the metallic sample to control the distance between aperture and sample.

Since this pioneering work many other schemes to form apertures have been put forward (e.g., Fischer, 1985; Morrison, 1988; Pool, 1988; Betzig *et al.*, 1988a,b; Pohl *et al.*, 1988a,b; Fischer *et al.*, 1988, 1989; Fischer and Pohl, 1989; Guerra, 1989, 1990; Moyer *et al.*, 1990; Girard and Spajer, 1990; Labani and Girard, 1990; Paesler *et al.*, 1990; Reddick *et al.*, 1990; Ferrell *et al.*, 1991; Qian and Wessels, 1991a,b; Denk and Pohl, 1991. One particularly appealing way is to use evanescent waves (see Fig. 1.45, left). Light is reflected at the glass–air interface from the glass side. Just outside the glass, an exponentially dampened optical wave running parallel to the glass is present. Any probe made of glass will convert part of the evanescent wave to a propagating wave again. The amount of light coupled into the glass decreases exponentially with increasing separation between the probe and the sample. Hence the phenomena is also called *optical tunneling*. If the size of the glass probe is considerably smaller than the wave-

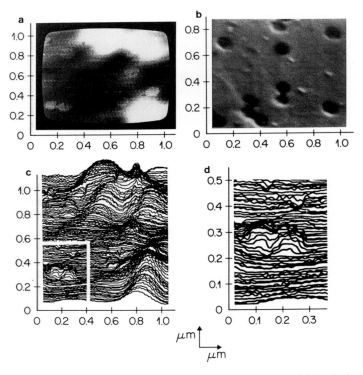

FIGURE 1.44 Near field optical microscopy results. a: High-resolution near field optical micrograph of a 30 nm Ta film with 100 nm holes. b: A scanning electron micrograph of the same sample as in a. c: A line scan record of part a. d: A magnification of the inset in c. (a–c are from Pohl et al., 1985, with permission; d is from Dürig et al., 1986a, with permission.)

length λ of the light, it will only collect light from its vicinity, which is, depending on the probe size, as small as $\lambda/20$. Hence this microscope exhibits super-resolution. Very sharp glass tips are conveniently manufactured from optical fibers. The fiber is first drawn apart over an oxygen–hydrogen torch. The already sharp tips are further etched in HF.

As an example, an image of a grating is shown in Fig. 1.46a. Surface features considerably smaller than the optical wavelength λ and spaced by similar distances can create a wide variety of standing wave patterns when illuminated by light in total internal reflection. Unlike far field standing waves, which have nodes spaced by $\lambda/2$, these patterns can have nodes separated by smaller distances. Since these patterns exist only in the near field they are not observable from a distance, for instance, by an optical microscope. The standing wave patterns are difficult to interpret. Only a careful analysis can separate true topographical features from the structure of the light in the near field. Easier to interpret are images measured by illuminating the sample surface through the glass probe and where the same glass probe collects the reflected light (see Fig. 1.45, right). This reflected light SNOM (Fischer et al., 1988) gives unambiguous data on the surface topography. However, experimental work has to confirm the theoretical prediction that the resolution stays the same. Figure 2.46b shows as an example of a topograph of an optical grating in reflection mode. The light source in this case was a laser diode emitting at 670 nm.

This kind of microscope could also be used to obtain spectra from very small areas. Assuming a microscope with a resolution of $\lambda/20$ and a wavelength $\lambda = 500$ nm, spectra from areas as small as 625 nm^2 could be measured. Dürig et al. (1986b) envisaged the use of this kind of microscopy to study fingerprints of biological functional units. First spectra obtained by near field optical methods have been published by Moyer et al. (1990) and by Qian and Wessels (1991a,b). A very versatile instrument could be built by combining the evanescent wave type microscope with the reflection type microscope. Liecerman et al. (1990) propose a light source smaller than the wavelength of the emitted light.

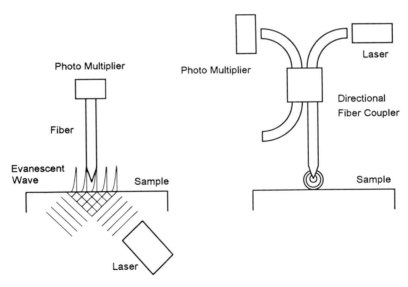

FIGURE 1.45 Principle of scanning near field optical microscopy. **Left:** The optical tunneling configuration. Light from the dense medium is totally reflected at the boundary. The evanescent waves are collected by a small aperture (here a sharpened fiber tip) and converted to a current by the photomultiplier. **Right:** A reflected light type near-field optical microscope. The same fiber is used to illuminate the sample and to collect the light. Two two light paths are joined in a directional fiber coupler.

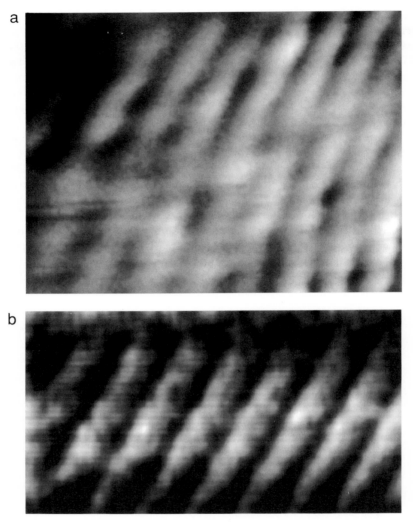

FIGURE 1.46 Scanning tunneling optical microscopy and scanning near field optical microscopy. The images are optical images of the grating shown in Figure 2.41. The optical microscope in both cases resolves height differences as small as 4 nm. **a**: Transmission type scanning tunneling optical microscope image. **b**: A reflection type scanning near field optical microscope image.

1.5 ELECTROCHEMISTRY WITH SXM

Electrochemistry is a very important tool in producing daily goods. Much is known about the experimental procedures to obtain the desired results. The comparison of the experimental results with theoretical predictions is often hampered by the inhomogeneous nature of the electrode surface. The liquid environment poses problems to the application of classic surface science tools to analyze the structure of the interface to the electrolyte. A number of groups have started to use STM and SFM to characterize the surfaces of electrodes (e.g., Siegenthaler and Christoph, 1990). Recently, SFM have been applied to this problem (Manne et al., 1991). Besides its technical applications, electrochemistry might become an essential tool with which to enhance bonding of organic molecules to surfaces or to prepare these surfaces (see Lindsay and Barris, 1987; see also Chapter 5, this volume). SFM in electrochemical environments offers a simpler arrangement of potentials than STM. SFM tips are less intrusive to electrochemistry. On the other hand, the tunneling tip is an additional electrode, which might be used not only to image surface topography, and to do local spectroscopy but also to trigger chemical reactions by electrical fields.

A comprehensive introduction to the electrochemical STM can be found in the Chapter 5, this volume. Usually the sample is used as the working electrode in the electrochemical set-up. The ion current flows between a counterelectrode and the sample. A currentless reference electrode is used to define a reference potential. All potentials are referred to this reference electrode. Besides the desired ion currents flowing to or from the sample there exist also ion currents to the tip. However, the potentials can be chosen such that the ionic current to the tip becomes negligible. The system tip sample behaves like a capacitor, provided the right combination of electrolyte and tip materials are used (Chapter 5, this volume; Siegenthaler and Christoph, 1990).

With such an electrochemical STM, cyclic voltammograms are used to deposit a controlled amount of material to the surface (down to monolayers or less) or to remove other, equally well-controlled quantities of material. Covering a sample surface with a monolayer and then removing the monolayer reveals, for instance, the locations where the chemical reaction has not been reversible.

The tip of an STM in an electrochemical environment is an additional disturbance in the distribution of potentials. The customary set-up of electrochemistry does not include such an electrode. Using an SFM as an imaging tool, one can dispose of nonconducting tips. These tips do not attract the flow of ions in the solution and hence are a much smaller disturbance. However, one cannot use the SFM to probe local potentials or local densities of states. A first report of the possibilities of SFM in an electrochemical environment is given by Manne *et al.* (1990).

Additional information on electrochemical experiments can be obtained from the following authors: Sonnenfeld and Handsma (1986), Arvia (1987), Morita *et al.* (1987, 1989), Lustenberger *et al.* (1988), Green *et al.* (1988), Robinson (1988b), Wiechers *et al.* (1988), Twomey *et al.* (1988), Hüsser *et al.* (1989), Itaya and Tomita (1989), Tomita *et al.* (1990), Vazquez (1989), Siegenthaler and Christoph (1990), Cataldi *et al.* (1990), Thundat *et al.* (1990a,b), Salvarezza *et al.* (1990), Gao *et al.* (1991), Robinson *et al.* (1991c), Carrejo *et al.* (1991), Sakamaki *et al.* (1991), and Binggeli *et al.* (1991).

1.6 LOCAL EXPERIMENTS

The SXM provides information on the topography of such various quantities as the surfaces of constant local density of states, of constant local charge density, of constant local temperatures, and so forth, on a very local scale. Furthermore, by applying an external stimulus they can provide the experimenter with information on the response of a small area. The probe itself can then be used to raster scan the surface, giving the experimenter a view of the sample surface with the same resolution as that obtainable in the local experiment.

The techniques outlined in the succeeding chapters in this section are known as *local experiments*. As an example of these techniques, I discuss the mixing of light with different frequencies in the tunneling junction of an STM.

1.6.1 Light Mixing

Arnold *et al.* (1987) demonstrated the use of an STM for generating difference frequencies between laser light from two lasers oscillating at different frequencies. Difference frequencies up to 5 THz were observed by Krieger *et al.* (1989, 1990) and by Völcker *et al.* (1991a,b). The same authors observed rectified currents from the light of a single CO_2 laser.

The generation of difference frequencies has been observed before in metal–insulator–metal point contact diodes (see Daniel *et al.*, 1981, and references therein). The STM is conceptually simpler than the metal–insulator–metal diodes, because the oxide layer, normally not as well characterized, is replaced by vacuum or air and because the characteristics of the junction are easily controlled over the normal STM electronics.

I discuss here the experiments of Krieger *et al.* (1990). The tunneling junction of the STM is brought into the focal spot of one or two CO_2 lasers. The CO_2 lasers can be individually tuned to give difference frequencies from a few megaHertz to 5 THz. The authors concluded that, at difference frequencies of about 10 MHz, the thermal expansion of the tip and sample due to the time-varying intensity of the two light waves is responsible for the modulation of the tunneling current (see also Amer *et al.*, 1986; Grafström *et al.*, 1991). The thermal time constant of the end of the tip can be smaller than a microsecond, thus explaining the fast response.

At difference frequencies of over a gigahertz a different effect takes place. The long, narrow tip of an STM acts like an antenna for the laser light focused on its end. The coupling of the light to the tip is greatest if the tip, the electrical field vector, and the direction of propagation of the light all are in the same plane. Hence the time-varying field of the light wave induces an electrical current in the tip. This alternating current will flow through the tunneling junction. Since the tunneling junction is nonlinear, we can expand the response of the tunnel junction in a Taylor series

$$I = I(V_b + V_I) = I(V_b) + \frac{\partial I}{\partial V} V_i + \frac{1}{2} \frac{\partial^2 I}{\partial V^2} V_i^2 + \cdots, \qquad (1.91)$$

where V_b is the static bias voltage of the tunneling junction, and V_i is the induced momentary voltage from the laser light. All of the partial derivatives are to be evaluated at $V = V_b$. The substitution $V_i = v_i \cos(\omega_i t)$; ($i = 1, 2$) into Eq. 1.91 gives, to second order,

$$\begin{aligned} I = {} & I(V_b) + \frac{\partial I}{\partial V} v_1 \cos(\omega_1 t) v_2 \cos(\omega_2 t) \\ & + \frac{1}{2} \frac{\partial^2 I}{\partial V^2}[v_1 \cos(\omega_1 t) v_2 \cos(\omega_2 t)]^2 + \cdots \\ \approx {} & I(V_b) + \frac{1}{4} \frac{\partial^2 I}{\partial V^2}(v_1 + v_2) \\ & + \cos(\omega_1 t) \frac{\partial I}{\partial V} v_1 \\ & + \cos(\omega_2 t) \frac{\partial I}{\partial V} v_2 \\ & + \cos(2\omega_1 t) \frac{1}{4} \frac{\partial^2 I}{\partial V^2} v_1^2 \\ & + \cos(2\omega_2 t) \frac{1}{4} \frac{\partial^2 I}{\partial V^2} v_2^2 \\ & + \cos(\omega_1 t + \omega_2 t) \frac{1}{2} \frac{\partial^2 I}{\partial V^2} v_1 v_2 \\ & + \cos(\omega_1 t - \omega_2 t) \frac{1}{2} \frac{\partial^2 I}{\partial V^2} v_1 v_2. \end{aligned} \quad (1.92)$$

Equation 1.92 shows the coefficients of the rectified current and the sum, difference, and doubled frequencies generated in the STM junction. Note that the sum, difference, and doubled frequencies are all proportional to $\partial^2 I/\partial V^2$. The question is whether this static current–voltage characteristic is valid at the infrared frequencies emitted by the CO_2 lasers. Krieger et al. (1990) conclude that the static current–voltage characteristics could explain their results. This implies that the tunneling time for electrons has to be $\leq 10^{-13}$ seconds. The agreement of the experimental data of Krieger et al. (1990) and this theory, developed therein, is excellent. Figure 1.47a shows an STM measurement of a graphite surface, and Figure 1.47b shows the variation of the mixing product over the same spot. Both images resolve the graphite structure.

Krieger et al. (1990) suggested the use of this technique for a new kind of spectroscopy. Adsorbed molecules on the sample surface would be illuminated by light tuned to one of their resonances. The light field in the vicinity of the molecule should then be influenced by the constant absorption and emission of photons. The locally modified electric field amplitude of the light wave would then change the amplitude of the difference signal or, alternatively, the rectified

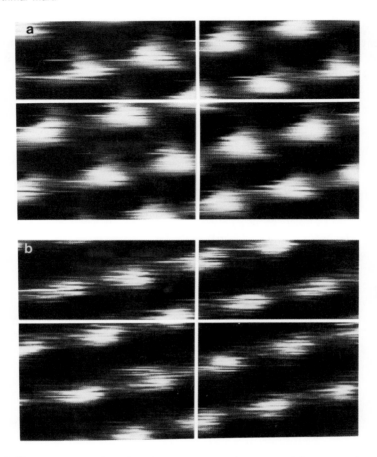

FIGURE 1.47 a: STM image of graphite. b: The mixing product image of the same surface spot as shown in a. (Courtesy of R. Völcker.)

current. It is also possible to tune a difference frequency to an absorption band of an adsorbate molecule. Again, the rectified current or the generated difference frequency should be modified in their amplitude. By selecting appropriate light frequencies, one can envisage an instrument for identifying specific groups on an organic molecule.

1.6.2 Additional References

Rectification of the tunneling current by a single molecule has been observed by Aviram *et al.* (1988). Cutler *et al.* (1987), Nguyen *et al.* (1989), and Huang *et al.* (1990a) propose the use of frequency-dependent mixing characteristics of tunneling junctions to measure the transit time of the electrons in the tunneling junction. Photon emission from an STM has been investigated by Coombs *et al.* (1988), Gimzewski *et al.* (1988), Smol'yaninov *et al.* (1990a,b), Kuk *et al.* (1990), Ushioda (1990), Takeuchi *et al.* (1991), and Berndt *et al.* (1991). Johansson *et al.* (1990) published an article on the theory of light emission by an STM.

1.7 NEW DEVELOPMENTS

Scanning probe microscopy is a developing field. Some techniques, such as scanning noise microscopy, are in their early stages (see Möller *et al.*, 1989); others, such as STM, are routinely applied. STM are on the market, with instruments optimized for various tasks, such as surface science under UHV conditions, imaging large areas under ambient conditions, and imaging electrochemical processes and biological samples. These instruments are computer controlled, with ever more sophisticated control electronics. Features to enhance and simplify the operation are constantly added. One example discussed is the miniaturization of STM.

1.7.1 Miniaturization of the STM

One way to improve the performance of an STM is the miniaturization of the instrument. The smaller the size, the better the frequency response because of the increased resonance frequency of the STM structure. A good signal-to-noise ratio for the tunneling current requires the placement of the preamplifiers close to the tip. Sizes smaller than 1 cm are difficult to obtain with conventional machining. The STM as an instrument is, in certain aspects, similar to pressure gauges microfabricated with silicon technology. Akamine *et al.* (1900a,b) and Albrecht *et al.* (1990) designed an STM on a chip (Fig. 1.48a). The piezoelectric structure for the scanning was made of ZnO films. The piezostructure with a length of 1 mm had a total scanning range of 15 μm, with a driving voltage of 30 V. The instrument is capable of resolving graphite on an atomic scale (see Fig. 1.48b). Clearly, this first design can be reduced further to submillimeter sizes. The design of Akamine *et al.* (1990a) does not yet incorporate integrated preamplifiers. If required, these preamplifiers could easily be incorporated into the silicon chip and would lessen many of the noise pick-up problems. The authors envisage the use of many of these STM in parallel to increase the recording density of storage devices (see also Thomson, 1988).

1.7.2 Other SXM

There are quite a few other SXM techniques that are not discussed in this chapter. For those interested in these techniques, a few references are given.

Scanning probe microscopes have been extensively used to modify surfaces (see Packard *et al.*, 1988; Schneir *et al.* 1988; Dijkkamp *et al.*, 1989; Hüsser *et al.*, 1989; Staufer *et al.*, 1989; Marrian and Colton, 1990; van Loenen *et al.*, 1990;

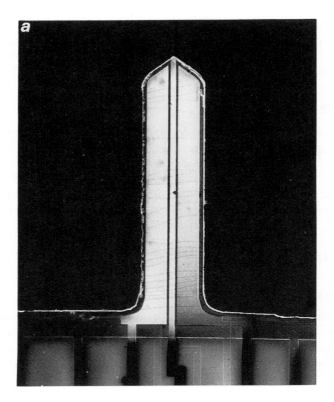
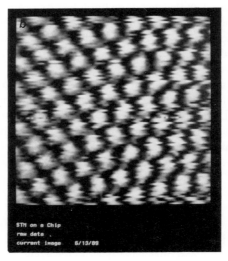

FIGURE 1.48 The piezoscanner of the STM on a chip. **a**: The STM on a chip. **b**: An image of the graphite surface taken by this microscope. (From Akamine *et al.*, 1990, with permission.)

Terashima *et al.*, 1990; Nagahara *et al.*, 1990; Hoffmann-Millack *et al.*, 1990; Yau *et al.*, 1991). A particularly elegant experiment placing xenon atoms was performed by Eigler and Schweizer (1990). Christen *et al.* (1991) describe a scanning cathodoluminescence microscope. Güthner *et al.* (1989) invented a scanning near field acoustic microscope. A tunneling acoustic microscope is described by Takata *et al.* (1989). Rohrbeck *et al.* (1991) detected surface acoustic waves by STM. Williams and Wickramasinghe (1986, 1988) and Dransfeld and Xu (1988) published details of a thermal microscope. A capacitance microscope has been described by Bugg and King (1988), Kleinknecht *et al.* (1988), Matey (1988), Kleinknecht and Meier (1989), and Williams *et al.* (1989). Baumgartner and Liess (1989) and Nonnenmacher *et al.* (1991) published a microkelvin probe. Williams and Wickramasinghe (1990) measured chemical potential variations. Hansma *et al.* (1989) and Prater *et al.* (1990) wrote about a scanning ion conductance microscope.

ACKNOWLEDGMENTS

I thank Heinrich Rohrer, Gerd Binnig, and Christoph Gerber for introducing me to the field of scanning tunneling microscopy. They show a continuing interest in my work and are always there for interesting and enlightening discussions. I enjoyed working with Erich Stoll, the IBM Rüschlikon image processing wizard, Alexis Baratoff, and Nicolas Garcia.

Paul Hansma gave me the chance to learn about the scanning force microscope, almost from the beginning of its history. I enjoyed the years in his group and the friendly and stimulating atmosphere he creates. I gratefully remember the collaboration with Richard Sonnenfeld, Jason Schneir, Scot Gould, Albrecht Weisenhorn, Barney Drake, and Sam Alexander.

I then joined the laboratory of Jürgen Mlynek. He gave me the opportunity to expand my activities beyond scanning force microscopy. The move to Konstanz with Jürgen Mlynek was supported by the members of the physics faculty and the Sonderforschungsbereich 306. Many discussions with and questions by Jürgen Mlynek, Rolf Möller, Jaime Colchero, Michael Hipp, Hartmut Bielefeldt, Heinz Gross, Mathias Amrein, W. Krieger, M. Völcker, Alain Humbert, and H. J. Güntherodt helped me to understand scanning probe microscopy better. I also want to acknowledge the help of the bibliographies prepared by P. J. Bryant and his group to find out about all the papers cited in the reference section.

This paper would not have been possible without the loving support of my wife, Gabriela, and my daughters, Anna and Irene.

APPENDIX A: BACKGROUND PLANE REMOVAL

Assume that the image affected by a background plane has n rows with m data points each. The background plane is defined by solving

$$a_i + b_j + c = z_{ij} \tag{1.94}$$

in a least-squares approximation. i and j are the indices for the points and run from 1 to m and 1 to n, respectively (Fig. 1.49). The $n \times m$ equations for all points can be combined to form the matrix equation

$$A\vec{a} = \vec{z}, \tag{1.95}$$

where

$$A = \begin{pmatrix} 1 & 1 & 1 \\ 2 & 1 & 1 \\ \vdots & \vdots & \vdots \\ m & 1 & 1 \\ 1 & 2 & 1 \\ 2 & 2 & 1 \\ \vdots & \vdots & \vdots \\ m & 2 & 1 \\ \vdots & \vdots & \vdots \\ m & n & 1 \end{pmatrix}, \quad \vec{a} = \begin{pmatrix} a \\ b \\ c \end{pmatrix}, \quad \text{and} \quad \vec{z} = \begin{pmatrix} z_{11} \\ z_{21} \\ \vdots \\ z_{m1} \\ z_{12} \\ z_{22} \\ \vdots \\ z_{m2} \\ \vdots \\ z_{mn} \end{pmatrix}.$$

The overdetermined system of Eq. 1.95 can be solved in terms of a least-squares fit by multiplying it from the left by A^T.

$$A^T A \vec{a} = A^T \vec{z}. \tag{1.96}$$

Written in components, Eq. 1.96 becomes

$$\begin{pmatrix} n \sum_{i=1}^{m} i^2 & \sum_{i=1}^{m}\sum_{j=1}^{n} ij & n \sum_{i=1}^{m} i \\ \sum_{j=1}^{n} ij & m \sum_{j=1}^{n} j^2 & m \sum_{j=1}^{n} \\ n \sum_{i=1}^{m} i & m \sum_{j=1}^{n} & n^2 \end{pmatrix} \begin{pmatrix} a \\ b \\ c \end{pmatrix}$$

$$= \begin{pmatrix} n\dfrac{m(m+1)(2m+1)}{6} & \dfrac{m(m+1)\,n(n+1)}{4} & n\dfrac{m(m+1)}{2} \\ \dfrac{m(m+1)\,n(n+1)}{4} & m\dfrac{n(n+1)(2n+1)}{6} & m\dfrac{n(n+1)}{2} \\ n\dfrac{m(m+1)}{2} & m\dfrac{n(n+1)}{2} & n^2 \end{pmatrix} \begin{pmatrix} a \\ b \\ c \end{pmatrix}$$

$$= \begin{pmatrix} \sum_{i=1}^{m}\sum_{j=1}^{n} i z_{ij} \\ \sum_{i=1}^{m}\sum_{j=1}^{n} j z_{ij} \\ \sum_{i=1}^{m}\sum_{j=1}^{n} z_{ij} \end{pmatrix}. \tag{1.97}$$

	1	2	3	...	m
1	z_{11}	z_{21}	z_{31}	...	z_{m1}
2	z_{12}	z_{22}	z_{32}	...	z_{m2}
3	z_{13}	z_{23}	z_{33}	...	z_{m3}
...	...	...	...	...	...
n	z_{1n}	z_{2n}	z_{3n}	...	z_{mn}

FIGURE 1.49 The labeling of the points in a data set for the background correction.

This system of three coupled equations can be solved with any standard numerical method (Press et al., 1989), such as the Gauss algorithm.

APPENDIX B: CORRECTION OF LINEAR DISTORTIONS IN TWO AND THREE DIMENSIONS

I discuss how to remove distortions in a plane. Figure 1.50 shows the coordinate system used. The Cartesian coordinate system defined by the unit vectors $\hat{e}_x$ and $\hat{e}_y$ is the real undistorted coordinate system. The piezo moves in a system defined by $\hat{e}'_x$ and $\hat{e}'_y$, which is not necessarily orthogonal and in which the two unit vectors, seen from the unprimed system, do not have the same length.

Without loss of generality we can assume that $\hat{e}_x = \hat{e}'_x$. $\hat{e}'_y$ is defined by the relation $\hat{e}'_y = a\hat{e}_x + b\hat{e}_y$. Now take a point $P(x, y)$ in the unprimed system and $P(x', y')$ in the piezo system and calculate a relation between the (x, y) and the (x', y').

$$x\hat{e}_x + y\hat{e}_y = x'\hat{e}'_x + y'\hat{e}'_y = x'\hat{e}_x + y'(a\hat{e}_x + b\hat{e}_y). \qquad (1.98)$$

By equating the components, we obtain

$$x = x' + ay'$$
$$y = by'. \qquad (1.99)$$

This relation allows us to remove the distortion in the image. For computational purposes, however, it is not ideal. To get to the undistorted image, we would have to interpolate both the x- and the y-components. By defining the basis

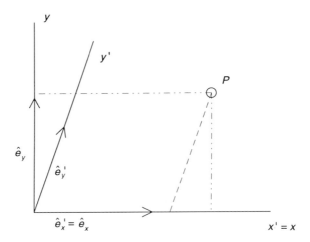

FIGURE 1.50 The coordinate system used for the linear distortion removal.

vectors differently, we can deduce a much more elegant equation in terms of computational efficiency. We set $\hat{e}'_x = a\hat{e}_x$ and $\hat{e}_y = b\hat{e}_x + \hat{e}_y$. We then obtain

$$x = ax' + by'$$
$$y = y'. \qquad (1.100)$$

Data are usually stored in computers in row order form, one row after the other. If we label the position of the points within a row by x and number the rows by y, then we can treat each row separately. We can in addition speed up the process by noting that for every row y the same values ax' will occur. These values can be calculated once and then stored in an array. The other term of the sum, $by' = by$, changes only from row to row. Hence the total computational effort per point is an array look-up and and addition, compared with an addition and an interpolation consisting of two multiplications and one addition in the first case. The three-dimensional correction is analogous.

Figure 1.51 gives the definition of the coordinate system. This time we make the two basis vectors $\hat{e}_z$ and $\hat{e}_{z'}$ collinear. $\hat{e}_x$ and $\hat{e}_{x'}$ form a plane whose normal direction is parallel to $\hat{e}_z$. As a last restriction, we set the $\hat{e}_y$-component of $\hat{e}_{y'}$ to 1. Doing the same calculations as in the two-dimensional case, we obtain

$$x = Ax' + By'$$
$$y = y'$$
$$z = Cx' + Dy' + Ez'. \qquad (1.101)$$

The computational effort for the xy plane is the same as for the two-dimensional case. In addition, we have two array look-ups (Cx' and Dy'), two addi-

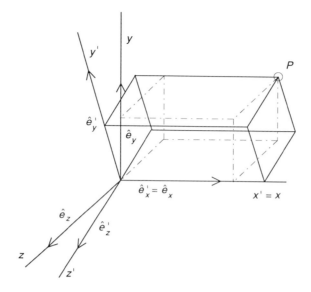

FIGURE 1.51 The coordinate system used for three-dimensional distortion removal.

tions, and one multiplication. The last three equations allow a complete removal of any linear distortion. I warn the reader that the coefficients A and B or $A-E$ have to be determined by independent means, because, with some clever choice, one can for instance change a hexagonal pattern to a cubic one. Measurements of angles and distances in surface unit cells would not be reliable.

REFERENCES

Abraham, D. W., and McDonald, F. A. (1990). Theory of magnetic force microscope images. *Appl. Phys. Lett.* 56, 1181.
Abraham, D. W., Williams, C. C., and Wickramasinghe, H. K. (1988a). High-resolution force microscopy of in-plane magnetization. *J. Microsc.* 152, 863.
Abraham, D. W., Williams, C. C., and Wickramasinghe, H. K. (1988b). Differential STM. *J. Microsc.* 152, 599.
Abraham, F. F., and Batra, I. P. (1989). Theoretical interpretation of atomic-force-microscope images of graphite. *Surf. Sci.* 209, L125–L132.
Abraham, F. F., Batra, I. P., and Çiraci, S. (1988). Effect of tip profile on atomic-force microscope images: a model study. *Phys. Rev. Lett.* 60, 1314.
Aers, G. C., and Leavens, C. R. (1988). The importance of current density "focusing" in STM image resolution. *J. Microsc.* 152, 65.
Aguilar, M., Pascual, P. J., and Santisteban, A. (1986). STM automation. *IBM J. Res. Dev.* 30, 525.
Aguilar, M., García, A., Pascual, P. J., Presa, J., and Santisteban, A. (1987). Computer system for scanning tunneling microscope automation. *Surf. Sci.* 181, 191.
Akama, Y., Nishimura, E., Sakai, A., and Murakami, H. (1990). New STM tip for measuring surface topography. *J. Vac. Sci. Technol.* A 8, 429.

Akamine, S., Albrecht, T. R., Zdeblick, M. J., and Quate, C. F. (1990a). A planar process for microfabrication of a scanning tunneling microscope. *Sens. Actuators* A21–A23, 964-970.

Akamine, S., Barrett, R. C., and Quate, C. F. (1990b). Improved atomic force microscope images using microcantilevers with sharp tips. *Appl. Phys. Lett.* 57, 316.

Albrecht, T. R., and Quate, C. F. (1987). Atomic resolution imaging of a nonconductor by atomic force microscopy. *J. Appl. Phys.* 62, 2599.

Albrecht, T. R., Quate C. F., Weisenhorn, A. L., and Hansma, P. K. (1989). Forces in atomic force microscopy in air and water. *Appl. Phys. Lett.* 54, 2651.

Albrecht, T. R., Akamine, S., Zdeblick, M. J., and Quate, C. F. (1990). Microfabrication of integrated STM. *J. Vac. Sci. Technol.* A 8, 317.

Albrektsen, O., Madsen, L. L., Mygind, J., and Morch, K. A. (1989). A compact STM with thermal compensation. *J. Phys.* 22, 39.

Alexander, S., Hellemans, L., Marti, O., Schneir, J., Elings, V., Hansma, P. K., Longmire, M., and Gurley, J. (1989). An atomic-resolution atomic-force microscope implemented using an optical lever. *J. Appl. Phys.* 65, 164.

Amer, N. M., Skumanich, A., and Ripple, D. (1986). Photothermal modulation of the gap distance in STM. *Appl. Phys. Lett.* 49, 137.

Anders, M., and Heiden, C. (1988). Imaging of tip-sample compliance in STM. *J. Microsc.* 152, 643.

Anders, M., Thaer, H., and Heiden, C. (1987). Simple micropositioning devices for STM. *Surf. Sci.* 181, 176.

Anders, M., Muck, M., and Heiden, C. (1990). Potentiometry for thin-film structures using atomic force microscopy. *J. Vac. Sci. Technol.* A 8, 394.

Anderson, H. L. (ed.) (1981). *AIP 50th Anniversary Physics Vade Mecum*. American Institute of Physics, Washington, DC, 41.

Arnold, L., Krieger, W., and Walther, H. (1987). Laser-frequency mixing in the junction of a STM. *Appl. Phys. Lett.* 51, 786–788.

Arvia, A. J. (1987). Scanning tunneling microscopy and electrochemistry. *Surf. Sci.* 181, 78.

Ash, E. A., and Nichols, G. (1972). Super resolution aperture scanning microscope. *Nature* 237, 510.

Ashcroft, N. W., and Mermin, N. D. (1976). *Solid State Physics*. Holt, Rinehart, and Winston, New York.

Aviram, A., Joachim, C., and Pomerantz, M. (1988). Evidence of switching and rectification by a single molecule effected by STM. *Chem. Phys. Lett.* 146, 490.

Banerjea, A., Smith, J. R., and Ferrante, J. (1990). Universal aspects of adhesion and atomic force microscopy. *J. Phys. Cond. Matter* 2, 8841.

Baratoff, A. (1983). Theory of scanning tunneling microscopy (STM). *Europhys. Conf. Abstracts* 7b, 364–367.

Baratoff, A. (1984). Theory of scanning tunneling microscopy—methods and approximations. *Physica* 127B, 143–150.

Baratoff, A., Binnig, G., Fuchs, H., Salvan, F., and Stoll, E. (1986). Tunneling microscopy and spectroscopy of semiconductor surfaces and interfaces. *Surf. Sci.* 168, 734.

Bardeen, J. (1961). Tunneling from a many-particle point of view. *Phys. Rev. Lett.* 6, 57.

Barniol, N., Perez, F., and Aymerich, X. (1991). On the three-dimensional scanning tunneling microscopy formalism. *J. Vac. Sci. Technol.* B9, 483.

Barrett, R. C., and Quate, C. F. (1990). Imaging polished sapphire with AFM. *J. Vac. Sci. Technol.* A 8, 394.

Barrett, R. C., and Quate, C. F. (1991a). High-speed, large-scale imaging with the atomic force microscope. *J. Vac. Sci. Technol.* B9, 302.

Barrett, R. C., and Quate, C. F. (1991b). Optical scan-correction system applied to atomic force microscopy. *Rev. Sci. Instrum.* 62, 1393.

Baski, A. A., Nogami, J., and Quate, C. F. (1991). Indium-induced reconstruction of the Si(100) surface. *Phys. Rev.* B43, 9316.

Batra, I. P., and Çiraci, S. (1988). Theoretical scanning tunneling microscopy/atomic force microscopy study of graphite including tip-sample interactions. *J. Vac. Sci. Technol.* 6, 313.

Batra, I. P., García, N., Rohrer, H., Salemink, H., Stoll, E., and Çiraci, S. (1987). A study of graphite surface with STM and electronic structure calculation. *Surf. Sci.* 181, 126.

Baumeister, T., and Marks, L. S. (1967). *Standard Handbook for Mechanical Engineers.* Ed. 7, pp. 5-29–5-53. McGraw-Hill, New York.

Baumgartner, H., and Liess, H. D. (1989). Micro Kelvin probe for local work-function measurements. *Rev. Sci. Instrum.* 59, 802.

Baym, G. (1969). *Lectures on Quantum Mechanics.* Benjamin/Cummings, Reading, MA.

Becker, J. (1987). Scanning tunneling microscope computer automation. *Surf. Sci.* 181, 200.

Bell, L. D., and Kaiser, W. J. (1988). Spatially resolved ballistic electron spectroscopy of subsurface interfaces. *J. Microsc.* 152, 605.

Bell, L. D., Hecht, M. H., Kaiser, W. J., and Davis, L. C. (1990). Direct spectroscopy of electron and hole scattering. *Phys. Rev. Lett.* 64, 2679.

Bendre, S., and Dharmadhikari, C. (1988). Design, construction and calibration of a PZT micromanipulator for STM. *J. Optics* 17, 67.

Berndt, R., Schlittler, R. R., and Gimzewski, J. K. (1991). Photon emission scanning tunneling microscope. *J. Vac. Sci. Technol.* B9, 573.

Berthe, R., Hartmann, U., and Heiden, C. (1988). Spatially resolved low-temperature spectroscopy on niobium bulk samples. *J. Microsc.* 152, 831.

Besocke, K. (1987). An easily operable scanning tunneling microscope. *Surf. Sci.* 181, 139.

Betzig, E., Barshatzky, H., Lewis, A., Isaacson, M., and Lin, K. (1988a). Super-resolution imaging with near field scanning optical microscopy (NSOM). *Ultramicroscopy* 25, 155–163.

Betzig, E., Isaacson, M., Barshatzky, H., Lewis, A., and Lin, K. (1988b). Near-field scanning optical microscopy (NSOM). *Proc. SPIE* 897, 91.

Biegelsen, D. K., Ponce, F. A., and Tramontana, J. C. (1989). Simple ion milling preparation of (111) tungsten tips. *Appl. Phys. Lett.* 54, 1223.

Binggeli, M., Carnal, D., Nyffenegger, R., Siegenthaler, H., Christoph, R., and Rohrer, H. (1991). Electrolytic scanning tunneling microscopy and point contact studies at electrochemically polished Au(111) substrates with or without Pb adsorbates. *J. Vac. Sci. Technol.* B9, 1985.

Binh, V. T. (1988). In situ fabrication of microtips for STM. *J. Microsc.* 152, 355.

Binh, V. T., and García, N. (1991). Atomic metallic ion emission, field surface melting and scanning tunneling microscopy tips. *J. Phys. I (Paris)* 1, 605.

Binnig, G., and Rohrer, H. (1982). Scanning tunneling microscopy. *Helv. Phys. Acta* 55, 726.

Binnig, G., and Rohrer, H. (1986). Scanning tunneling microscopy. *IBM J. Res. Deve.* 30, 355.

Binnig, G., and Smith, D. P. E. (1986). Single-tube three-dimensional scanner for scanning tunneling microscopy. *Rev. Sci. Instrum.* 57, 1688.

Binnig, G., Rohrer, H., Gerber, C., and Weibel, E. (1982a). Vacuum tunneling. *Physica* 109 & 110B, 2075.

Binnig, G., Rohrer, H., Gerber, C., and Weibel, E. (1982b). Tunneling through a controllable vacuum gap. *Appl. Phys. Lett.* 40, 178.

Binnig, G., Rohrer, H., Gerber, C., and Weibel, E. (1983). 7 × 7 Reconstruction on Si(111) resolved in real space. *Phys. Rev. Lett.* 50, 120.

Binnig, G., Frank, K. H., Fuchs, H., García, N., Reihl, B., Salvan, F., and Williams, A. R. (1985a). Tunneling spectroscopy and inverse photoemission: image and field states. *Phys. Rev. Lett.* 55, 991–994.

Binnig, G., García, N., and Rohrer, H. (1985b). Conductivity sensitivity of inelastic scanning tunneling microscopy. *Phys. Rev.* B32, 1336.

Binnig, G., Quate, C. F., and Gerber, C. (1986). Atomic force microscope. *Phys. Rev. Lett.* 56, 930.

Binnig, G., Gerber, C., Stoll, E., Albrecht, T. R., and Quate, C. F. (1987). Atomic resolution with the atomic force microscope. *Europhys. Lett.* 3, 1281–1286.

Blackford, B. L., and Jericho, M. H. (1990). Simple two-dimensional piezoelectric micropositioner for STM. *Rev. Sci. Instrum.* 61, 182.

Blackford, B. L., Dahn, D. C., and Jericho, M. H. (1987). High-stability bimorph scanning tunneling microscope. *Rev. Sci. Instrum.* 58, 1343.

Blackman, G. S., Mate, C. M., and Philpott, M. R. (1990). Interaction forces of a sharp tungsten tip with molecular films on silicon surfaces. *Phys. Rev. Lett.* 65, 2270.

Bono, J., and Good, R. H., Jr. (1985). Theoretical discussions of the scanning tunneling microscope. *Surf. Sci.* 151, 543.

Brinkman, W. F., Dynes, R. C., and Rowell, J. M. (1969). Tunneling conductance of asymmetrical barriers. *J. Appl. Phys.* 41, 1915–1921.

Brown, A., and Cline, R. W. (1990). A low cost, high performance imaging system for scanning tunneling microscopy. *Rev. Sci. Instrum.* 61, 1484.

Bryant, A., Smith, D. P. E., and Quate, C. F. (1986). Imaging in real time with the tunneling microscope. *Appl. Phys. Lett.* 48, 832.

Bryant, P. J., Kim, H. S., Zheng, Y. C., and Yang, R. (1987). Technique for sharpening STM tips. *Rev. Sci. Instrum.* 58, 1115.

Bryant, P. J., Miller, R. G., Deeken, R., Yang, R., and Zheng, Y. C. (1988). Scanning tunneling and AFM performed with the same probe in one unit. *J. Microsc.* 152, 871.

Bryant, P. J., Kim, H. S., Deeken, R. H., and Cheng, Y. C. (1990). Surface force measurements on picometer and piconewton scales. *J. Vac. Sci. Technol.* A8, 3502.

Bugg, C. D., and King, P. J. (1988). Scanning capacitance microscopy. *J. Phys.* E 21, 147.

Burnham, N. A., and Colton, R. J. (1989). Measuring the nanomechanical properties and surface forces of materials using an atomic force microscope. *J. Vac. Sci. Technol.* A7, 2906.

Burnham, N. A., Dominguez, D. D., Mowery, R. L., and Colton, R. J. (1990). Probing the surface forces of monolayer films with an atomic-force microscope. *Phys. Rev. Lett.* 64, 1931.

Burnham, N. A., Colton, R. J., and Pollock, H. M. (1991). Interpretation issues in force microscopy. *J. Vac. Sci. Technol.* A9, 2548.

Büttiker, M., and Landauer, R. (1986). Traversal time for tunneling. *IBM J. Res. Dev.* 30, 451.

Cahill, D. G., and Hamers, R. J. (1991). Surface photovoltage of Ag on Si(111)-(7 × 7) by scanning tunneling microscopy. *Phys. Rev.* B44, 1387.

Carlos, W. E., and Cole, M. W. (1980). Interaction between a He atom and a graphite surface. *Surf. Sci.* 91, 339.

Carr, R. G. (1988). Finite element analysis of PZT tube scanner motion for STM. *J. Microsc.* 152, 379.

Carrejo, J. P., Thundat, T., Nagahara, L. A., Lindsay, S. M., and Majumdar, A. (1991). Scanning tunneling microscopy investigations of polysilicon films under solution. *J. Vac. Sci. Technol.* B9, 955.

Cataldi, T. R. I., Blackham, I. G., Briggs, G. A. D., Pethica, J. B., Allen, H., and Hill, O. (1990). In situ scanning tunneling microscopy: new insight for electrochemical electrode-surface investigations. *J. Electroanal. Chem. Interf.* 290, 1.

Chalmers, S. A., Gossard, A. C., Weisenhorn, A. L., Gould, S. A. C., Drake, B., and Hansma, P. K. (1989). Determination of tilted superlattice structure by atomic force microscopy. *Appl. Phys. Lett.* 55, 2491–2493.

Chen, C. J. (1990). Tunneling matrix elements in three-dimensional space: the derivative rule and the sum rule. *Phys. Rev.* B42, 8841.

Chen, C. J. (1991a). Attractive interatomic force as a tunneling phenomenon. *J. Phys. Cond. Matter* 3, 1227.

Chen, C. J. (1991b). Microscopic view of scanning tunneling microscopy. *J. Vac. Sci. Technol.* A9, 44.

Chen, Y., Xu, W., and Huang, J. (1989). A simple new technique for preparing STM tips. *J. Phys.* E22, 455.

Chiang, S., and Wilson, R. J. (1986). Construction of a UHV STM. *IBM J. Res. Dev.* 30, 515.

Chornik, B., Aravena, R., Grahmann, C., Venegas, R., and Gaete, L. (1991). Automatic rough approximation system for a scanning tunneling microscope. *Rev. Sci. Instrum.* 62, 1863.

Christen, J., Grundmann, M., and Bimberg, D. (1991). Scanning cathodoluminescence microscopy: a unique approach to atomic-scale characterization of heterointerfaces and imaging of semiconductor inhomogeneities. *J. Vac. Sci. Technol.* B9, 2358.

Chu, C. S., and Sorbello, R. S. (1990). Phase-sensitive scanning tunneling potentiometry and the local transport field in mesoscopic systems. *Phys. Rev.* B42, 4928.

Chung, M. S., Feuchtwang, T. E., and Cutler, P. H. (1987). Spherical tip model in the theory of the scanning tunneling microscope. *Surf. Sci.* 181, 412.

Çiraci, S., and Batra, I. P. (1987). Scanning-tunneling microscopy at small tip-to-surface distances. *Phys. Rev.* B36, 6194.

Çiraci, S., and Tekman, E. (1989). Atomic theory of STM. *Phys. Rev.* B40, 10286.

Çiraci, S., Baratoff, A., and Batra, I. P. (1990a). Tip–sample interaction effects in scanning tunneling and AFM. *Phys. Rev.* B41, 2763.

Çiraci, S., Baratoff, A., and Batra, I. P. (1990b). Site-dependent electronic effects, forces, and deformations in scanning tunneling microscopy of flat metal surfaces. *Phys. Rev.* B42, 7618.

Colton, R. J., Baker, S. M., Baldeschwieler, J. D., and Kaiser, W. J. (1987). "Oxide-free" tip for scanning tunneling microscopy. *Appl. Phys. Lett.* 51, 305.

Cooley, J. W., and Tukey, J. W. (1965). An algorithm for machine calculation of complex Fourier series. *Math. Computation* 19, 297–301.

Coombs, J. H., Gimzewski, J. K., Reihl, B., Sass, J. K., and Schlittler, R. R. (1988). Photon emission experiments with the STM. *J. Microsc.* 152, 325.

Corb, B. W., Ringger, M., and Güntherodt, H.-J. (1985). An electromagnetic microscopic positioning device for the STM. *J. Appl. Phys.* 58, 3947.

Cortager, R., Beauvillain, J., Ajustron, F., Lacaze, J. C., and Tremollieres, C. (1991). A stage for submicron displacements using electromagnetic coils and its application to scanning tunneling microscopy. *Rev. Sci. Instrum.* 62, 830.

Cox, M. P., Heppell, T., and Hanrieder, W. (1989). A new UHV system with integrated STM for industrial applications. *J. Phys.* E22, 788.

Cricenti, A., Generosi, A., Gori, E., Chiarotti, G., and Selci, S. (1989a). New air operating STM. *Surf. Sci.* 211/212, 143.

Cricenti, A., Selci, S., Generosi, R., Gori, E., and Chiarotti, G. (1989b). Shaping of tungsten tips for STM. *Solid State Commun.* 70, 897.

Cutkosky, R. D. (1990). Versatile scan generator and data collector for scanning tunneling microscopes. *Rev. Sci. Instrum.* 61, 960.

Cutler, P. H., Feuchtwang, T. E., Tsong, T. T., Kuk, Y., Nguyen, H., and Lucas, A. A. (1987). Proposed use of the STM tunnel junction for the measurement of a tunneling time. *Phys. Rev.* B35, 7774.

Daniel, H.-U., Steiner, M., and Walther, H. (1981). Response of metal-insulator-metal point contact diodes to visible laser light. *Appl. Phys.* 25, 7.

Danielson, G. C., and Lanezos, C. (1942). Some improvements in practical Fourier analysis and their application to X-ray scattering from liquids. *J. Franklin Inst.* 233, 365–380; 435–452.

Das, B., and Mahanty, J. (1987). Spatial distribution of tunnel current and application to STM: a semiclassical treatment. *Phys. Rev.* B36, 898.

Davis, L. C., Everson, M. P., Jaklevic, R. C., and Shen, W. (1991). Theory of the local density of states on a metal: comparison with scanning tunneling spectroscopy of a Au(111) surface. *Phys. Rev.* B43, 3821.

DeKoven, B. M., and Meyers, G. F. (1991). Friction studies in ultra high vacuum of Fe surfaces with thin films from exposure to perfluorodiethylether. *J. Vac. Sci. Technol.* A9, 2570.

Demuth, J. E., Hamers, R. J., Tromp, R. M., and Welland, M. E. (1986). A simplified STM for surface science studies. *J. Vac. Sci. Technol.* A4, 1320.

den Boef, A. J. (1990). Preparation of magnetic tips for a scanning force microscope. *Appl. Phys. Lett.* 56, 2045.

den Boef, A. J. (1991). The influence of lateral forces in scanning force microscopy. *Rev. Sci. Instrum.* 62, 88.

Denk, W., and Pohl, D. W. (1991). Near-field optics: microscopy with nanometer-size fields. *J. Vac. Sci. Technol.* B9, 510.

Dijkkamp, D., Hoeven, A. J., Lenssinck, J. M., and Dielman, J. (1989). Direct writing in Si with a STM. *Appl. Phys. Lett.* 55, 1312.

DiLella, D. P., Wandass, J. H., Colton, R. J., and Marrian, C. R. K. (1989). Control systems for STM with tube scanners. *Rev. Sci. Instrum.* 60, 997.

Doyen, G., Koetter, E., Vigneron, J. P., and Scheffler, M. (1990). Theory of scanning tunneling microscopy. *Appl. Phys.* A51, 281.

Drake, B., Sonnenfeld, R., Schneir, J., Hansma, P. K., Slough, G., and Coleman, R. V. (1986). Tunneling microscope for operation in air or fluids. *Rev. Sci. Instrum.* 57, 441–445.

Drake, B., Sonnenfeld, R., Schneir, J., and Hansma, P. K. (1987). Scanning tunneling microscopy of processes at liquid–solid interfaces. *Surf. Sci.* 181, 92.

Drake, B., Prater, C. B., Weisenhorn, A. L., Gould, S. A. C., Albrecht, T. R., Quate, C. F., Cannell, D. S., Hansma, H. G., and Hansma, P. K. (1989). Imaging crystals, polymers, and processes in water with the AFM. *Science* 343, 1586.

Dransfeld, K., and Xu, J. (1988). The heat transfer between a heated tip and a substrate: fast thermal microscopy. *J. Microsc.* 152, 35.

Ducker, W. A., Cook, R. F., and Clarke, D. R. (1990). Force measurement using an AC atomic force microscope. *J. Appl. Phys.* 67, 4045.

Duke, C. B. (1969). Tunneling in Solids. Academic Press, New York.

Dürig, U., Pohl, D. W., and Rohner, F. (1986a). Near-field optical-scanning microscopy. *J. Appl. Phys.* 59, 3318–3327.

Dürig, U., Pohl, D. W., and Rohner, F. (1986b). Near-field optical scanning microscopy with tunnel-distance regulation. *IBM J. Res. Dev.* 30, 478–483.

Dürig, U., Züger, O., and Pohl, D. W. (1988). Force sensing in STM: observation of adhesion forces on clean metal surfaces. *J. Microsc.* 152, 259.

Edel'man, V. S., Trayanovskii, A. M., Khaikin, M. S., Stepanyan, G. A., and Volodin, A. P. (1991). The scanning tunneling microscopy combined with the scanning electron microscopy—a tool for the nanometry. *J. Vac. Sci. Technol.* B9, 618.

Eigler, D. M., and Schweizer, E. K. (1990). Positioning single atoms with a STM. *Nature* 344, 524.

Elrod, S. A., de Lozanne, A. L., and Quate, C. F. (1984a). Low temperature vacuum tunneling microscope. *Appl. Phys. Lett.* 45, 1240.

Elrod, S. A., de Lozanne, A. L., and Quate, C. F. (1984b). *J. Appl. Phys.* 55, 3544.

Elrod, S. A., Bryant, A., de Lozanne, A. L., Park, S., Smith, D., and Quate, C. F. (1986). Tunneling microscopy from 300 to 4.2 °K. *IBM J. Res. Dev.* 30, 387–395.

Emch, R., Descoutes, P., and Niedermann, P. (1988). A small STM with a large scan range for biological studies. *J. Microsc.* 152, 85.

Erlandson, R., Hadziioannou, G., Mate, C. M., McClelland, G. M., and Chiang, S. (1988). Atomic scale friction between the muscovite mica cleavage plane and a tungsten tip. *J. Chem. Phys.* 89, 5190.

Ezaki, T., Enomoto, H., Baba, M., Ikeda, Y., Ogura, S., Kuramochi, E., and Ozaki, H. (1990). Angle-resolved tunneling between two atomic planes. *J. Vac. Sci. Technol.* A8, 182.

Feenstra, R. M., and Mårtensson, P. (1988). *Phys. Rev. Lett.* 61, 447.

Fein, A. P., Kirtley, J. R., and Feenstra, R. M. (1987). STM for low temperature, high magnetic field, and spatially resolved spectroscopy. *Rev. Sci. Instrum.* 58, 1806.

Ferrell, T. L., Goundonnet, J. P., Reddick, R. C., Sharp, S. L., and Warmack, R. (1991). The photon scanning tunneling microscope. *J. Vac. Sci. Technol.* B9, 525.

Feuchtwang, T. E., Cutler, P. H., and Miskovsky, N. M. (1983). A theory of vacuum tunneling microscopy. *Phys. Lett.* 99A, 167.

Fiering, J. O., and Ellis, F. M. (1990). Versatile apparatus for etching scanning tunneling microscope tips. *Rev. Sci. Instrum.* 61, 3911.

Fink, H. W. (1986). Mono-atomic tips for STM. *IBM J. Res. Dev.* 30, 460.

Fischer, U. C. (1985). Optical characteristics of 0.1 μm circular apertures in a metal film as light sources for ultramicroscopy. *J. Vac. Sci. Technol.* 33, 386.

Fischer, U. C., and Pohl, D. W. (1989). Observation of single-particle plasmons by near-field optical microscopy. *Phys. Rev. Lett.* 62, 458–461.

Fischer, U. C., Dürig, U. T., and Pohl, D. W. (1988). Near-field optical-scanning microscopy in reflection. *Appl. Phys. Lett.* 52, 249–251.

Fischer, U. C., Dürig, U. T., and Pohl, D. W. (1989). Scanning near-field optical microscopy (SNOM) in reflection or scanning optical tunnelling microscopy (SOTM). *Scan. Microsc.* 3, 1–7.

Frohn, J., Wolf, J. F., Besocke, K., and Teske, M. (1989). Coarse tip distance adjustment and positioner for STM. *Rev. Sci. Instrum.* 60, 1200.

Fuchs, H., Eustachi, W., and Seifert, R. (1989). A data acquisition and image processing system for STM. *Scanning* 11, 139.

Fujita, M., Nagayoshi, H., and Yoshimori, A. (1990). Electronic-structure theory of the dimers adatoms and stacking fault model on Si(111) reconstructed surface — comparison with scanning tunneling spectroscopy. *J. Vac. Sci. Technol.* 8, 166.

Gao, X., Hamlin, A., and Weaver, M. J. (1991). Potential-dependent reconstruction at ordered Au(100)–aqueous interfaces as probed by atomic-resolution scanning tunneling microscopy. *Phys. Rev. Lett.* 67, 648.

García, N. (1986). Theory of STM and spectroscopy: resolution, image and field states, and thin oxide layers. *IBM J. Res. Dev.* 30, 533.

García, N., and Flores, F. (1984). Theoretical studies for STM. *Physica* 127B, 137.

García, N., Ocal, C., and Flores, F. (1983). Model theory for scanning tunneling microscopy. *Phys. Rev. Lett.* 50, 2002.

García, R., Sáenz, J. J., Mosler, J., and García, N. (1986). Distance-voltage characteristics in STM. *J. Phys.* C19, L131.

García, R., Sáenz, J. J., Soler, J. M., and García, N. (1987). Tunneling current through localized surface states. *Surf. Sci.* 181, 69–77.

García-Cantú, R., and Huerta-García, M. A. (1987). Inductoscanner tunneling microscope. *Surf. Sci.* 181, 216.

García-Cantú, R., and Huerta-Garnica, M. A. (1989). Direct tip structures determination by STM. *Colloq. Phys.* C-8, 40.

García-Cantú, R., and Huerta-Garnica, M. A. (1990). Long-scan imaging by STM. *J. Vac. Sci. Technol.* A8, 354.

García-García, R., and García, N. (1991). Geometry induced quantum states in scanning tunneling spectroscopy. *Surf. Sci.* 251–252, 408.

García-García, R., and Sáenz, J. J. (1991). Is scanning tunneling potentiometry a useful tool for probing the surface potential? *Surf. Sci.* 251–252, 223.

Garnaes, J., Kragh, F., Morch, K. A., and Tholen, A. R. (1990). Transmission electron microscopy of scanning tunneling tips. *J. Vac. Sci. Technol.* A8, 441.

Gerber, C., and Marti, O. (1985). Magnetostrictive positioner. *IBM Technol. Discl. Bull.* 27, 6373.

Gerber, C., Binnig, G., Fuchs, H., Marti, O., and Rohrer, H. (1986). Scanning tunneling microscope combined with a scanning electron microscope. *Rev. Sci. Instrum.* 57, 221–224.

Gerritsen, J. W., van Kempen, H., and Wyder, P. (1985). High stability scanning tunneling microscopy. *Rev. Sci. Instrum.* 56, 1573.

Gewirth, A. A., Craston, D. H., and Bard, A. J. (1989). Fabrication and characterization of microtips for in situ STM. *J. Electroanal. Chem.* 261, 477–482.

Giaever, I. (1960). Energy gap in superconductors measured by electron tunneling. *Phys. Rev. Lett.* 5, 147–148.

Gibson, G. A., and Schultz, S. (1991). A high-sensitivity alternating-gradient magnetometer for use in quantifying magnetic force microscopy. *J. Appl. Phys.* 69, 5880.

Giessibl, F. J., Gerber, C., and Binnig, G. (1991). A low-temperature atomic force/scanning tunneling microscope for ultrahigh vacuum. *J. Vac. Sci. Technol.* B9, 984.

Gimzewski, J. K., Reihl, B., Coombs, J. H., and Schlittler, R. R. (1988). Photon emission with the STM. *Z. Phys.* B72, 497.

Girard, C. (1991). Theoretical atomic-force-microscopy study of a stepped surface: nonlocal effects in the probe. *Phys. Rev.* B43, 8822.

Girard, C., and Spajer, M. (1990). Model for reflection near field optical microscopy. *Appl. Optics* 29, 3726.

Girard, C., van Labeke, C., and Vigoureux, J. M. (1989). Van der Waals force between a spherical tip and a solid surface. *Phys. Rev.* B40, 12133.

Girard, C., Maghezzi, S., and van Labeke, D. (1990). Interaction between a spherical probe and an atomic lattice: implication for atomic force microscopy on graphite and diamond. *Surf. Sci.* 234, 181.

Gleyzes, P., Kuo, P. K., and Boccara, A. C. (1991). Bistable behavior of a vibrating tip near a solid surface. *Appl. Phys. Lett.* 58, 2989.

Göddenhenrich, T., Hartmann, U., Anders, M., and Heiden, C. (1988). Investigation of Bloch wall fine structures by magnetic force microscopy. *J. Microsc.* 152, 527.

Göddenhenrich, T., Lemke, H., Hartmann, U., and Heiden, C. (1990a). Force microscope with capacitive displacement detection. *J. Vac. Sci. Technol.* A8, 383.

Göddenhenrich, T., Lemke, H., Hartmann, U., and Heiden, C. (1990b). Magnetic force microscopy of domain wall stray fields on single-crystal iron whiskers. *Appl. Phys. Lett.* 56, 2578.

Göddenhenrich, T., Lemke, H., Muck, M., Hartmann, U., and Heiden, C. (1990c). Probe calibration in magnetic force microscopy. *Appl. Phys. Lett.* 57, 2612.

Golubok, A. O., and Tarasov, N. A. (1990). The effect of geometric surface nonuniformities on the measurement of the local work function by scanning tunneling microscopy. *Sov. Tech. Phys. Lett.* 16, 418.

Goodman, F. O., and García, N. (1991). Roles of the attractive and repulsive forces in atomic force microscopy. *Phys. Rev.* B43, 4728.

Gould, S. A. C., Burke, K., and Hansma, P. K. (1989). Simple theory for the atomic-force microscope with a comparison of theoretical and experimental images of graphite. *Phys. Rev.* B40, 5363-5366.

Grafström, S., Kowalski, J., Neumann, R., Probst, O., and Wörtge, M. (1990). A compact STM control and data acquisition system based on a Macintosh II workstation. *J. Vac. Sci. Technol.* A8, 357.

Grafström, S., Kowalski, J., Neumann, R., Probst, O., and Wörtge, M. (1991). Analysis and compensation of thermal effects in laser-assisted scanning tunneling microscopy. *J. Vac. Sci. Technol.* B9, 568.

Green, M. P., Richter, M., Xing, X., Scherson, D., Hanson, K. J., Ross, P. N., Jr., Carr, R., and Lindau, I. (1988). In situ STM studies of electrochemical underpotential deposition of Pb on Au(111). *J. Micros.* 152, 823.

Griffith, J. E., Miller, G. L., Green, C. A., Grigg, D. A., and Russell, P. E. (1990). A scanning tunneling microscope with a capacitance based position monitor. *J. Vac. Sci. Technol.* B8, 2023-2027.

Grütter, P., Wadas, A., Meyer, E., Hidber, H.-R., and Güntherodt, H.-J. (1989). Magnetic force microscopy of a CoCr thin film. *J. Appl. Phys.* 66, 6001-6006.

Grütter, P., Rugar, D., Mamin, H. J., Castillo, G., Lambert, S. E., Lin, C.-J., Valletta, R. M., Wolter, O., Bayer, T., and Greschner, J. (1990a). Batch fabricated sensors for magnetic force microscopy. *Appl. Phys. Lett.* 57, 1820.

Grütter, P., Wadas, A., Meyer, E., Heinzelmann, H., Hidber, H.-R., and Güntherodt, H.-J. (1990b). High resolution magnetic force microscopy. *J. Vac. Sci. Technol.* A8, 406.

Grütter, P., Jung, T., Heinzelmann, H., Wadas, A., Meyer, E., Hidber, H.-R., and Güntherodt, H.-J. (1990c). 10-nm Resolution by magnetic force microscopy on FeNdB. *J. Appl. Phys.* 67, 1437.

Guckenberger, R., Wiegrabe, W., Hillebrand, A., Hartmann, T., and Wang, Z. (1989). STM of hydrated bacterial surface protein. *Ultramicroscopy* 31, 327.

Guerra, J. M. (1989). Photon tunneling microscopy. *Proc. SPIE* 1009, 254.

Guerra, J. M. (1990). Photon tunneling microscopy. *Appl. Optics* 29, 3741.

Guinea, F., and García, N. (1990). Scanning tunneling microscopy, resonant tunneling, and counting electrons: a quantum standard of current. *Phys. Rev. Lett.* 65, 281.
Gundlach, K. H. (1966). Zur Berechnung des Tunnelstroms durch eine trapezförmige Potentialstufe. *Solid State Electronics* 9, 949–957.
Güntherodt, H.-J., and Wiesendanger, R. (eds.) (1992). Scanning Tunneling Microscopy. Vols. I and II. Springer-Verlag, New York.
Güthner, P., Fischer, U. C., and Dransfeld, K. (1989). Scanning near-field acoustic microscope. *Appl. Phys.* 48, 89.
Haase, O., Borbonus, M., Muralt, P., Koch, R., and Rieder, K. H. (1990). A novel ultrahigh vacuum scanning tunneling microscope for surface science studies. *Rev. Sci. Instrum.* 61, 1480.
Hamers, R. J., and Cahill, D. G. (1991). Ultrafast time resolution in scanned probe microscopies: surface photovoltage on Si(111)–(7 × 7). *J. Vac. Sci. Technol.* B9, 514.
Hamers, R. J., and Köhler, I. K. (1989). Determination of the local electronic structure of atomic-sized defects on Si(001) by tunneling spectroscopy. *J. Vac. Sci. Technol.* A7, 2854.
Hamers, R. J., and Markert, K. (1990a). Surface photovoltage on Si(111)–(7 × 7) probed by optically pumped scanning tunneling microscopy. *J. Vac. Sci. Technol.* A8, 3524.
Hamers, R. J., and Markert, K. (1990b). Atomically resolved carrier recombination at Si(111)–7 × 7 surfaces. *Phys. Rev. Lett.* 64, 1150.
Hamers, R. J., Tromp, R. M., and Demuth, J. E. (1986). Surface electronic structure of Si(111)–7 × 7 resolved in real space. *Phys. Rev. Lett.* 56, 1972.
Hansma, P. K., Drake, B., Marti, O., Gould, S. A. C., and Prater, C. B. (1989). The scanning ion conductance microscope. *Science* 243, 641–643.
Hao, H. W., Baro, A., and Sáenz, J. J. (1991). Electrostatic and contact forces in force microscopy. *J. Vac. Sci. Technol.* B9, 1323.
Hartman, T. E. (1984). Tunneling through asymmetric barriers. *J. Appl. Phys.* 35, 3283.
Hartmann, U. (1989a). The point dipole approximation in magnetic force microscopy. *Phys. Lett.* A137, 475.
Hartmann, U. (1989b). Analysis of Bloch-wall fine structures by magnetic force microscopy. *Phys. Rev.* B40, 7421.
Hartmann, U. (1989c). Bit analysis of magnetic recording media by force microscopy. *Phys. Status Solidi* A115, 285.
Hartmann, U. (1990a). Magnetic microfield analysis by force microscopy. *J. Magetism Magnetic Mater.* 83, 545.
Hartmann, U. (1990b). Theory of magnetic force microscopy. *J. Vac. Sci. Technol.* A8, 411.
Hartmann, U. (1990c). Manifestation of zero-point quantum fluctuations in atomic force microscopy. *Phys. Rev.* B42, 1541.
Hartmann, U. (1991a). Theory of van der Waals microscopy. *J. Vac. Sci. Technol.* B9, 465.
Hartmann, U. (1991b). van der Waals interactions between sharp probes and flat sample surfaces. *Phys. Rev.* B43, 2404.
Hartmann, U., and Heiden, C. (1988). Calculation of the Bloch wall contrast in magnetic force microscopy. *J. Microsc.* 152, 281.
Hartmann, U., Göddenhenrich, T., Lemke, H., and Heiden, C. (1990). Domain-wall imaging by magnetic force microscopy. *IEEE Transm. Magn.* 26, 1512.
Hasegawa, Y., Kuk, Y., Tung, R. T., Silverman, P. J., and Sakurai, T. (1991). Ballistic electron emission in silicide–silicon interfaces. *J. Vac. Sci. Technol.* B9, 578.
Hashizume, T., Kamiya, I., Hasegawa, Y., Sano, N., Sakurai, T., and Pickering, H. W. (1988). A role of a tip geometry on STM images. *J. Microsc.* 152, 347.
Heben, M. J., Dovek, M. M., Lewis, N. S., Penner, R. M., and Quate, C. F. (1988). Preparation of STM tips for in situ characterization of electrode surfaces. *J. Microsc.* 152, 651.
Hecht, M. H., Bell, L. D., and Kaiser, W. J. (1989). Ballistic-electron-emission microscopy of subsurface defects at the Au–GaAs(100) interface. *Appl. Surf. Sci.* 41–42, 17.
Hecht, M. H., Bell, L. D., Kaiser, W. J., and Davis, L. C. (1990). Ballistic-hole spectroscopy of interfaces. *Phys. Rev.* B42, 7663.

Hellemans, L., Waeyaert, K., Hennau, F., Stockman, L., Heyvaert, I., and van Haesendonck, C. (1991). Can atomic force microscopy tips be inspected by atomic force microscopy? *J. Vac. Sci. Technol.* B9, 1309.

Hermsen, J. G. H., van Kempen, H., Nelissen, B. J., Soethout, L. L., van der Walle, G. F. A., Weijs, P. J. W., and Wyder, P. (1987). New mechanical constructions for the scanning tunneling microscope. *Surf. Sci.* 181, 183.

Hipps, K. W., Fried, G., and Fried, D. (1990). A scanning tunneling microscope with a wide sampling range. *Rev. Sci. Instrum.* 61, 1869.

Hobbs, P. C. D., Abraham, D. W., and Wickramasinghe, H. K. (1989). Magnetic force microscopy with 25 nm resolution. *Appl. Phys. Lett.* 55, 2357.

Hoeven, A. J., van Loenen, E. J., van Hooft, P. J. G., and Oostveen, K. (1990). A multiprocessor data acquisition and analysis system for scanning tunneling microscopy. *Rev. Sci. Instrum.* 61, 1668.

Hoffmann-Millack, B., Roberts, C. J., and Steer, W. S. (1990). Surface modification and atomic resolution on a vacuum-annealed gold foil in air by STM. *J. Appl. Phys.* 67, 1749.

Horie, C., and Miyazaki, H. (1990). Atomic-force-microscopy images of graphite due to van der Waals interactions. *Phys. Rev.* B42, 11757.

Hosaka, S., Hasegawa, T., Hosoki, S., and Takata, K. (1990). Fast scanning tunneling microscope for dynamic observation. *Rev. Sci. Instrum.* 61, 1342.

Huang, Z. H., Cutler, P. H., Feuchtwang, T. E., and Kazes, E. (1989). A multi-dimensional tunneling theory with application to STM. *Colloq. Phys.* C-8, 6.

Huang, Z. H., Cutler, P. H., Feuchtwang, T. E., Kazes, E., Nguyen, H. Q., and Sullivan, T. E. (1990a). Model studies of tunneling time. *J. Vac. Sci. Technol.* A8, 186.

Huang, Z. H., Feuchtwang, T. E., Cutler, P. H., and Kazes, E. (1990b). The Wentzel-Kramers-Brilloin method in multidimensional tunneling: application to STM. *J. Vac. Sci. Technol.* A8, 177.

Huang, Z.-H., Weimer, M., and Allen, R. E. (1991). The image potential in scanning tunneling microscopy of semiconductor surfaces. *J. Vac. Sci. Technol.* B9, 2402.

Hüsser, O. E., Craston, D. H., and Bard, A. J. (1989). Scanning electrochemical microscopy: high resolution deposition and etching of metals. *J. Electrochem. Soc.* 136, 3222.

Ibe, J. P., Bey, P. P., Jr., Brandow, S. L., Brizzolara, R. A., Burnham, N. A., DiLella, C. P., Marrian, C. R. K., and Colton, R. J. (1990). On the electrochemical etching of tips for scanning tunneling microscopy. *J. Vac. Sci. Technol.* A8, 3570.

Israelachvili, J. N. (1985). Intermolecular and Surface Forces with Applications to Colloidal and Biological Systems. Academic Press, New York.

Isshiki, N., Kobayashi, K., and Tsukada, M. (1991). First-principle simulation of scanning tunneling microscopy/spectroscopy with cluster models of W, Pt, TiC, and impurity adsorbed. *J. Vac. Sci. Technol.* B9, 475.

Itaya, K., and Tomita, E. (1989). In-situ STM of semiconductor (n-TiO$_2$)/liquid interfaces: a role of band bending in semiconductors. *Chem. Lett.*, pp. 285–288.

Jeon, D., and Willis, R. F. (1991a). Inchworm controller for fine approach in a scanning tunneling microscope. *J. Vac. Sci. Technol.* A9, 2418.

Jeon, D., and Willis, R. F. (1991b). Feedback system response in a scanning tunneling microscope. *Rev. Sci. Instrum.* 62, 1650.

Jericevic, Z., Benson, D. M., Bryan, J., and Smith, L. C. (1988). Geometric correction of digital images using orthonormal decomposition. *J. Microsc.* 149, 233.

Jericho, M. H. (1989). STM imaging technique for weakly bonded surface deposits. *J. Appl. Phys.* 65, 5237.

Jericho, M. H., Dahn, D. C., and Blackford, B. L. (1987). Scanning tunneling microscope with micrometer approach and thermal compensation. *Rev. Sci. Instrum.* 58, 1349.

Johansson, P., Monreal, R., and Apell, P. (1990). Theory for light emission from a scanning tunneling microscope. *Phys. Rev.* B42, 9210.

Jona, F., Strozier, J. A., and Yang, W. S. (1982). Low-energy electron diffraction for surface structure analysis. *Rep. Prog. Phys.* 45, 527–585.

Joyce, S. A., and Houston, J. E. (1991). A new force sensor incorporating force-fedback control for interfacial force microscopy. *Rev. Sci. Instrum.* 62, 710.

Kaiser, W. J., Bell, L. D., Hecht, M. H., and Grunthaner, F. J. (1989). Ballistic electron emission microscopy and spectroscopy of Au/GaAs interfaces. *J. Vac. Sci. Technol.* B7, 945.

Kaneko, R. (1988). A frictional force microscope controlled with an electromagnet. *J. Micros.* 152, 363.

Kato, T., and Tanaka, I. (1990). A scanning tunneling microscopy/spectroscopy system for cross-sectional observations of epitaxial layers of semiconductors. *Rev. Sci. Instrum.* 61, 1664.

Kato, T., Osaka, F., Tanaka, I., and Ohkouchi, S. (1991). A scanning tunneling microscope using dual-axis inchworms for the observation of a cleaved semiconductor surface. *J. Vac. Sci. Technol.* B9, 1981.

Kawakatsu, H., Hoshi, Y., Higuchi, T., and Kitano, H. (1991). Crystalline lattice for metrological applications and positioning control by a dual tunneling-unit scanning tunneling microscope. *J. Vac. Sci. Technol.* B9, 651.

Kenny, T. W., Waltman, S. B., Reynolds, J. K., and Kaiser, W. J. (1991). Micromachined silicon tunnel sensor for motion detection. *Appl. Phys. Lett.* 58, 100.

Khaikin, M. S. (1989). STM with wide field of vision. *Instrum. Exp. Tech. (USSR)* 32, 188.

Kirk, M. D., Albrecht, T. R., and Quate, C. F. (1988). Low-temperature atomic force microscope. *Rev. Sci. Instrum.* 59, 833–835.

Kirtley, J. (1978). Theoretical interpretation of IETS Data. In *Inelastic Electron Tunneling Spectroscopy*. Springer Series in Solid-State Science, Vol. 4. (T. Wolfram, ed.). Springer-Verlag, New York, pp. 80–91.

Kleinknecht, H. P., and Meier, H. (1989). The capacitance microscope: a non-contacting stylus technique for the investigation of semiconductor surfaces. *Diffus. Defect Data Solid State Data* B6, 411.

Kleinknecht, H. P., Sandercock, J. R., and Meier, H. (1988). An experimental scanning capacitance microscope. *Scan. Microsc.* 2, 1839.

Klitsner, T., Becker, R. S., and Vickers, J. S. (1990). Observation of the effect of tip electronic states on tunnel spectra acquired with the scanning tunneling microscope. *Phys. Rev.* B41, 3837.

Kobayashi, K., and Tsukada, M. (1990). Simulation of STM image based on electronic states of surface/tip system. *J. Vac. Sci. Technol.* A8, 170.

Kobayashi, K., Isshiki, N., and Tsukada, M. (1990). Simulation of scanning tunneling microscopy/spectroscopy based on electronic states theory. *Solid State Phys.* 25, 39.

Kochanski, G. P. (1989). Nonlinear alternating-current tunneling microscopy. *Phys. Rev. Lett.* 62, 2285.

Krieger, W., Kopperman, H., Suzuki, T., and Walter, H. (1989). The generaiton of laser difference frequencies using the STM. *IEEE Trans. Instrum. Meas.* 24, 905.

Krieger, W., Suzuki, T., Völcker, M., and Walther, H. (1990). Generation of microwave radiation in the tunneling junction of a scanning tunneling microscope. *Phys. Rev.* B41, 10229.

Kubby, J. A., Wang, Y. R., and Greene, W. J. (1991). Fabry-Perot transmission resonances in tunneling spectroscopy. *Phys. Rev.* B43, 9346.

Kuk, Y., and Silverman, P. J. (1986). Role of tip structure in scanning tunneling microscopy. *Appl. Phys. Lett.* 48, 1597.

Kuk, Y., and Silverman, P. J. (1990). Scanning tunneling spectroscopy of metal surfaces. *J. Vac. Sci. Technol.* A8, 289.

Kuk, Y., Becker, R. S., Silverman, P. J., and Kochanski, G. P. (1990). Optical interactions in the junction of a scanning tunneling microscope. *Phys. Rev. Lett.* 65, 456.

Kuwabara, M., Lo, W., and Spence, J. C. H. (1989). Reflection electron microscope imaging of an operating STM. *J. Vac. Sci. Technol.* A7, 2745.

Labani, B., and Girard, C. (1990). Optical interaction between a dielectric tip and a nanometric lattice: implications for near-field microscopy. *J. Optic Soc. Am.* B7, 936.

Laegsgaard, E., Besenbacher, F., Mortensen, K., and Stensgaard, I. (1988). A fully automated "thimble-size" STM. *J. Microsc.* 152, 663.

Laiho, R., Levola, T., and Snellman, H. (1987). A scanning tunneling microscope constructed in a rigid square frame. *Surf. Sci.* 181, 370.

Laloyaux, T., Lucas, A. A., Vigneron, J.-P., Lambin, P., and Morawitz, H. (1988). Lateral resolution of the STM. *J. Micros.* 152, 53.

Landman, U., and Luedtke, W. O. (1991). Nanomechanics and dynamics of tip-substrate interactions. *J. Vac. Sci. Technol.* B9, 414–423.

Lang, C. A., Dovek, M. M., and Quate, C. F. (1989). Low-temperature ultrahigh-vacuum STM. *Rev. Sci. Instrum.* 60, 3109.

Lang, N. D. (1985). Vacuum tunneling current from an adsorbed atom. *Phys. Rev. Lett.* 55, 230.

Lang, N. D. (1986a). Theory of single-atom imaging in the scanning tunneling microscope. *Phys. Rev. Lett.* 56, 1164–1167.

Lang, N. D. (1986b). Electronic structure and tunneling current for chemisorbed atoms. *IBM J. Res. Dev.* 30, 374–379.

Lang, N. D. (1986c). Spectroscopy of single atoms in the STM. *Phys. Rev.* B34, 5947.

Lang, N. D. (1987a). Apparent size of an atom in the scanning tunneling microscope as a function of bias. *Phys. Rev. Lett.* 58, 2579.

Lang, N. D. (1987b). Resistance of a one-atom contact in the scanning tunneling microscope. *Phys. Rev.* B36, 8173.

Leavens, C. R., and Aers, G. C. (1986). On calibration of tunneling microscope vacuum gaps using electron interferometry. *Solid State Commun.* 59, 285.

Leavens, C. R., and Aers, G. C. (1987). Effect of lattice vibrations in scanning tunneling microscopy. *Surf. Sci.* 181, 250.

Le Duc, H. G., Kaiser, W. J., and Stern, J. A. (1986). GaAs *pn*-junction studied by scanning tunneling potentiometry. *Appl. Phys. Lett.* 49, 1441.

Lemke, H., Göddenhenrich, T., Bochem, H. P., Hartmann, U., and Heiden, C. (1990). Improved microtips for scanning probe microscopy. *Rev. Sci. Instrum.* 61, 2538.

Libioulle, L., Ronda, A., Taborelli, M., and Gilles, J. M. (1991). Deformations and nonlinearity in scanning tunneling microscope images. *J. Vac. Sci. Technol.* B9, 655.

Liecerman, K., Harush, S., Lewis, A., and Kopelman, R. (1990). A light source smaller than the optical wavelength. *Science* 247, 59.

Lindsay, S. M., and Barris, B. (1987). Imaging deoxyribose nucleic acid molecules on a metal surface under water by scanning tunneling microscopy. *J. Vac. Sci. Technol.* A6, 544.

Louis, E., and Flores, F. (1987). Correlation between charge and current corrugations in scanning tunneling microscopy. *Phys. Rev.* B35, 1433.

Louis, E., Flores, F., and Echenique, P. M. (1986). Current saturation through image surface states in STM. *Solid State Commun.* 59, 453.

Ludeke, R., Prietsch, M., and Samsavar, A. (1991). Ballistic electron emission spectroscopy of metals on GaP(110). *J. Vac. Sci. Technol.* B9, 2342.

Lustenberger, P., Rohrer, H., Christoph, R., and Siegentaler, H. (1988). STM at potentially controlled electrode surfaces in electrolytic environments. *J. Electroanal. Chem.* 243, 225.

Lyding, J. W., Skala, S., Hubacek, J. S., Brockenborough, R., and Ganmmie, G. (1988). Design and operation of variable temperature STM. *J. Micros.* 152, 371.

Maghezzi, S., Girard, C., and van Labeke, D. (1991). Model for a van der Waals interaction between a metallic probe and a dielectric surface: implication for atomic force microscopy. *J. Phys. I (Paris)* 1, 289.

Mamin, H. J., Abraham, D. W., Ganz, E., and Clarke, J. (1985). Two-dimensional remote micropositioner for a scanning tunneling microscope. *Rev. Sci. Instrum.* 56, 2168.

Mamin, H., Ganz, E., Abraham, D. W., Thomson, R. E., and Clarke, J. (1986). Contamination-mediated deformation of graphite by STM. *Phys. Rev.* B34, 9015.

Mamin, H. J., Rugar, D., Stern, J. E., Fontana, Jr., R. E., and Kasiraj, P. (1989). Magnetic force microscopy of thin Permalloy films. *Appl. Phys. Lett.* 55, 318.

Manassen, Y., Hamers, R. J., Demuth, J. E., and Castellano, Jr., A. J. (1989). Direct observation of the precession of individual paramagnetic spins on oxidized slicon surfaces. *Phys. Rev. Lett.* 62, 2531.

Manne, S., Massie, J., Elings, V. B., Hansma, P. K., and Gewirth, A. A. (1991). Electrochemistry on a gold surface observed with the atomic force microscope. *J. Vac. Sci. Technol.* B9, 950.

Mansuripur, M. (1989). Computation of fields and forces in magnetic force microscopy. *IEEE Trans. Magn.* 25, 3467.

Maps, J. (1991). Simple raster generator for use with scanning tunneling microscopes. *Rev. Sci. Instrum.* 62, 357.

Marcus, R. B., Ravi, T. S., Gmitter, T., Chin, K., Liu, D., Orvis, W. J., Ciarlo, D. R., Hunt, C. E., and Trujillo, J. (1990). Formation of silicon tips with <1 nm radius. *Appl. Phys. Lett.* 56, 236.

Marrian, C. R. K., and Colton, R. J. (1990). Low-voltage electron beam lithography with a STM. *Appl. Phys. Lett.* 56, 775.

Marti, O. (1986). *Scanning Tunneling Microscope at Low Temperatures.* Dissertation ETH, Zürich No. 8595 (unpublished).

Marti, O., Binnig, G., Rohrer, H., and Salemink, H. (1986). Low-temperature scanning tunneling microscope. *Surf. Sci.* 181, 230.

Marti, O., Drake, B., and Hansma, P. K. (1987). Atomic force microscopy of liquid-covered surfaces: atomic resolution images. *Appl. Phys. Lett.* 51, 484.

Marti, O., Gould, S. A. C., and Hansma, P. K. (1988a). Control electronics for atomic force microscopy. *Rev. Sci. Instrum.* 59, 836–839.

Marti, O., Ribi, H. O., Drake, B., Albrecht, T., Quate, C. F., and Hansma, P. K. (1988b). Atomic force microscopy of an organic monolayer. *Science* 239, 50.

Marti, O., Colchero, J., and Mlynek, J. (1990). combined scanning force and friction microscopy of mica. *Nanotechnology* 1, 141.

Martin, Y., Abraham, D. W., and Wickramasinghe, K. (1988). High resolution capacitance measurement and potentiometry by force microscopy. *Appl. Phys. Lett.* 52, 1103.

Martin, Y., and Wickramasinghe, H. K. (1987). Magnetic imaging by "force microscopy" with 1000 Å resolution. *Appl. Phys. Lett.* 50, 1455.

Martin-Rodero, A., Ferrer, J., and Flores, F. (1988). Contact resistance and saturation effects in the STM: the resistance quantum unit. *J. Microsc.* 152, 317.

Mate, C. M., McClelland, G. M., Erlandsson, R., and Chiang, S. (1987). Atomic-scale friction of a tungsten tip on a graphite surface. *Phys. Rev. Lett.* 59, 1942.

Matey, J. R. (1988). Scanning capacitance microscopy. *Proc. SPIE* 897, 110.

Matey, J. R., Crandall, R. S., and Bryck, B. (1987). Bimorph driven x-y-z translation stage for scanned image microscopy. *Rev. Sci. Instrum.* 58, 567.

McClelland, G. M., Erlandsson, R., and Chiang, S. (1987). Atomic force microscopy: general principles and a new implementation. *Rev. Progr. Quant. Non-Destruc. Eval.* 6, 1307.

McCord, M. A. (1991). An x-y-z stage for scanning proximity microscopes using elastic elements. *Rev. Sci. Instrum.* 62, 530.

Meepagala, S. C., Real, F., Reyes, C. B., Novoselskaya, A., Rong, Z., and Wolf, E. L. (1990). Compact scanning tunneling microscope with easy-to-construct $X-Z$ inertial sample translation stage. *J. Vac. Sci. Technol.* A8, 3555.

Meepagala, S. C., Real, F., and Reyes, C. B. (1991). Tip–sample interaction forces in scanning tunneling microscopy: effects of contaminants. *J. Vac. Sci. Technol.* B9, 1340.

Melmed, A. J. (1991). The art and science and other aspects of making sharp tips. *J. Vac. Sci. Technol.* B9, 601.

Meyer, G., and Amer, N. M. (1988). Novel optical approach to atomic force microscopy. *Appl. Phys. Lett.* 53, 1045.

Meyer, G., and Amer, N. M. (1990). Simultaneous measurement of lateral and normal forces with an optical-beam-deflection atomic force microscope. *Appl. Phys. Lett.* 57, 2089.

Meyer, E., Wiesendanger, R., Anselmetti, D., Hidber, H. R., Güntherodt, H.-J., Levy, F., and Berger, H. (1990). Different response of AFM and STM to charge density. *J. Vac. Sci. Technol.* A8, 495.

Michel, B., and Travaglini, G. (1988). An STM for biological applications: bioscope. *J. Microsc.* 152, 681.

Miller, R. G., and Bryant, P. J. (1989). Atomic force microscopy of layered compounds. *J. Vac. Sci. Technol.* A7, 2879.

Miller, G. L., Griffith, J. F., Wagner, E. R., and Grigg, D. A. (1991). A rocking beam electrostatic balance for the measurement of small forces. *Rev. Sci. Instrum.* 62, 705.

Miyamoto, I., Ezawa, T., and Itabashi, K. (1991). Ion beam fabrication of diamond probes for a scanning tunneling microscope. *Nanotechnology* 2, 52.

Mizes, H. A., Park, S., and Harrison, W. A. (1987). Multiple-tip interpretation of anomalous scanning-tunneling-microscopy images of layered materials. *Phys. Rev.* B36, 4491.

Möller, R., Esslinger, A., and Koslowski, B. (1989). Noise in vacuum tunneling: application for a novel scanning microscope. *Appl. Phys. Lett.* 55, 2360.

Möller, R., Esslinger, A., and Koslowski, B. (1990a). Thermal noise in vacuum STM at zero bias voltage. *J. Vac. Sci. Technol.* A8, 590.

Möller, R., Esslinger, A., and Rauscher, M. (1990b). Tunneling tips imaged by STM. *J. Vac. Sci. Technol.* A8, 434.

Möller, R., Baur, C., Esslinger, A., and Kürz, P. (1991). Scanning noise potentiometry. *J. Vac. Sci. Technol.* B9, 609.

Moreland, J., and Rice, P. (1990). High-resolution, tunneling-stabilized magnetic imaging and recording. *Appl. Phys. Lett.* 57, 310.

Morita, S., Okada, T., Ishigame, Y., Sato, C., and Mikoshiba, N. (1986). Voltage-dependence of STM on a titanium surface in air. *Jpn. J. Appl. Phys.* 25, L516.

Morita, S., Otsuka, I., Okada, T., Yokoyama, H., Iwasaki, T., and Mikoshiba, N. (1987). Construction of a STM for electrochemical studies. *Jpn. J. Appl. Phys.* 26, L1853.

Morita, S., Okada, T., and Mokoshiba, N. (1989). Effect of potentiostatic control on in situ images of Ag and Au electrodes in 0.1 M KCl solution. *Jpn. J. Appl. Phys.* 28, 535.

Morrison, G. H. (1989). Near-field scanning optical microscopy. *Anal. Chem.* 61, 1075.

Moyer, P. J., Jahncke, C. L., Paesler, M. A., Reddick, R. C., and Warmack, R. J. (1990). Spectroscopy in the evanescent field with an analytical photon scanning tunneling microscope. *Phys. Lett.* A145, 343.

Mulhern, P. J., Hubbard, T., Arnold, C. S., Blackford, B. L., and Jericho, M. (1991). A scanning force microscope with a fiber-optic-interferometer displacement sensor. *Rev. Sci. Instrum.* 62, 1280.

Muralt, P., and Pohl, D. (1986). Scanning tunneling potentiometry. *Appl. Phys. Lett.* 48, 514.

Muralt, P., Meier, H., Pohl, D. W., and Salemink, H. W. M. (1987). STM and potentiometry on a semiconductor heterojunction. *Appl. Phys. Lett.* 50, 1352.

Musselman, I. H., and Russell, P. E. (1990). Platinum/iridium tips with controlled geometry for scanning tunneling microscopy. *J. Vac. Sci. Technol.* A8, 3558.

Nagahara, L. A., Thundat, T., and Lindsay, S. M. (1989). Preparation and characterization of STM tips for electrochemical studies. *Rev. Sci. Instrum.* 60, 3128.

Nagahara, L. A., Thundat, T., and Lindsay, S. M. (1990). Nanolithography on semiconductor surfaces under an etching solution. *Appl. Phys. Lett.* 57, 270.

Neddermeyer, H., and Drechsler, M. (1988). Electric-field induced changes of W(110) and W(111) tips. *J. Microsc.* 152, 459.

Neubauer, G., Cohen, S., McClelland, G. M., Horne, D., and Mate, C. M. (1990). Force microscopy with a bidirectional capacitance sensor. *Rev. Sci. Instrum.* 61, 2296.

Nguyen, H. Q., Cutler, P. H., Feuchtwang, T. E., Huang, Z.-H., Kuk, Y., Silverman, P. J., Lucas, A. A., and Sullivan, T. E. (1989). Mechanisms of current rectification in an STM tunnel junction and the measurement of an operational tunneling time. *IEEE Trans. Electron Devices* 36, 2671.

Nishikawa, O., Tomitori, M., Iwawaki, F., and Katsuki, F. (1989). Image quality of STM and apex profile of scanning tip. *Colloq. Phys.* C-8, 22.

Nishikawa, O., Tomitori, M., and Minakuchi, A. (1987). Piezoelectric and electrostrictive ceramics for STM. *Surf. Sci.* 181, 210.

Nishikawa, O., Tomitori, T., Iwawaki, F., and Hirano, N. (1990). Correlation between STM/spectroscopy images and apex profiles of scanning tips. *J. Vac. Sci. Technol.* A8, 421.

Nishimura, K. (1991). A spring-guided micropositioner with linearized subnanometer resolution. *Rev. Sci. Instrum.* 62, 2004.

Nogami, J., Park, S.-I., and Quate, C. F. (1989). Structure of submonolayers of tin on Si(111) studied by STM. *J. Vac. Sci. Technol.* A7, 1919.

Noguera, C. (1989). Validity of the transfer Hamiltonian approach: application to the STM spectroscopic mode. *J. Phys.* 50, 2587.

Noguera, C. (1990). Theoretical approach to the scanning tunneling microscope. *Phys. Rev.* B42, 1629.

Nonnenmacher, M., O'Boyle, M. P., and Wickramasinghe, H. K. (1991). Kelvin probe force microscopy. *Appl. Phys. Lett.* 58, 2921.

Ohnishi, S., and Tsukada, M. (1989). Molecular orbital theory for the STM. *Solid State Commun.* 71, 391.

Ohnishi, S., and Tsukada, M. (1990). Effect of microscopy electronic states of the tip on STM image. *J. Vac. Sci. Technol.* A8, 174.

Okayama, S., Bando, H., Tokumoto, H., and Kajimura, K. (1985). Piezoelectric actuator for STM. *Jpn. J. Appl. Phys.* 24, 152.

Okumura, A., and Goshi, Y. (1990). A method of scanning tunneling microscope tip approach and several means to improve S/N ratio. *Solid State Phys.* 25, 434.

Oppenheimer, J. R. (1928). Three notes on the quantum theory of aperiodic effects. *Phys. Rev.* 31, 66.

Orosz, L., and Balazs, E. (1986). Calculation of a quasi-self-consistent barrier for metal–vacuum–metal junctions. *Surf. Sci.* 177, 444.

Ott, H. W. (1976). Noise Reduction Techniques in Electronic Systems. John Wiley & Sons, New York.

Overney, G., Zhong, W., and Tomanek, D. (1991). Theory of elastic tip–surface interactions in atomic force microscopy. *J. Vac. Sci. Technol.* B9, 479.

Packard, W. E., Liang, Y., Dai, N., Dow, J. D., Nicolaides, R., Jaklevic, R. C., and Kaiser, W. J. (1988). Nano-machining of gold and semiconductor surfaces. *J. Microsc.* 152, 715.

Paesler, M. A., Moyer, P. J., Jahncke, C. J., Johnson, C. E., Reddick, R. C., Warmack, R. J., and Ferrell, T. (1990). Analytical photon scanning tunneling microscopy. *Phys. Rev.* B42, 6750.

Pancorbo, M., Anguiano, E., Diaspro, A., and Aguilar, M. (1990). A Wiener filter with circular-aperture-like point spread function to restore scanning tunneling microscopy (STM) images. *Pattern Recognition Lett.* 11, 553.

Pancorbo, M., Aguilar, M., Anguiano, E., and Diaspro, A. (1991). New filtering techniques to restore scanning tunneling microscopy images. *Surf. Sci.* 251–252, 418.

Park, S.-I., and Quate, C. F. (1986). Tunneling microscopy of graphite in air. *Appl. Phys. Lett.* 48, 112.

Park, S.-I., and Quate, C. F. (1987). Theories of the feedback and vibrational isolation systems for the STM. *Rev. Sci. Instrum.* 58, 2004.

Park, S.-I., and Quate, C. F. (1987a). Digital filtering of STM images. *J. Appl. Phys.* 62, 312.

Park, S.-I., Nogami, J., and Quate, C. F. (1987). Effect of tip morphology on images obtained by STM. *Phys. Rev.* B36, 2863.

Park, S.-I., Nogami, J., and Quate, C. F. (1988). Metal-induced reconstruction of silicon(111) surface. *J. Microsc.* 152, 727.

Park, K.-S., Huh, Y.-S., Jeon, I.-C., and Kim, S. (1991). Scanning tunneling microscope with novel coarse sample positioning technique. *J. Vac. Sci. Technol.* B9, 636.

Pelz, J. P. (1991). Tip-related artifacts in scanning tunneling spectroscopy. *Phys. Rev.* B43, 6746.

Pelz, J. P., and Koch, R. H. (1990). Tip-related artifacts in scanning tunneling potentiometry. *Phys. Rev.* B41, 1212.

Person, B. N. J., and Baratoff, A. (1987). Inelastic electron tunneling from a metal tip: the contribution from resonant processes. *Phys. Rev. Lett.* 58, 2575.

Piner, R., and Reifenberger, R. (1989). Computer control of the tunnel barrier width for the STM. *Rev. Sci. Instrum.* 60, 3123.

Pitarke, J. M., Echenique, P. M., and Flores, F. (1989). Apparent barrier height for tunneling electrons in STM. *Surf. Sci.* 217, 267.

Pitarke, J. M., Flores, F., and Echenique, P. M. (1990). Tunneling spectroscopy: surface geometry and interface potential effects. *Surf. Sci.* 234, 1.

Pitsch, M., Metz, O., Kohler, H.-H., Heckmann, K., and Strnad, J. (1989). Atomic resolution with a new atomic force tip. *Thin Solid Films* 175, 81.

Pohl, D. W. (1986). Some design criteria in STM. *IBM J. Res. Dev.* 30, 417.

Pohl, D. W. (1987a). Sawtooth nanometer slider: a versatile low voltage piezoelectric translation device. *Surf. Sci.* 181, 174.

Pohl, D. W. (1987b). Dynamic piezoelectric translation devices. *Rev. Sci. Instrum.* 58, 54.

Pohl, D. W., Denk, W., and Dürig, U. (1985). Optical stethoscopy: imaging with $\lambda/20$. *Proc. SPIE* 565, 56-61.

Pohl, D. W., Fischer, U. Ch., and Dürig, U. T. (1988a). Scanning near-field optical microscopy (SNOM). *J. Microsc.* 152, 853-861.

Pohl, D. W., Fischer, U. Ch., and Dürig, U. T. (1988b). Scanning near-field optical microscopy (SNOM): basic principles and some recent developments. *Proc. SPIE* 897, 84.

Poirier, G. E., and White, J. M. (1989). A new ultra-high vacuum STM design for surface science studies. *Rev. Sci. Instrum.* 60, 3113.

Poirier, G. E., and White, J. M. (1990). Diffraction grating calibration of scanning tunneling microscope piezoscanners. *Rev. Sci. Instrum.* 61, 3917.

Pollmann, J., Druger, P., and Mazur, A. (1987). Self-consistent electronic structure of semi-infinite Si(001) (2 × 1) and Ge(001) (2 × 1) with model calculations for STM. *J. Vac. Sci. Technol.* B5, 945.

Pool, R. (1988). Near-field microscopes beat the wavelength limit. *Science* 241, 25.

Prater, C. B., Drake, B., Gould, S. A. C., Hansma, H. G., and Hansma, P. K. (1990). Scanning ion-conduction microscope and atomic force microscope. *Scanning* 12, 50.

Press, W. H., Flannery, B. P., Teukolsky, S. A., and Vetterling, W. T. (1989). Numerical Recipes in Pascal: the Art of Scientific Computing. Cambridge University Press, New York.

Probst, O., Grafström, S., Kowalski, J., Neumann, R., and Wörtge, M. (1991). A tunneling atomic force microscope with inertial tip-to-sensor approach. *J. Vac. Sci. Technol.* B9, 626.

Qian, L. Q., and Wessels, B. W. (1991a). Scanning tunneling optical spectroscopy of semiconductor quantum well structures. *Appl. Phys. Lett.* 58, 253.

Qian, L. Q., and Wessels, B. W. (1991b). Scanning tunneling optical spectroscopy of semiconductors. *Appl. Phys. Lett.* 58, 1295.

Reddick, R. C., Warmack, R. J., Chilcott, D. W., Sharp, S. L., and Ferrell, T. L. (1990). Photon scanning tunneling microscopy. *Rev. Sci. Instrum.* 61, 3669.

Reiss, G., Vancea, J., Wittman, H., Zweck, J., and Hoffmann, H. (1990a). STM on rough surfaces, tip-shape-limited resolution. *J. Appl. Phys.* 67, 1156.

Reiss, G., Schneider, F., Vancea, J., and Hoffmann, H. (1990b). Scanning tunneling microscopy on rough surfaces: deconvolution of constant current images. *Appl. Phys. Lett.* 57, 867.

Renner, C., Niedermann, P., Kent, A. D., and Fischer, O. (1990a). A versatile low-temperature STM. *J. Vac. Sci. Technol.* A8, 330.

Renner, C., Niedermann, P., Kent, A. D., and Fischer, O. (1990b). A vertical piezoelectric inertial slider. *Rev. Sci. Instrum.* 61, 965.

Riis, E., Simonsen, H., Worm, T., Nielsen, U., and Besenbacher, F. (1989). Calibration of the electrical response of piezoelectric elements at low voltage using laser interferometry. *Appl. Phys. Lett.* 54, 2530.

Robinson, R. S. (1988a). Increasing the scanning speed of STM. *J. Microsc.* 152, 387.

Robinson, R. S. (1988b). Real-time STM of surfaces under active electrochemical control. *J. Microsc.* 152, 541.

Robinson, R. S. (1990). Interactive computer correction of piezoelectric creep in scanning tunneling microscopy images. *J. Comput. Assisted Microsc.* 2, 53.

Robinson, R. S., Kimsey, T. H., and Kimsey, R. (1991a). Desktop computer-based management of images and digital electronics for scanning tunneling microscopy. *J. Vac. Sci. Technol.* B9, 631.

Robinson, R. S., Kimsey, T. H., and Kimsey, R. (1991b). A digital integrator and scan generator coupled with dynamic scanning for scanning tunneling microscopy. *Rev. Sci. Instrum.* 62, 1772.

Robinson, R. S., Sternitzke, K., McDermott, M. T., and McCreery, R. L. (1991c). Morphology and electrochemical effects of defects on highly oriented pyrolytic graphite. *J. Electrochem. Soc.* 138, 2412.

Rohrbeck, W., Chilla, E., Fröhlich, H.-J., and Riedel, J. (1991). Detection of surface acoustic waves by scanning tunneling microscopy. *Appl. Phys.* A52, 344.

Rugar, D., and Hansma, P. K. (1990). Atomic force microscopy. *Phys. Today* 43, 23.

Rugar, D., Mamin, H. J., and Güthner, P. (1989). Improved fiber-optic interferometer for atomic force microscopy. *Appl. Phys. Lett.* 55, 2588–2590.

Rugar, D., Mamin, H. J., Güthner, P., Lambert, S. E., Stern, J. E., McFadyen, I., and Yogi, T. (1990). Magnetic force microscopy: general principles and application to longitudinal recording media. *J. Appl. Phys.* 68, 1169.

Rugar, D., and Grütter, P. (1991). Mechanical parametric amplification and thermomechanical noise squeezing. *Phys. Rev. Lett.* 67, 699.

Sacks, W., and Noguera, C. (1991a). Beyond Tersoff and Hamann: a generalized expression for the tunneling current. *J. Vac. Sci. Technol.* B9, 488.

Sacks, W., and Noguera, C. (1991b). Generalized expression for the tunneling current in scanning tunneling microscopy. *Phys. Rev.* B43, 11612.

Sacks, W., Gauthier, S., Rousset, S., Klein, J., and Esrick, M. (1987). Surface topography in STM: a free-electron model. *Phys. Rev.* B36, 961.

Sáenz, J. J., García, N., Grütter, P., Meyer, E., Heinzelmann, H., Wiesendanger, R., Rosenthaler, L., Hidber, H. R., and Güntherodt, H.-J. (1987). Observation of magnetic forces by the atomic force microscope. *J. Appl. Phys.* 62, 4293–4295.

Sakamaki, K., Hinokuma, K., Fujishima, A. (1991). Photoelectrochemical in situ observation of n-MoS$_2$ in aqueous solutions using a scanning tunneling microscope. *J. Vac. Sci. Technol.* B9, 944.

Salmeron, M., Ogletree, D. F., Ocal, C., Wang, H.-C., Neubauer, G., Kolbe, W., and Meyers, G. (1991). Tip-surface forces during imaging by scanning tunneling microscopy. *J. Vac. Sci. Technol.* B9, 1347.

Salvarezza, R. C., Alonso, C. A., Vara, J. M., Albano, E., Martin, H. O., and Arvia, A. J. (1990). Monte Carlo simulation applicable to the growth of rough metal overlayers: parametric relationship related to the electrochemical roughening. *Phys. Rev.* B41, 2502.

Sarid, D. (1991). Scanning Force Microscopy. Oxford University Press, New York.

Sarid, D., and Elings, V. (1991). Review of scanning force microscopy. *J. Vac. Sci. Technol.* B9, 431.

Sarid, D., Weissenberger, V., Iams, D. A., and Ingle, J. T. (1989). Theory of the laser diode interaction in scanning force microscopy. *IEEE J. Quantum Electron.* 25, 1968.

Sarid, D., Iams, D. A., Ingle, J. T., Weissenberger, V., and Ploetz, J. (1990). Performance of a scanning force microscope using a laser diode. *J. Vac. Sci. Technol.* A8, 378.

Saurenbach, F., and Terris, B. D. (1990). Imaging of ferroelectric domain walls by force microscopy. *Appl. Phys. Lett.* 56, 1703.

Scheinfein, M. R., Unguris, J., Cellota, R. J., and Pierce, D. T. (1989). Influence of the surface on magnetic domain-wall microstructure. *Phys. Rev. Lett.* 63, 668.

Schmidt, H., Heil, J., Wesner, J., and Grill, W. (1990). Atomic force sensors constructed from carbon and quartz fibers. *J. Vac. Sci. Technol.* A8, 388.

Schmid, A. K., and Kirschner, J. (1991). An ultrahigh vacuum-compatible scanning tunneling microscope head mounted on a 2-3/4 in. outer diameter flange. *J. Vac. Sci. Technol.* B9, 648.

Schneir, J., Hansma, P. K., Slough, G., and Coleman, R. V. (1986a). Tunneling microscope for operation in air or fluids. *Rev. Sci. Instrum.* 57, 441.

Schneir, J., Sonnenfeld, R., Hansma, P. K., and Tersoff, J. (1986b). Tunneling microscopy study of the graphite surface in air and water. *Phys. Rev.* B34, 4979.

Schneir, J., Sonnenfeld, R., Marti, O., Hansma, P. K., Demuth, J. E., and Hamers, R. J. (1988). Tunneling microscopy, lithography, and surface diffusion on an easily prepared, atomically flat gold surface. *J. Appl. Phys.* 63, 717–721.

Schönenberger, C., Alvarado, S. F. (1989). A differential interferometer for force microscopy. *Rev. Sci. Instrum.* 60, 3131–3134.

Schönenberger, C., and Alvarado, S. F. (1990a). Probing single charges by scanning force microscopy. *Mod. Phys. Lett.* B5, 871.

Schönenberger, C., and Alvarado, S. F. (1990b). Understanding magnetic force microscopy. *Z. Phys.* B80, 373.

Schönenberger, C., and Alvarado, S. F. (1990c). Observation of single charge carriers by force microscopy. *Phys. Rev. Lett.* 65, 3162.

Schönenberger, C., Alvarado, S. F., Lambert, S. F., and Sanders, I. L. (1990). Separation of magnetic and topographic effects in force microscopy. *J. Appl. Phys.* 67, 7278.

Schowalter, L. J., and Lee, E. Y. (1991). Role of elastic scattering in ballistic-electron-emission microscopy of Au/Si(001) and Au/Si(111) interfaces. *Phys. Rev.* B43, 9308.

Schroer, P. H., and Becker, J. (1986). Computer automation for STM. *IBM J. Res. Dev.* 30, 543.

Schummers, A., Halling, H., Besocke, K. H., and Cox, G. (1991). Controls and software for tunneling spectroscopy. *J. Vac. Sci. Technol.* B9, 615.

Schwoebel, P. R. (1987). In-situ morphological control of STM probes by helium cathode bombardment. *Surf. Sci.* 181, 154.

Selloni, A., Carnevali, P., Tosatti, E., and Chen, C. D. (1985). Voltage dependent scanning tunneling microscopy of a crystal surface: graphite. *Phys. Rev.* B31, 2602.

Shang, G. Y., Yao, J. E., and He, J. (1991). A new scanning tunneling microscope with large field of view and atomic resolution. *J. Vac. Sci. Technol.* B9, 612.

Shen, Y. R. (1984). The Principles of Nonlinear Optics. John Wiley & Sons, New York.

Shimizu, N., Kimura, T., Nakamura, T., and Umebu, I. (1990). An ultrahigh vacuum STM with a new inchworm mechanism. *J. Vac. Sci. Technol.* A8, 333.

Siegenthaler, H., and Christoph, R. (1990). In-situ scanning tunneling microscopy in electrochemistry. *Nato ASI Ser. E Appl. Sci.* 184, 242–267.

Simmons, J. G. (1963). Generalized formula for the electric tunnel effect between similar electrodes separated by a thin insulating film. *J. Appl. Phys.* 34, 1739.

Simmons, J. G. (1964). Generalized thermal $J-V$ characteristic for the electric tunnel effect. *J. Appl. Phys.* 35, 2655.

Simpson, A. M., and Wolfs, W. (1987). Thermal expansion and piezoelectric response of PZT channel 5800 for use in low-temperature scanning tunneling microscopy designs. *Rev. Sci. Instrum.* 58, 2193.

Smith, D. P. E., and Binnig, G. (1986). Ultrasmall scanning tunneling microscope for use in a liquid-helium storage dewar. *Rev. Sci. Instrum.* 57, 2630.

Smith, D. P. E., and Elrod, S. A. (1985). Magnetically driven micropositioners. *Rev. Sci. Instrum.* 56, 1970.

Smol'yaninov, I., Khakin, M. S., and Edel'man, V. S. (1990a). Emission of light by the tunnel junction of a scanning tunneling microscope. *JETP Lett.* 52, 201.

Smol'yaninov, I. I., Khaikin, M. S., and Edelman, V. S. (1990b). Light emission from the tunneling junction of the scanning tunneling microscope. *Phys. Lett.* A149, 410.

Snyder, E. J., Eklund, E. A., and Williams, R. S. (1990). Effects of tip size and asymmetry on scanning tunneling microscope topographs. *Surf. Sci.* 239, L487.

Soethout, L. L., Gerritsen, J. W., Groeneveld, P. P. M. C., Nelissen, B. J., and van Kempen, H. (1988). STM measurements on graphite using correlation averaging of the data. *J. Microsc.* 152, 251.

Soler, J. M., Baro, A. M., García, N., and Rohrer, H. (1986). Interatomic forces in STM: giant corrugation of the graphite surface. *Phys. Rev. Lett.* 57, 444.

Sonnenfeld, R., and Hansma, P. K. (1986). Atomic-resolution microscopy in water. *Science* 232, 211.

Sonnenfeld, R., and Schardt, B. C. (1986). Tunneling microscopy in an electrochemical cell. *Appl. Phys. Lett.* 49, 1172.

Staufer, U., Scandella, L., and Wiesendanger, R. (1989). Direct writing of nanometer scale structures on glassy metals by the STM. *Z. Phys.* B77, 281.

Stedman, M. (1988). Limits of topographic measurement by scanning tunneling and atomic force microscopy. *J. Microsc.* 152, 611–618.

Stilinger, F. H., and Weber, T. (1985). Computer simulation of local order in condensed phases of silicon. *Phys. Rev.* B31, 5262.

Stoll, E. (1984). Resolution of the scanning tunneling microscopy. *Surf. Sci.* 143, L411.

Stoll, E. P. (1988). Why do "dirty" tips produce higher-resolution images when graphite is scanned in a STM? *J. Phys.* C21, L921.

Stoll, E., and Gimzewski, J. K. (1991). Fundamental and practical aspects of differential scanning tunneling microscopy. *J. Vac. Sci. Technol.* B9, 643.

Stoll, E., and Marti, O. (1986). Restoration of scanning-tunneling-microscope data blurred by limited resolution and hampered by 1/f-like noise. *Surf. Sci.* 181, 222–229.

Stoll, E., Baratoff, A., Selloni, A., and Carnevali, P. (1984). Current distribution in the scanning vacuum tunnel microscope: free electron model. *J. Phys.* C17, 3073.

Stroscio, J. A., Feenstra, R. M., and Fein, A. P. (1986). Electronic structure of the Si(111)–2 × 1 surface by scanning tunneling microscopy. *Phys. Rev. Lett.* 57, 2579.

Stupian, G. W., and Leung, M. S. (1989). The use of a linear piezoelectric actuator for coarse motion in a vacuum compatible STM. *J. Vac. Sci. Technol.* A7, 2895.

Sueoka, K., Okuda, K., Matsubara, N., and Sai, F. (1991). Study of tip magnetization behavior in magnetic force microscope. *J. Vac. Sci. Technol.* B9, 1313.

Sugawara, Y., Ishizaka, T., and Morita, S. (1990a). Simultaneous imaging of a graphite surface with atomic force/scanning tunneling microscope (AFM/STM). *Jpn. J. Appl. Phys.* 29, 1539.

Sugawara, Y., Ishizaka, T., Morita, S., Imai, S., and Mikoshiba, M. (1990b). Simultaneous observation of atomically resolved AFM/STM images of a graphite surface. *Jpn. J. Appl. Phys.* 29, 296.

Sugihara, K. (1990). An ultrahigh vacuum tip transfer system for the STM field ion microscopy. *Rev. Sci. Instrum.* 61, 81.

Sumetskii, M. Y. (1988). Penetrability of asymmetric tunnel microjunction and quasiclassical theory of the STM. *JETP* 67, 438.

Takata, K., Hasegawa, T., Hosaka, S., and Hosoki, S. (1989). Tunneling acoustic microscope. *Appl. Phys. Lett.* 55, 1718.

Takeuchi, K., Uehara, Y., Ushioda, S., and Morita, S. (1991). Prism-coupled light emission from a scanning tunneling microscope. *J. Vac. Sci. Technol.* B9, 557.

Teague, E. C. (1986). Room temperature gold-vacuum-gold tunneling experiments. *J. Res. Nat. Bur. Stand.* 91, 171–233.

Terashima, K., Kondoh, M., and Yoshida, T. (1990). Fabrication of nucleation sites for nanometer size selective deposition by STM. *J. Vac. Sci. Technol.* A8, 581.

Terris, B. D., Stern, J. E., Rugar, D., and Mamin, H. J. (1989). Contact electrification using force microscopy. *Phys. Rev. Lett.* 63, 2669–2672.

Terris, B. D., Stern, J. E., Rugar, D., and Mamin, H. J. (1990). Localized charge force microscopy. *J. Vac. Sci. Technol.* A8, 374.

Tersoff, J. (1986). Anomalous corrugations in scanning tunneling microscopy: imaging of individual states. *Phys. Rev. Lett.* 57, 440–443.

Tersoff, J. (1989). Method for the calculation of STM images and spectra. *Phys. Rev.* B40, 11990.

Tersoff, J. (1990a). Role of tip electronic structure in STM images. *Phys. Rev.* B41, 1235.

Tersoff, J. (1990b). Theory of scanning tunneling microscopy and spectroscopy. *Nato ASI Ser. E Appl. Sci.* 184, 77–95.

Tersoff, J., and Lang, N. D. (1990). Tip-dependent corrugation of graphite in scanning tunneling microscopy. *Phys. Rev. Lett.* 65, 1132.

Tersoff, J., and Hamann, D. R. (1983). Theory and application for the scanning tunneling microscope. *Phys. Rev. Lett.* 50, 1998–2001.

Tersoff, J., and Hamann, D. R. (1985). Theory of the STM. *Phys. Rev.* B31, 805.
Thomson, D. J. (1988a). The STM as an information storage device. *J. Microsc.* 152, 627.
Thomson, W. T. (1988b). Theory of Vibration with Applications. Unwin Hyman Ltd., London.
Thundat, T., Nagahara, L. A., and Lindsay, S. M. (1990a). STM studies of semiconductor electrochemistry. *J. Vac. Sci. Technol.* A8, 539.
Thundat, T., Nagahara, L. A., Oden, P., and Lindsay, S. M. (1990b). Direct observation of bioelectrochemical processes by STM. *J. Vac. Sci. Technol.* A8, 645.
Tietze, U., and Schenk, C. (1988). Advanced Electronic Circuits. Ed. 9. Springer Verlag, New York.
Tomanek, D., and Zhong, W. (1991). Palladium–graphite interaction potentials based on first-principles calculations. *Phys. Rev.* B43, 12623.
Tomanek, D., Overney, G., Miyazaki, H., Mahanti, S. D., and Güntherodt, H.-J. (1989). Theory for the AFM of deformable surfaces. *Phys. Rev. Lett.* 63, 876.
Tomanek, D., Zhong, W., and Thomas, H. (1991). Calculation of an atomically modulated friction force in atomic-force microscopy. *Europhys. Lett.* 15, 887.
Tomita, E., Matsuda, N., and Itaya, K. (1990). Surface electronic structure of semiconductor (p- and n-Si) electrodes in electrolytes solution. *J. Vac. Sci. Technol.* A8, 534.
Tomitori, M., Iwawaki, F., Hirano, N., Katsuki, F., and Nishikawa, O. (1990). Corrugation of Si surfaces and profiles of tip apexes. *J. Vac. Sci. Technol.* A8, 222.
Tromp, R. M., Hamers, R. J., and Demuth, J. E. (1986). Quantum states and atomic structure of silicon surface. *Science* 234, 304.
Troyanovskii, A. M. (1989). Feedback control system for STM. *Instrum. Exp. Tech. (USSR)* 32, 188.
Tsukada, M., Kobayashi, K., and Ohnishi, S. (1990). First-principles theory of the STM simulation. *J. Vac. Sci. Technol.* A8, 160.
Tsukada, M., Kobayashi, K., and Isshiki, N. (1991). Effect of tip atomic and electronic structure on scanning tunneling microscopy/spectroscopy. *Surf. Sci.* 242, 12.
Tsukamoto, S., Siu, B., and Nakagiri, N. (1991). Twin-probe scanning tunneling microscope. *Rev. Sci. Instrum.* 62, 1767.
Twomey, T., Wiechers, J., Kolb, D. M., and Behm, R. J. (1988). In situ, atomic scale observation of electrode topography and reactions. *J. Microsc.* 152, 537.
Umeda, N., Ishizaki, S., and Uwai, H. (1991). Scanning attractive force microscope using photothermal vibration. *J. Vac. Sci. Technol.* B9, 1318.
Ushioda, S. (1990). Light emission associated with tunneling phenomena. *J. Lumin.* 47, 131.
Valdes, J., Kohanoff, J. J., Lobbe, E. E., Lopez Bancalari, R., Porfiri, M. E., and García-Cantú, R. (1988). Battery operated STM. *J. Microsc.* 152, 675.
van de Leemput, L. E. C., Rongen, P. H. H., Timmerman, B. H., and van Kempen, H. (1991). Calibration and characterization of piezoelectric elements as used in scanning tunneling microscopy. *Rev. Sci. Instrum.* 62, 989.
van Loenen, E. J., Dijkkamp, D., and Hoeven, A. J. (1988). Clean and metal-contaminated Si(110) surfaces studied by RHEED, XPS, and STM. *J. Microsc.* 152, 487.
van Loenen, E. J., Dijkkamp, D., Hoeven, A. J., Lenssinck, J. M., and Dieleman, J. (1990). Nanometer scale structuring of Si by direct indentation. *J. Vac. Sci. Technol.* A8, 574.
Van de Walle, G., Van Kempen, H., and Wyder, P. (1986). Tip structure determination by STM. *Surf. Sci.* 167, L219.
van Kempen, H. (1990). Spectroscopy using conduction electrons. *Nato ASI Ser. E Appl. Sci.* 184, 242–267.
van Kempen, H., and van de Walle, G. F. A. (1986). Applications of a high-stability STM. *IBM J. Res. Dev.* 30, 509–514.
Vasile, M. J., Grigg, D. A., Griffith, J. E., Fitzgerald, J. E., and Russel, P. (1991). Scanning probe tips formed by focused ion beams. *Rev. Sci. Instrum.* 62, 2167.
Vazquez, L. (1989). STM–SEM combination study on the electrochemical growth mechanism and structure of gold overlayers: a quantitative approach to electrochemical surface roughening. *Surf. Sci.* 215, 171.

Vieira, S. (1986). The behaviour and calibration of some piezoelectric ceramics used in the STM. *IBM J. Res. Dev.* 30, 553.

Vieira, S., Ramos, M. A., Hortal, M. A., and Buendia, A. (1987). A new design of the scanning tunneling microscope unit. *Surf. Sci.* 181, 376.

Völcker, M., Krieger, W., and Walther, H. (1991a). Laser-driven scanning tunneling microscope. *Phys. Rev. Lett.* 66, 1717.

Völcker, M., Krieger, W., Suzuki, T., and Walther, H. (1991b). Laser-assisted scanning tunneling microscopy. *J. Vac. Sci. Technol.* B9, 541.

Wadas, A. (1988). The theoretical aspect of AFM used for magnetic materials. *J. Magnetism Magnetic Mater.* 71, 147.

Wadas, A. (1989). Description of magnetic imaging in AFM. *J. Magnetism Magnetic Mater.* 78, 263.

Wadas, A., and Grütter, P. (1989). Theoretical approach to magnetic force microscopy. *Phys. Rev.* B39, 12013.

Wadas, A., and Güntherodt, H.-J. (1990a). The topography effect on magnetic images in magnetic force microscopy. *J. Appl. Phys.* 68, 4767.

Wadas, A., and Güntherodt, H.-J. (1990b). Lateral resolution in magnetic force microscopy. *Phys. Lett.* A146, 227.

Wandass, J. H., Murday, J. S., and Colton, R. J. (1989). Magnetic field sensing with magnetostrictive materials using a tunneling tip detector. *Sens. Actuators* 19, 211.

Wang, X. S., Phameuf, R. J., and Williams, E. D. (1988). Comparison of LEED and STM measurements of vicinal Si(111). *J. Microsc.* 152, 473.

Watanabe, M. O., Tanaka, K., and Sakai, A. (1990). High-temperature STM. *J. Vac. Sci. Technol.* A8, 327.

Weaver, J. M. R., and Abraham, D. W. (1991). High resolution atomic force microscopy potentiometry. *J. Vac. Sci. Technol.* B9, 1559.

Weaver, J. M. R., and Wickramasinghe, H. K. (1991). Semiconductor characterization by scanning force microscope surface photovoltage. *J. Vac. Sci. Technol.* B9, 1562.

Wengelnik, H., and Neddermeyer, H. (1990). Oxygen-induced sharpening process of W(111) tips for STM use. *J. Vac. Sci. Technol.* A8, 438.

Wetsel, G. C., Jr., McBride, S. E., Warmack, R. J., and van de Sande, B. (1989). Calibration of STM transducers using optical beam deflection. *Appl. Phys. Lett.* 55, 528.

Whitehouse, D. J. (1976). Stylus techniques. In *Characterization of Solid Surfaces*. (P. E. Kane and G. P. Larrabee, eds.), p. 49. Plenum, New York.

Wickramasinghe, H. K. (1989). Scanned-probe microscopes. *Sci. Am.* 261, 74.

Wiechers, J., Twomey, T., Kolb, D. M., and Behm, R. J. (1988). An in situ STM study of Au(111) with atomic scale resolution. *J. Electroanal. Chem.* 248, 451.

Wiesendanger, R., Güntherodt, H.-J., Güntherodt, G., Gambino, R. J., and Ruf, R. (1990). Observation of vacuum tunneling of spin-polarized electrons with the scanning tunneling microscope. *Phys. Rev. Lett.* 65, 247.

Williams, C. C., and Wickramasinghe, H. K. (1986). Scanning thermal profiler. *Appl. Phys. Lett.* 49, 1587.

Williams, C. C., and Wickramasinghe, H. K. (1988). Thermal and photo thermal imaging on a sub 100 nanometer scale. *Proc. SPIE* 897, 129.

Williams, C. C., and Wickramasinghe, H. K. (1990). Microscopy of chemical-potential variations on an atomic scale. *Nature (London)* 344, 317.

Williams, C. C., Hough, W. P., and Rishton, S. A. (1989). Scanning capacitance microscopy on a 25 nm scale. *Appl. Phys. Lett.* 55, 203.

Wolf, E. L. (1985). Principles of Electron Tunneling Spectroscopy. *Int. Ser. Monogr. Phys.* 71.

Wolter, O., Bayer, T., and Gerschner, J. (1991). Micromachined silicon sensors for scanning force microscopy. *J. Vac. Sci. Technol.* B9, 1353.

Xiao, G. L., Liu, H. F., Chi, S. D., Xue, Z. Q., Li, X., and Tsong, T. T. (1988). Development of an STM-LEED-FIM combination system. *J. Phys. Colloq.* 49, 37.

Yamagata, Y., Higuchi, T., Saeki, H., and Ishimaru, H. (1990). Ultrahigh vacuum precise positioning device utilizing rapid deformations of piezoelectric elements. *J. Vac. Sci. Technol.* A8, 4098.

Yang, R., Miller, R., and Bryant, P. J. (1988). Atomic force profiling utilizing contact force. *J. Appl. Phys.* 63, 570.

Yao, J. E., Shang, G. Y., Jiao, Y. K., Yi, Y., Bai, C. L., He, J., Zhong, J. C., and Rong, D. N. (1988). A STM for operation in air *J. Microsc.* 152, 671.

Yasutake, M., and Miyata, C. (1990). STM combined with optical microscope for large sample measurement. *J. Vac. Sci. Technol.* A8, 350.

Yata, M., Ozaki, M., Sakata, S., Yamada, T., Kohno, A., and Aono, M. (1989). Titanium carbite single-cristal tips for high-resolution STM. *Jpn. J. Appl. Phys.* 28, L885.

Yau, S.-T., Saltz, D., and Nayfeh, M. H. (1991). Laser-assisted deposition of nanometer structures using a scanning tunneling microscope. *Appl. Phys. Lett.* 57, 2913.

Yokoyama, K., Hashizume, T., Tanaka, H., Sumita, I., Takao, M., and Sakurai, T. (1991). In situ tip exchange mechanism for the Demuth-type scanning tunneling microscope. *J. Vac. Sci. Technol.* B9, 623.

Young, R., Ward, J., and Scire, F. (1971). Observation of metal–vacuum–metal tunneling, field emission, and the transition region. *Phys. Rev. Lett.* 27, 922.

Young, R., Ward, J., and Scire, F. (1972). The topographiner: an instrument for measuring surface microtopography. *Rev. Sci. Instrum.* 43, 999.

Yuan, J.-Y., Shao, Z., and Gao, C. (1991). Alternative method of imaging surface topologies of nonconducting bulk specimens by scanning tunneling microscopy. *Phys. Rev. Lett.* 67, 863.

Zeglinski, D. M., Ogletree, D. F., Beebe, T. P., Jr., Hwang, R. Q., Somorjai, G. A., and Salmeron, M. B. (1990). An ultrahigh vacuum scanning tunneling microscope for surface science studies. *Rev. Sci. Instrum.* 61, 3769.

Zheng, N. J., and Tsong, I. S. T. (1990). Resonant-tunneling theory of imaging close-packed metal surfaces by STM. *Phys. Rev.* B41, 2671.

Zhong, W., Overney, G., and Tomanek, D. (1991). Limits of resolution in atomic force microscopy images of graphite. *Europhys. Lett.* 15, 49.

STM in Biology

CHAPTER 2

STM of Proteins and Membranes

M. Amrein
Institute for Medical Physics and Biophysics
University of Münster, Münster, Germany

H. Gross
Institute for Cell Biology
Swiss Federal Institute of Technology
ETH-Hönggerberg, Zürich, Switzerland

R. Guckenberger
MPI for Biochemistry
Martinsried, Germany

2.1 Introduction
2.2 STM for Biological Applications
2.3 Tunneling Tips
    2.3.1 Tip-Sample Interaction
    2.3.2 Tip Fabrication
    2.3.3 Tip Conditioning
2.4 Sample Preparation
    2.4.1 Substrates
    2.4.2 Sample Deposition
    2.4.3 Dehydration of the Objects
    2.4.4 Metal Coating
    2.4.5 Metal Replicas
2.5 STM of Uncoated Specimens
    2.5.1 STM Imaging under Conventional Tunneling Conditions
    2.5.2 STM Imaging at a Tip-Sample Voltage ≥ 5 V
2.6 STM of Metal-Coated Specimens
    2.6.1 Conductively Coated Specimens
    2.6.2 Metal Replicas
2.7 Image Processing and Quantitative Analysis of STM Data
    2.7.1 Distortion Correction and Appropriate Display of STM Images

2.7.2 Digital Image Averaging
2.7.3 Height Measurement
2.7.4 Comparison of TEM and STM data
2.8 Conclusions and Prospects
2.8.1 STM Imaging of Biological Macromolecular Structures
2.8.2 STM in Molecular Structure Research
References

2.1 INTRODUCTION

The capability of scanning tunneling microscopy (STM) to reveal the three-dimensional structure of a surface with atomic resolution under various conditions, as well as to probe electronic and elastic properties of a truly local nature, makes it attractive in biological molecular structure research. Direct observations of structural and even dynamic aspects of macromolecules in their natural environment appear to be possible. However, STM of biological materials is hampered due to their poor intrinsic electrical conductivity.

Resorting to conductive coating films (e.g., Travaglini *et al.*, 1987; Amrein *et al.*, 1988, 1989; Guckenberger *et al.*, 1988; Garcia *et al.*, 1989) or metal replicas (Zasadzinski *et al.*, 1988; Blackford *et al.*, 1991) makes it relatively easy to obtain reproducible, trustworthy images that are not influenced by the low conductivity of biological macromolecular structures. When using a fine-grain, stable coating film, the method permits direct three-dimensional visualization of structural details at least at the level of individual subunits in protein complexes.

On the other hand, there is an increasing number of reports that prove that STM of uncoated biological macromolecules and macromolecular structures is feasible, although these bulky organic structures were initially considered not to be sufficiently conductive (e.g., Baró *et al.*, 1985; Welland *et al.*, 1989; Voelker *et al.*, 1988; Stemmer *et al.*, 1989; Amrein *et al.*, 1989; Guckenberger *et al.*, 1989; Miles *et al.*, 1990; see also Chapters 3, 5, and 6, this volume). Usually the contrast in these STM images cannot solely be attributed to the relief of the structures under investigation. The electronic and elastic properties of the biomaterials contribute to the image formation in a way not yet fully elucidated.

Scanning tunneling microscopy can be performed in modes other than the imaging mode. For example, spectroscopic data on a protein have been collected by ramping the bias voltage and recording the corresponding tunneling current (Welland *et al.*, 1989; see also Chapter 3, this volume). Images of dI/ds, the change in tunneling current I on fast modulation of the tip-sample separation s, taken simultaneously with the STM images, may reveal elastic properties of the investigated structures (e.g., Travaglini *et al.*, 1988).

Scanning tunneling microscopy of uncoated biological specimens quite often suffers from poor reproducibility. Nevertheless, by using an unconventionally high tunneling voltage ($V_T \geq 5$ V) and a very low tunneling current ($I_T \leq 1$ pA,) it has been possible to resolve a two-dimensional protein crystal, the HPI layer, and a bacterial membrane with good reproducibility (Guckenberger *et al.*, 1989, 1991). However, the large tip-sample separation that is encountered under these imaging conditions reduces the lateral resolution.

Furthermore, high-resolution data of uncoated polypeptides and proteins have been obtained reproducibly from molecules arranged in ordered arrays on a graphite substrate (e.g., Miles *et al.*, 1990; Hörber *et al.*, 1991; see also Chapter 3, this volume).

The following section is concerned with STM instrumentation suitable for biological application, including tip fabrication. Then sample preparation is detailed. Some general aspects of STM imaging such as the influence of the tunneling tip geometry on STM topographical images are then discussed. It has to be mentioned here that all STM images presented in this Chapter have been acquired by operating the STM in the constant current mode, i.e., the imaging mode, where the tip-sample distance is kept constant in first instance. Only under these conditions are truly three-dimensional data of surface reliefs obtained. Scanning tunneling microscopy imaging of both uncoated and freeze-dried, metal-coated biological membranes and regular protein structures are described. We then discuss digital image processing of raw STM data. It is performed in order to compensate for image distortions, to enhance the visibility of the structures of interest for display, and to analyze the structural data quantitatively. Methods of digital image "filtration" to optimize the signal-to-noise ratio in STM topographical images of ordered protein arrays are presented.

The biological structures included as examples have all been well characterized by means other than STM, which allows quantitative comparison of the data. This is important for the interpretation of STM images, considering that STM is a new technique in molecular structure research. We attempt to point out how a combination of STM and transmission electron microscopy (TEM) data may provide a better understanding of the three-dimensional structure and the functional interactions of biological macromolecules.

2.2 STM FOR BIOLOGICAL APPLICATIONS

We now describe the basic requirements of scanning tunneling microscopes suited for biological applications. A number of designs have been published (e.g., Emch *et al.*, 1988; Michel and Travaglini, 1988; Guckenberger *et al.*, 1988), and there is now a number of commercially available instruments that are well suited for macromolecular structure research (e.g., Nanoscope II, Digital Instruments, Inc.; STM-SA1, Park Scientific Instruments; Micro-STM, Omicron Vakuumphysik GmbH). These microscopes are usually operated in air.

The relatively small maximum field of view and the slow scan speed of most STM make searching for biological objects a tedious task when they are not densely and homogeneously distributed on the support. This problem can be partially overcome by the combination of a large scan range ($\geq 2\mu m^2$) and a reproducible coarse positioning system. To facilitate the identification and the positioning of sample areas of interest, the STM can be combined with an auxiliary microscope that allows low magnification. Electron microscopes cover the range of magnifications from almost macroscopic to molecular dimensions. However, they require vacuum and therefore are only suitable as auxiliary microscopes for applications in which vacuum conditions are not disturbing and the objects are not sensitive to electron irradiation (e.g., metal-coated biological specimens). In fact, STM have been incorporated into scanning electron microscopes (SEM) (Gerber *et al.*, 1986; Ichinokawa *et al.*, 1987; Anders *et al.*, 1988), SEM, extended for transmission mode [S(T)EM] (Stemmer *et al.*, 1988), and TEM (Panitz, 1987; Spence, 1988). However, both complete dehydration and electron irradiation disrupt the native conformation of uncoated biological molecules. Thus, a light microscope may be used as an auxiliary device (e.g., Emch, 1988; Guckenberger, 1988). Fluorescent labeling allows detection of objects that are too small to be resolved directly. In addition, sample regions of interest can quite often be localized in the light microscope by macroscopic traces. For example, well-spread molecules that have been adsorbed by spraying techniques are always found near the border of macroscopic droplets.

Due to the small electrical conductivity of the biological macromolecules, tunneling currents as low as possible should be aimed for. With a special preamplifier such as that designed for the patch clamp technique (EPC-7 probe, HEKA Elektronik P. Schulze, D-6734 Lamprecht), it has been possible to achieve tunneling currents as low as 0.1 pA (Stemmer *et al.*, 1989; Guckenberger *et al.*, 1989). A preamplifier design that allows a similarly low current at an even higher bandwidth has been published by Michel *et al.* (1991).

Uncoated proteins and membranes need to be hydrated in order to keep their native conformation. The degree of hydration also influences the process of STM imaging. To be able to work under defined conditions, the microscope can be placed in a humidity-controlled box. A liquid cell that makes STM in water possible may be foreseen.

2.3 TUNNELING TIPS

2.3.1 Tip-Sample Interaction

The properties of the tunneling tips are decisive for the quality of the STM images. To resolve the surface features satisfactorily, stable tunneling is required, and the tips have to be sufficiently sharp (i.e., they should confine the tunneling current to as small an area as a few angstroms squared). The foremost part must be shaped needlelike over a distance that corresponds at least to the specimen corrugation so that it can dip into narrow pores and cavities of the biological structures. These extremely sharp tips should be based on low-aspect-ratio shanks in order to provide mechanical stability.

When scanning in the constant current mode, the tunneling tip follows a surface such that the tip-sample separation is kept constant. On strongly corrugated samples, the "effective" tip may become a function of the local corrugation, and hence the resulting STM topographical images will be strongly dependent on the tip geometry (Chicon *et al.*, 1987). Smooth images with no fine structures being resolved most likely indicate a rather "flat" tip. In some cases, specimen features are imaged twice or more times on the same topographical image due to the tunneling current jumping from one "local tip" to another. Such multiple tips frequently cause artifacts at steep and high structural features such as membrane borders and protein filaments. There is, for example, a number of early publications on STM of biological filamentous structures in which bundles are reported to be present that almost certainly are just multiple images of a single filament (e.g., Travaglini *et al.*, 1987). Figures 2.1 and 2.2 show tip artifacts. Figure 2.2 illustrates how the tip-sample interaction can give rise to such an artifact.

2.3.2 Tip Fabrication

Tunneling tip manufacture has generally been rather intuitive and tedious. We have been using tips electrochemically etched from gold, platinum/iridium, or tungsten wire. These materials result in tips of more or less similar qualities. Mechanically cut wires usually act as multiple tips when scanning over a strongly corrugated specimen and therefore are not recommended for most biological applications.

In the following, we describe how to make tungsten tips based on a method published by Bryant *et al.* (1987). A tungsten wire (0.25 mm in diameter) is

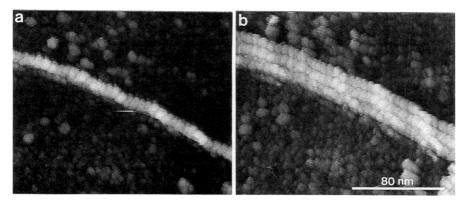

FIGURE 2.1 Filamentous structures often appear as "bundles" when imaged with STM. **a**: STM image of a single, metal-coated, native actin filament. **b**: After accidentally crashing the tip into the sample, the same filament appears as a bundle of three.

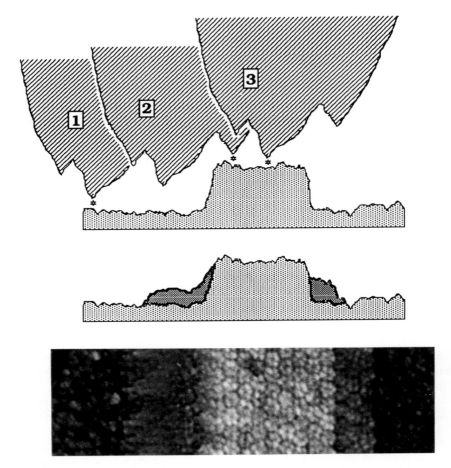

FIGURE 2.2 Illustration of possible image formation based on the example of the STM image of a bacteriophage T4 polyhead at the bottom of the figure (see also Section 2.6). The upper part of the figure shows the tunneling tip in three subsequent positions while scanning across a polyhead. The tunneling current between a local tip and the specimen surface is indicated by asterisks. The resulting topographical image is presented in the middle, with the artifacts caused by the particular tip geometry being indicated by the densely dotted area. In position 1, the long whisker of the tip is tunneling on the support, whereas in position 2 the short whisker starts to image the phage over a short distance giving rise to the artifactual border on the left side of the polyhead. Note that this border is "convoluted" with the local tip geometry in such a way that the regular surface structure, visible in the central part of the polyhead, is no longer resolved.

tapered electrochemically in an NaOH solution (10% w/w) by applying an alternating voltage of about 2 V between the tip and a Pt-electrode. The lower part of the wire is protected from etching by paraffin, which is stripped off afterwards. Etching proceeds just below the solution meniscus until the lower end of the wire falls down. It is recommended to use the drop-off end of the wire as a tip, because its disconnection from the electrode terminates etching instantaneously. Polishing off the finest asperities due to continued etching is thus avoided. Although the method works quite reliably, tip diameters smaller than 20 nm are only rarely obtained.

2.3.3 Tip Conditioning

To achieve good resolution, the tips must almost always be conditioned during scanning by applying pulses to the tunneling voltage (pulse height ≥ 3 V, pulse duration ≤ 1 μsec). Thus the actual tunneling tips arise during scanning. As a rule of thumb, the sharper a tip is the less stable it is. Therefore, for optimal resolution, the tip may have to be conditioned more than once during the acquisition of one single image. Thin Pt-C films (2–5 nm) evaporated onto glass or mica are well suited as samples for conditioning the tips before moving to the actual biological sample. The performance of the tip is judged by the appearance of the typical granular structure of these films (Fig. 2.3). Besides the topographic signal, dI/ds has proven to be a most valuable tool with which to detect quickly any contamination of the tip. In the case of a stiff sample, dI/ds corresponds to the local barrier height (Coombs *et al.*, 1986), and a very low apparent barrier height indicates that tunneling no longer occurs. If dirt cannot be removed by pulsing, the tip can be treated with sandpaper. Even though they do not appear sharp anymore, such tips quite often perform well. This is probably due to new "nano-tips" created in the course of this procedure.

2.4 SAMPLE PREPARATION

2.4.1 Substrates

Uncoated or metal-coated biological macromolecular structures to be studied with STM first have to be immobilized onto a solid support. Specimen supports for STM must allow a strong immobilization of the samples, and they have to be well conductive, at least when STM of uncoated specimens is intended. Substrate surfaces should be as smooth as possible so that they do not interfere with the structure of the biological specimen in the STM image. In addition, they must be mechanically stable to avoid an elastic response to the inevitable forces generated between tip and sample during scanning (Stemmer *et al.*, 1989).

The first results of STM in biological application were obtained with highly oriented pyrolytic graphic (HOPG) as the specimen support (e.g., Baró *et al.*, 1985; Travaglini *et al.*, 1987), and it is still widely used (e.g., Edstrom *et al.*, 1990;

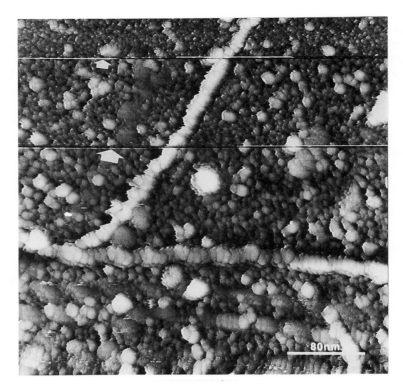

FIGURE 2.3 Tip shaping demonstrated on native actin filaments, adsorbed to a Pt-C support film and subsequently metal coated. The STM image was acquired from the bottom to the top. In the bottom part of the image the actin filaments appear at least twice as wide as expected from TEM data. After a first series of pulses (large arrow) an obviously sharper tip has imaged the structures more accurately. After applying pulses one more time (small arrow) the filament appears only slightly broadened, and the support film shows its typical grain size of about 2 nm.

Olk *et al.*, 1991; Haggerty *et al.*, 1991). Flakes of HOPG may be flat on an atomic scale over areas of micrometers, mechanically stable, and well conductive. However, the adsorption of biological specimens has proven to be very poor on graphite, causing the structures to aggregate. They are also easily moved by the tip (e.g., Travaglini *et al.*, 1987). Aggregates of the biomaterials on the graphite surface in some cases form regular arrays. Such arrays have withstood the mechanical stress caused by the scanning process and therefore could be reproducibly imaged (see Chapter 3, this volume). Defects and the fine structure of step edges of pure graphite can be very misleading when they resemble the expected structure of the specimen. When working with periodic molecular structures such as quasicrystalline Langmuir–Blodgett films, the intrinsic periodic structure of the graphite might be disturbing (e.g., Clemmer and Beebe, 1991; see also Fig. 2.4).

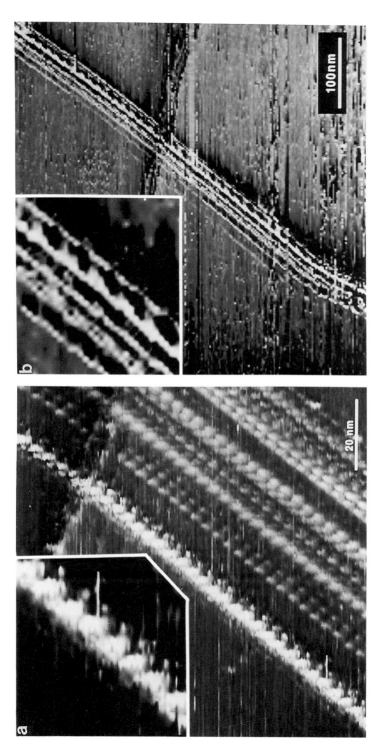

FIGURE 2.4 Structures observed on untreated HOPG. Such eigenstructures can be very misleading when using HOPG as a specimen support for STM imaging of biological macromolecules. Straight, filamentous structures with a helical appearance are often seen on HOPG. When imaged by a multiple tip they seem to be arranged in bundles. The filament on the left in a resembles very much a left-handed conformation of a double-stranded DNA. The structures in **b** could easily be interpreted as helical arrangements of a globular protein. Two times enlarged sections are displayed in the top left of **a** and **b**. A multiple tip image of a small graphite flake might have caused the pattern shown in **c**. Astonishingly, as shown in **d**, hexagonal patterns with a strong amplitude and lattice dimensions much larger than the atomic lattice of the graphite are sometimes observed on HOPG. *(Continues)*

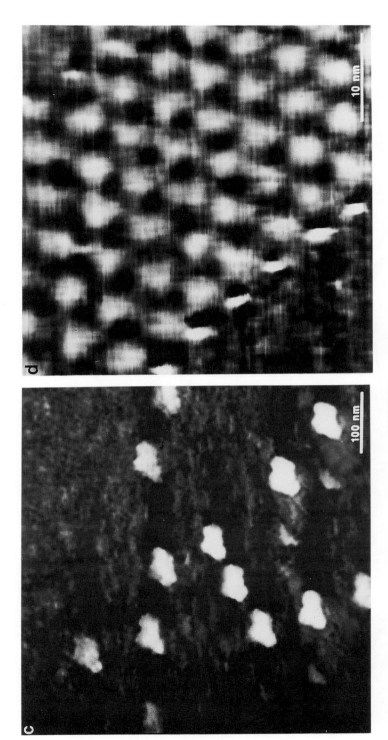

FIGURE 2.4 *(Continued)*

The surface of crystalline gold has been used for STM imaging of DNA (see Chapter 5, this volume). Its suitability as a support for membranes and proteins is currently being investigated (Häussling et al., 1991; see also Chapter 3, this volume). In any case, the possibility of binding biological matter strongly to the gold surface by chemisorption makes it an especially interesting support for STM measurements in solution.

Although less flat than HOPH and gold surfaces, thin evaporated films of Pt-C deposited onto either glass or mica have proven to be suitable supports for STM that do not display the problems that occur with HOPG. Adsorption of proteins and membranes onto Pt-C films is usually very good, and the films give very reproducible, high-quality pictures provided that the tunneling tip is in good condition. The maximum corrugation of Pt-C films of a nominal thickness of 5 nm amounts to approximately 2 nm and the standard deviation is 0.3 nm. The films are directly evaporated onto the freshly cleaved mica platelets or light microscope coverslips at room temperature by an electron beam–heated gun in a high-vacuum apparatus (e.g., Balzers BAF 400T).

Thin carbon films, commonly used in TEM, are marginally smoother than Pt-C films. However, they do not always allow stable tunneling when directly evaporated onto mica platelets or light microscope coverslips. This is probably due to the much higher electrical resistance of carbon films than Pt-C films of the same nominal thickness. On the other hand, the carbon films become sufficiently conducting when they are mounted on gold-coated, fenestrated, plastic films that are themselves attached to TEM grids (Stemmer et al., 1989). In addition, this allows the search for a suitable sample region first in an TEM or STEM before acquiring an STM image of the same area (Fig. 2.5). When the STM is combined with an auxiliary light microscope, the tunneling tip can be easily placed over the selected area by aid of the characteristic pattern of holes in the plastic film.

2.4.2 Sample Deposition

On depositing the sample, a homogeneous distribution of the biological specimens in as close to a native conformation as possible is aimed for. The forces between the objects and the support present a major problem. Strongly bound biological specimens are prevented from clustering on the support surface, and they may also better withstand the forces that arise during scanning between tip and sample. On the other hand, the structure can be substantially distorted by strong binding forces, well known fromTEM and STM of biological specimens (e.g., Wang et al., 1990; see also Section 2.7.3).

There are several methods with which to immobilize the biological specimens on a solid support, the most common of which is adsorption based on van der Waals forces and forces between dipoles and charges. The quality of the adsorption depends very strongly on the support used. There exists a wide

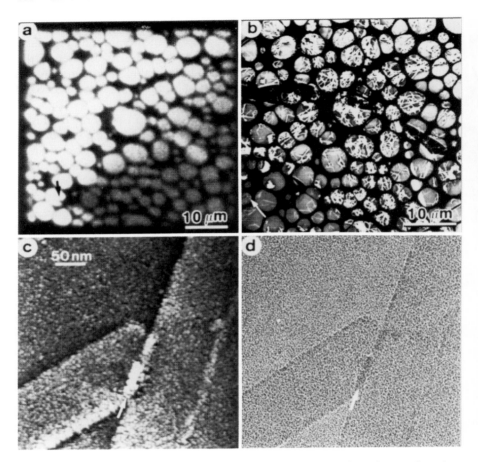

FIGURE 2.5 Carbon films mounted on gold-coated, fenestrated plastic films that are themselves attached to TEM grids. This specimen support allows the search for a suitable sample region first in a TEM or STEM before acquiring an STM image of the very same area. When the STM is combined with a light microscope, the tunneling tip may be easily placed over the selected area by aid of the characteristic pattern of holes in the plastic film. The tip positioning is demonstrated on the example of metal-coated bacteriophage T4 polyheads. a: Positioning of the tunneling tip (arrow) is observed in the integrated light microscope. b: The identical area as in a shown via STEM. Here a suitable sample region may be chosen (arrow). c, d: An STM image and a TEM micrograph, respectively, of the same specimen area [From Stemmer et al. (1989) Fig. 3.5, p. 268.]

experience with the adsorption of biological specimens to carbon films from TEM. Pt-C films have proven to behave very much the same beside a slightly stronger adsorption (Amrein et al., 1988; Wang et al., 1990).

For the adsorption of many biological specimens to carbon films or Pt-C films, the support films have to be rendered hydrophilic by glow discharge (e.g., in a Harrick Plasma cleaner). For good adsorption in some cases (e.g., recA – DNA complexes, purple membranes) bivalent cations (e.g., magnesium acetate ≤ 10 mM) are required in both the adsorption solution and the washing solution (Sogo et al., 1987). Some specimens adsorb better to hydrophobic surfaces. Obviously glow discharge must here be omitted. It is recommended to use carbon films or Pt-C films that have been stored for several weeks in air, since they become more hydrophobic with time. Sometimes it is worthwhile to use hydrophobic supports in order to prevent destructive binding forces, even when the objects adsorb much less densely (Henderson et al., 1990).

Filamentous structures that do not adsorb from solution in a well-spread manner, such as myosin, may be deposited by the spraying technique. Here the object solution is sprayed in very fine drops onto the substrate. Usually the solution must contain up to 50% glycerol in order to prevent clustering of the molecules. The sample may be rinsed afterwards.

Electrochemical methods for the immobilization of the objects can be used when STM measurements are carried out in fluids (see Chapter 5, this volume). The objects are adsorbed by applying appropriate potentials between the sample solution and the support. They are imaged while maintaining the potential. Chemical binding and chemisorption are deposition methods too. In the case of a gold support, for example, chemisorption may be accomplished via thiols and disulfides (Häussling et al., 1991).

To check the quality of the adsorption, it is very efficient to prepare a sample for TEM in parallel with the STM sample. In case of a Pt-C support film it is usually sufficient to check the actual preparation on a carbon-coated electron microscope grid prepared in parallel with the STM sample, since the adsorption properties of carbon films and Pt-C films are about the same.

2.4.3 Dehydration of the Objects

The natural environment of most biological macromolecules is an aqueous solution. Indeed, when STM imaging of the uncoated specimens is intended, dehydration can principally be avoided by tunneling in water or in a buffer solution (Häussling et al., 1991; Lindsay et al., 1989; see also Chapter 5, this volume). Here, however, a stable immobilization of the specimens may prove to be very difficult. On the other hand, dehydration by simple air-drying engenders dramatic structural alterations due to surface tension forces (Wildhaber and Gross, 1985).

When coating of the biological material is intended, the material must be dehydrated after adsorption in order to expose the surface for metal deposition. Here artifacts due to air-drying can be avoided by freezing and subsequent vacuum sublimation of the water phase (freeze-drying). Polysaccharides, nucleic acids, most proteins, and even lipid bilayers, however, are hydrophilic, that is, they are naturally surrounded by a shell of strongly bound, well-ordered water molecules. This hydration shell is essential in the establishment and stabilization of the native conformation of biological macromolecules (e.g., the correct folding of a polypeptide chain into a protein; for an introduction, see Kellenberger, 1987). It has been shown by means of TEM of metal coated protein lattices that the removal of this shell causes the protein structures to collapse partially even when they are freeze-dried (Gross, 1987).

When protein structures are freeze-dried at a temperature no higher than 193 K, however, a fraction of water remains bound that we refer to as the *water of the hydration shell*. It desorbs in distinct peaks when the specimens are heated above about 220 K (Fig. 2.6). These findings and those in many other studies (for a review, see Robards and Sleyter, 1985) show that freeze-drying of the adsorbed macromolecular structures at 193 K is completed within a reasonable time (1–3 hr) while the hydration shell is substantially maintained.

For freeze-drying, the samples are quickly frozen, mounted onto the specimen table under liquid nitrogen, and transferred onto the precooled cold-stage (193 K) of the freeze-etch unit (e.g., Balzers BAF 400T). Freeze-drying is carried out under high vacuum conditions ($p < 10^{-6}$ mbar) for 1–3 hr, depending on the object. After freeze-drying, the sample must be stabilized by coating before it can be warmed up to room temperature and withdrawn from the vacuum system.

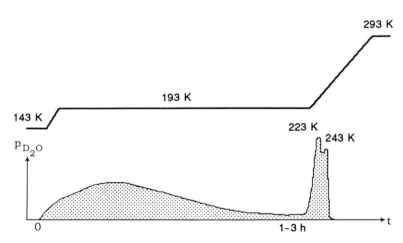

FIGURE 2.6 The course of the sublimating specimen water during the freeze-drying procedure. The specimen water was substituted with D_2O, allowing mass spectrometric differentiation between the specimen water (D_2O) and water vapor originating from elsewhere in the UHV freeze-etch unit. After drying at 193 K for 1–3 hours, depending on the biological sample, an equilibrium at a very low sublimation rate was reached. Warming to room temperature resulted in a further sublimation of the specimen water in at least two distinctive peaks for all specimens tested. The removal of this strongly bound water, which we refer to as the *water of the hydration shell*, caused the biological structures to collapse partially.

2.4.4 Metal Coating

Metal coating of macromolecular structures for STM is performed in order to render the surface uniformly conducting. The coating should cover the specimen homogeneously at a very small film thickness so that fine surface features are not blurred. When optimal structural resolution is aimed for, the sample has to be freeze-dried, and consequently, the coating film must stabilize the well-preserved protein conformation during the whole period of the transfer from the sample preparation unit to the microscope and STM imaging.

Films evaporated by an electron beam–heated gun have provided the best results with respect to the demands outlined above. The formation of thin metal films proceeds via thermal accommodation, surface diffusion, nucleation, and crystal coalescence (e.g., Venables *et al.*, 1984). Grains in the condensate arise due to the lateral mobility of the evaporated material on the specimen surface. If the exposed structures cause an unequal distribution of surface forces, nucleation and crystal growth occur preferentially at surface sites with higher binding energy. Therefore at higher resolution such preferential nucleation (decoration) can be responsible for revealing structural details, as well as the actual surface relief itself (Gross *et al.*, 1985).

Pt-C has been the most commonly used evaporation material in STM of biological specimens. It reveals well-conducting, very stable coating films when applied to air-dried biological specimens. However, it does not sufficiently stabilize well-preserved biological structures obtained by freeze-drying. Here the Pt-C films form large clusters, and the structures underneath collapse when the sample is warmed up and withdrawn from the vacuum system. In high-resolution shadowing for TEM, alterations of the metal films can be hindered by an additional stabilizing carbon film (Bachmann *et al.*, 1966). Such an additional carbon coat, however, blurs fine surface features when probed by an STM.

Pt-Ir-C films, on the other hand, have proven to remain three-dimensionally stable when transferred to atmospheric conditions after freeze-drying of the sample. They allow stable tunneling and yield a much finer granularity than Pt-C films in both TEM and STM (Amrein *et al.*, 1988). In addition, preferential nucleation can almost be neglected. However, the ratio of the components is crucial. The compositions of the films change as a function of the mass evaporated from the source electrode, a Pt-Ir cylinder (diameter 1.5 mm, 70% Ir) inserted into a graphite rod (diameter 2 mm) Fig. 2.7a) Wavelength dispersive x-ray (WDX) analysis of the films made in parallel with TEM and STM experiments have revealed that for three-dimensional stability a carbon content of at least 25% (w) is required (Wepf *et al.*, 1991). This is reproducibly achieved after pre-evaporating about 13.8 mg from the source (Fig. 2.7b). For stable tunneling the metal content of the films should be no less than 60%.

To achieve a homogeneous, coherent coating of the freeze-dried biological matter, Pt-Ir-C films can be rotary evaporated from an elevation angle of 65° to a calculated average film thickness of about 1.5 nm. Thus a coherent metal coat arises that mainly superelevates, as well as slightly broadens, the relief underneath.

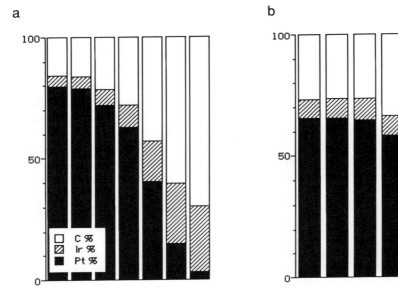

FIGURE 2.7 a: Wavelength dispersive X-ray analysis of Pt-Ir-C films, sequentially collected from a new evaporation source. The first film was collected within a range of 6–9 mg mass evaporated from the source electrode, the second from 9–12 mg, and so forth. The ratio of Pt, Ir, and C changes upon the evaporation of the mass from the source. The Pt content decreases continuously, while the Ir and C content increase. b: WDX analysis of Pt-Ir-C films after a preevaporation of about 13 mg from the Pt-Ir-C electrode. The carbon content of a least 25%, required for stable and fine granular films, is reproducibly attained after a preevaporation of 13.8 mg.

2.4.5 Metal Replicas

As an alternative to adsorption and coating, fluid, nonconductive biomaterials can be replaced by a rigid, highly conductive freeze-fracture replica (Zasadzinsky et al., 1988; Blackford et al., 1991). Metal replicas for STM are obtained by conventional freeze-fracture techniques. The sample is quickly frozen (e.g., in a Balzers FSU 010 propan jet), transferred to a freeze-etch unit, and fractured. The fractured surfaces can either be immediately replicated or freeze-etched prior to their being metal coated. Here Pt-C proves to be well suited as coating material because the metal film can be further stabilized with a carbon layer (30 nm). The carbon layer renders the replica sufficiently rigid. Finally, the biomaterial is removed from the replica so that the impression of the sample surface is now accessible to the STM measurements.

2.5 STM OF UNCOATED SPECIMENS

2.5.1 STM Imaging under Conventional Tunneling Conditions

There seems to be no generally applicable way to obtain STM images of uncoated membranes and proteins under conventional tunneling conditions. Mechanical stabilization via the formation of regular arrays has been successfully used (see Chapter 3, this volume). A variety of supports have been used. In some experiments the samples had to be kept wet or hydrated in order to make STM imaging possible (Amrein *et al.*, 1989; Guckenberger *et al.*, 1990). The tunneling currents and voltages reported in most of the experiments on uncoated organic material are ≥ 50 pA and ≤ 1 V, respectively. However, reducing the tunneling current and increasing the bias voltage seem to enhance the reproducibility of STM imaging of uncoated proteins and membranes (Welland *et al.*, 1989; Stemmer *et al.*, 1989). The following examples are intended to point out some of the characteristics of the approaches discussed thus far rather than to be a complete representation of the published work. More examples are given in Chapter 3, this volume.

2.5.1.1 PURPLE MEMBRANE

The light energy–transducing purple membrane of the *Halobacterium halobium* consists of the protein bacteriorhodopsin embedded in a lipid bilayer. It is one of the best-characterized natural membranes because of its intrinsic simplicity and because bacteriorhodopsin forms a perfect two-dimensional crystal of plane group p3 and a lattice constant of 6.2 nm. The three-dimensional structure of the protein has been determined at near atomic resolution by low-dose electron microscopy of tilted, unstained specimens and extended image processing (Henderson *et al.*, 1990).

For STM, purple membranes have been adsorbed to Pt-C–coated mica platelets. It has not been difficult to obtain images of the membrane at a tunneling voltage between 0.1 and 1 V (the tip biased with either polarization) and a tunneling current between 50 pA and 1 nA (Figs. 2.8, 2.9). Air-drying and adsorption cause cracks in the membranes along lattice lines. Therefore membrane borders quite often include an angle of 120° or 60° observed with both TEM and STM (Fig. 2.9). Compared with the corrugation of the Pt-C support film, the purple membranes seem to have an even smoother surface, and the soft corrugation on top of the membranes does not reflect the protein lattice. The latter may be due to air-drying, which causes the membranes to lose their order on the surface (Studer *et al.*, 1981). A thin layer of an adsorbate that obscures the surface could also be a reason. However, rows with a regular spacing of about 6 nm appear near steps. They might have been caused by a decay along lattice lines of the membrane at its border.

2.5.1.2 RecA–DNA COMPLEXES

RecA protein from *Escherichia coli* is involved in several crucial steps in homologous genetic recombination (Howard-Flanders *et al.*, 1984). Transmission electron microscope images of negatively stained complexes have revealed that recA-DNA complexes form right-handed helical filaments of about a 10 nm

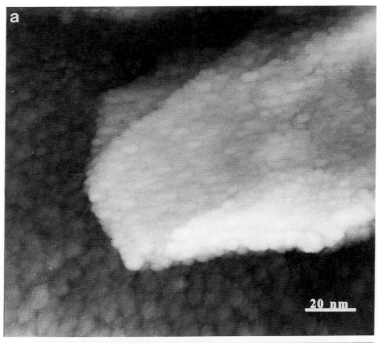

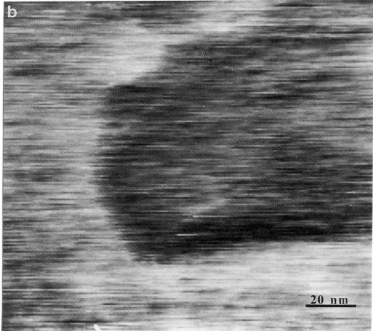

diameter in which about six recA protomers contribute to one helical repeat (Egelman and Stasiak, 1986)

Unlike purple membranes, recA–DNA complexes could not be imaged by STM when they were directly adsorbed to the Pt-C support. Scanning tunneling microscope images of such a sample merely showed the typical support structure with no indication of the presence of the protein. Subsequent metal coating of such a sample, on the other hand, made it easy to obtain filaments all over. Therefore, the uncoated complexes must have been immediately pushed away by the tip during scanning. This was most probably because these filaments had been less well contacted and mechanically fixed than the extended membrane structures. To make STM imaging of uncoated recA–DNA complexes possible, the Pt-C support had to be treated with magnesium chloride ($MgCl_2$) by putting it on a drop of a 5 mM $MgCl_2$ solution for 1 minute before adsorption (Amrein et al., 1989). After removing excess solution and drying, a thin layer of $MgCl_2$ remained bound. Adsorption on such support was very dense as determined by TEM. During tunneling the sample had to be kept humid to achieve a stable, not too noisy tunnel current. This was achieved by periodic wetting of the sample with double-distilled water.

Scanning tunneling miscroscopy images of the densely adsorbed complexes in some regions could barely be separated into individual filaments, whereas in other regions the complexes could easily be distinguished as characteristic right-handed helical filaments (Fig. 2.10). Quite often filaments had been moved over a short distance by the scanning tip. An individual filament is shown in Figure 2.11 in which every striation seems to be composed of three to four parts. Since this image represents only one-half of the surface of the helical filament, every helical turn of the complex is composed of about six such parts that are presumably the recA protomers.

2.5.2 STM Imaging at a Tip-Sample Voltage ≥ 5 V

With a tip-sample voltage of 5 V or higher it has been possible to resolve the surface of a two-dimensional protein crystal, the hexagonally packed intermediate (HPI) layer, clearly and with good reproducibility (Guckenberger et al., 1989). Images of the purple membrane acquired under these conditions some-

---

FIGURE 2.8 a: Scanning tunneling microscopic topographical image and b: dI/ds image of uncoated purple membrane adsorbed to a Pt-C support film. The membrane is not perfectly flatly adsorbed, and the edges are partially folded over, a situation usually encountered with such preparations (see also Figs. 2.15 and 2.16). The apparent height of the uncoated membranes lies between 3 and 4.5 nm, the higher value corresponding to approximately that expected for the height of the purple membrane. The lower value may be explained by a repulsive elastic response when imaging the membrane. The latter is also evident from the decrease in the apparent barrier height on the purple membrane compared with the support (b).

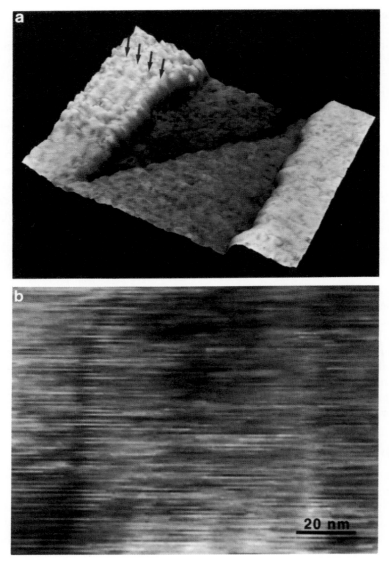

FIGURE 2.9 a: Surface view representation of an uncoated purple membrane directly adsorbed to the Pt-C support and a second layer on top of it on the right. The lower layer shows a crack typical for air-dried preparations. Rows with a regular spacing of about 6 nm appear near the edge of the crack (arrows). They might have been caused by decay, along lattice lines, of the membrane at its border. Here the membranes are measured with approximately the expected height of 4.5 nm. b: In fact, the dI/ds map shows a much less pronounced decrease of the apparent barrier height over the membrane when compared with Figure. 2.8.

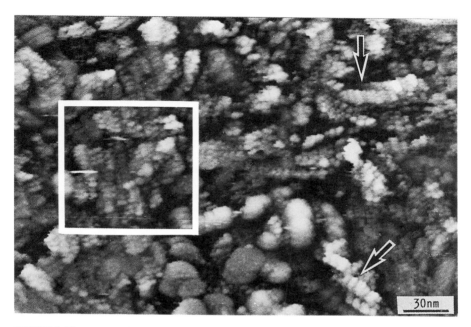

FIGURE 2.10 Low magnification STM image of unstained recA–DNA complexes densely adsorbed on an $MgCl_2$-treated Pt-C film. In some regions individual filaments cannot be separated (box), even though a striation caused by the helical repeat of the structure can easily be distinguished. Arrows point to single filaments. They show the characteristically right-handed helical structure. The image is a top view in gray-tone representation.

times showed positive, sometimes negative, contrast (Guckenberger *et al.*, 1991). The contrast, as well as the resolution, obtained on the uncoated specimens appears to depend mainly on the shape of the tunneling tip. To allow good imaging of uncoated biological material at these high voltages, the current has to be below 1 pA. The humidity plays a central role too.

When STM imaging is performed at high voltage, the work functions of platinum and tungsten are exceeded and "tunneling" has to be considered as field emission. However, the tip-sample distance has proven to be very large in such STM experiments (see Fig. 2.16), and an effective barrier height on the order of 1 eV has to be assumed. This small barrier height, being indicative that the tip or the substrate is not clean, makes Schottky emission an alternative emission mechanism.

2.5.2.1 HPI LAYER

The HPI layer is a regular protein monolayer found outermost in the cell wall of *Deinococcus radiodruans,* a bacterium famous for its exceptional radioresistence

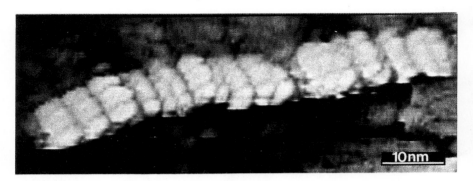

FIGURE 2.11 Scanning tunneling microscopic image of an individual, unstained recA–DNA complex. Every striation, corresponding to one helical turn, consists of a three to four partite structure. Such substructure seems to represent the RecA protomers.

(Baumeister *et al.*, 1982). Three-dimensional models of the HPI layer have been obtained by electron microscopy of both tilted, negatively stained specimens and freeze-dried, heavy metal–shadowed specimens (Baumeister *et al.*, 1986; Gross, 1985; Guckenberger, 1985). Compared with the purple membrane, the HPI layer possesses a prominent surface on both the outer and the more corrugated inner surface. It has p6 symmetry and a lattice constant of 18 nm.

For STM imaging of an uncoated HPI layer, the isolated specimens were directly adsorbed to Pt-C–coated coverslips out of a buffer solution (Tris HCl, pH 7.5) containing a detergent [0.1% litium dodecyl sulfate (LDS)]. The sample was then washed with double-distilled water.

As in the case of recA–DNA complexes, conventional STM imaging of the uncoated layers, directly adsorbed to the Pt-C support, was not possible. Here insufficient conductivity was overcome when applying a tip-sample voltage above 8 V for most of the time and selecting the current below 0.5 pA (Guckenberger *et al.*, 1989).

Figures 2.12 and 2.13 present STM images of an uncoated HPI layer scanned under these conditions. The periodic structure of the layer appears clearly, and in one case even the pore in the core is resolved (Fig. 2.13). At a voltage below 4.5 V or a current above 0.5 pA, the periodicity of the HPI layer smears out and the tip apparently dips into the surface or begins to oscillate. Clearly the relative humidity plays a central role in providing the conductivity needed. It had to be kept within a range of 30–45% in order to resolve the HPI layer satisfactorily. Below a relative humidity of 10% it was not possible to find any indications of the HPI layer on the support, whereas at a humidity above 45% the tip often began to oscillate vertically.

2.5.2.2 PURPLE MEMBRANE

Scanning tunneling microscopy images of uncoated purple membrane, directly adsorbed to Pt-C–coated mica, are easily obtained at tunneling currents lower

2 STM of Proteins and Membranes 149

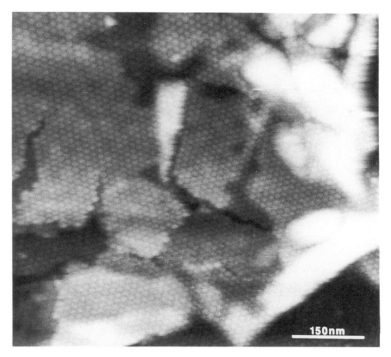

FIGURE 2.12 Scanning tunneling microscopic image of an uncoated HPI layer acquired at a bias voltage of −8 V, a "tunneling" current of 0.3 pA, and a relative humidity of 35%. The periodical structure of the HPI layer is clearly resolved. Locally several sheets are piled up on top of each other.

than 1 pA and tunneling voltages above 5 V (Guckenberger *et al.*, 1991). Surprisingly, the membranes appear sometimes with positive sometimes with negative contrast (Figs. 2.14, 2.15) on the same sample. The actual contrast seems to depend mainly on the tip used, and the polarity of the tunneling voltage does not apparently influence the results obtained by a certain tip. The apparent thickness of the membranes that have been measured range from +3.5 nm (maximal positive contrast) to −10 nm. One membrane on top of another has similar contrast as long as the tunneling voltage is high enough. However, images acquired with too small of a voltage (5–10 V, depending on the actual tip) do not reveal the second layer.

Usually the membranes are best resolved in images with positive contrast, and the poorest resolution is observed in the case of very strong negative contrast. However, the basic p3 lattice of the purple membrane could not be observed in either case. The contrast decreases noticeably when the same area is scanned several times or when the tunneling current is increased above 0.5 pA. The humidity of the ambient air affects the STM images as well. Positive contrast is increased at higher humidities, whereas the contrary is observed with tips that image the membranes with reversed contrast. Interestingly, the purple membranes

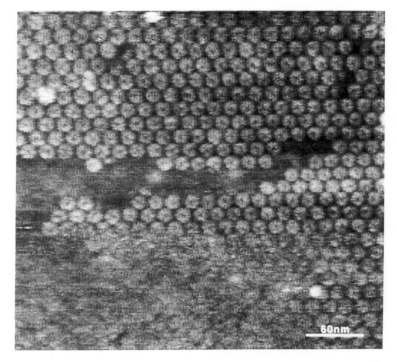

FIGURE 2.13 Scanning tunneling microscopic image of an uncoated HPI layer acquired at a bias voltage of 4.7 V and a "tunneling" current of 0.3 pA. In this high magnification image even the pores in the core are visible.

can still be imaged at relative humidities well below 10%, ambient conditions under which the HPI layer can no longer be observed.

To get an idea of the absolute tip-sample distance, the vertical tip position has been measured versus the tunneling voltage at a given, small tunneling current. Figure 2.16 shows a typical recording on a Pt-C film without any purple membrane adsorbed to it. The observed slope depends strongly on the tip used. Tips that provide strong negative contrast have proven to be far (up to 50 nm) from the sample. In any case, there is enough space for the tip to move toward the specimen without touching it.

2.5.2.3 THEORETICAL MODEL OF IMAGE FORMATION

In the following section, a simple theoretical model is presented that can account for the variable contrast observed in STM images of uncoated purple membranes by relating the contrast formation to the shape of the tunneling tip (Guckenberger et al., 1991). Here it is not important whether field emission or Schottky

2 STM of Proteins and Membranes   **151**

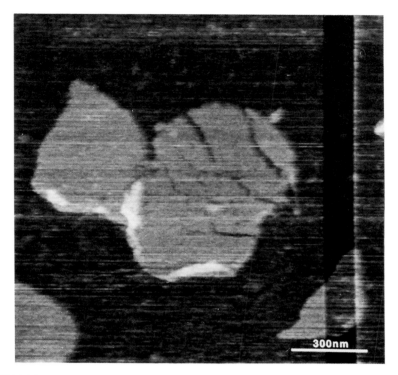

FIGURE 2.14 Scanning tunneling microscopic image of uncoated purple membrane adsorbed onto a Pt-C support. The data have been acquired at a bias voltage of 13 V, a "tunneling" current of 0.15 pA, and a relative humidity of 40%. In the vertical stripe on the right the bias voltage has been decreased to 11 V. Note that the membrane patches appear with positive contrast.

emission accounts for the imaging current. In both cases the current depends on the electric field at the negative electrode, as well as on the work function Φ and the area from where the electrons are emitted. Obviously for a blunt tip the emitting area will be larger than for a sharp tip, independent of the polarization of the tip chosen. Therefore, to obtain a certain current at a given voltage, the tip-sample separation will be larger for a blunt tip than for a sharp tip (Fig. 2.17a,c).

When the purple membrane is included in the model, surface states are assumed that define the Fermi level at the surface of the membrane that faces the tip. Depending on the polarization, the electrons are emitted from these surface states to the tip or they reach the surface states from the tip via inelastic processes at the surface (Fig. 2.17b,d). In both cases the membrane becomes charged. Finally, an electron transport through the membrane is assumed that depends on a potential difference, V_{PM}, between the membrane surface and the support film.

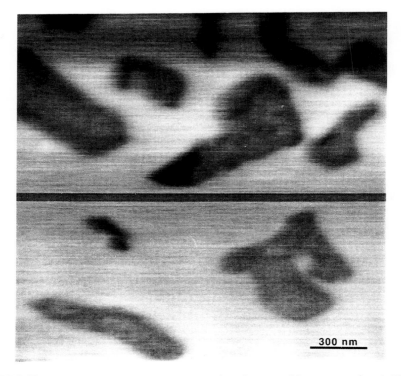

FIGURE 2.15 Scanning tunneling microscopic image of another area of the same sample as in Figure 2.14, using another tip. Here the membranes appear with negative contrast. The imaging parameters were 10 V bias voltage, 0.2 pA "tunneling current," and a relative humidity of 36%. In the horizontal stripe in the middle of the image the bias voltage was reduced to −8.5 V.

The characteristic potential slope of the tip can be steeper than the effective slope across the membrane. In this case the tip will retract when it is moved from over a substrate region to over a membrane in order to keep the tunneling current constant during STM imaging. This gives rise to the positive contrast observed in case of a sharp tip (Fig. 2.17b). In case of a blunt tip the slope at the tip is flatter than the slope across the membrane, and the tip will come closer to the sample over membrane patches (Fig. 2.17d). Note that the latter situation can only be explained when charging of the membrane surface is taken into account. Without charges the potential slope across the membrane would always be less steep, because it has a larger dielectric constant than air.

The differences in potential at the surfaces of the membrane and the support film, V_{PM}, can be determined from the measured slope of the potentials and the differences between the apparent height and the known height of the purple membrane (4.8 nm). V_{PM} observed thus far ranges from 2 to 5 V, with most

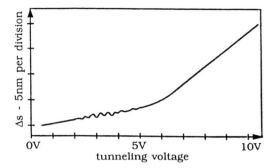

FIGURE 2.16 Typical plot of the relative vertical tip position versus the tunneling voltage on Pt/C. Tunneling current It = 0.6 pA. Exchanging the tip can result in large changes in the slope of the right part of the curve.

values around 3.5 V. V_{PM} seems to be independent of the tunneling current as long as it does not exceed 1 pA. On the other hand, it becomes smaller when the humidity of the ambient air is increased.

2.6 STM OF METAL-COATED SPECIMENS

2.6.1 Conductively Coated Specimens

Insufficient conductivity for STM imaging of membranes and proteins can be overcome by metal coating the samples. This not only makes imaging relatively easy but also provides STM topographies that are not affected by the intrinsic conductivity properties of the biological materials. Sample preparation is derived from a well-established method used in TEM to enhance contrast of the biological specimens. The biological specimens are adsorbed to the substrate from an aqueous solution, dehydrated, and coated with a conducting and stabilizing film (see Section 2.4). The sample, without any further processing, is now ready for STM imaging.

2.6.1.1 RecA–DNA COMPLEXES

The recA–DNA complexes were formed under the same conditions as for STM of the uncoated specimens. They were directly adsorbed to the Pt-C support, freeze-dried, and rotary shadowed with about 1.5 nm Pt-Ir-C from an elevation of 65° (Amrein *et al.*, 1988). Figure 2.18 shows an STM image of the complexes. The thick filaments show a pronounced striation caused by the 10 nm pitch of their helical structure. Note that uncomplexed DNA extending from complexed regions is also clearly visible as free thin filaments.

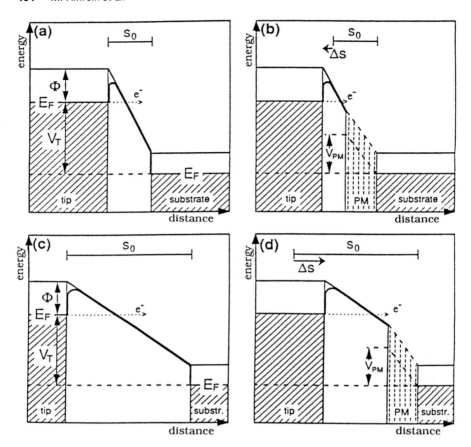

FIGURE 2.17 Schematic drawing explaining positive and negative contrast in images of purple membranes assuming field emission. It shows the potential between tip and sample for a sharp tip (a,b) and a blunt tip (c,d) at sample regions without (a,c) and with (b,d) purple membranes. The surface of the purple membranes becomes charged by the field-emitted electrons. A voltage-dependent conduction mechanism within the purple membranes is assumed to keep the voltage drop V_{PM} across the purple membranes constant in the first approximation. The movement of the tip in constant current mode over purple membranes regions with respect to areas without purple membranes renders positive (b) and negative (d) contrast in the purple membranes images.

2 STM of Proteins and Membranes 155

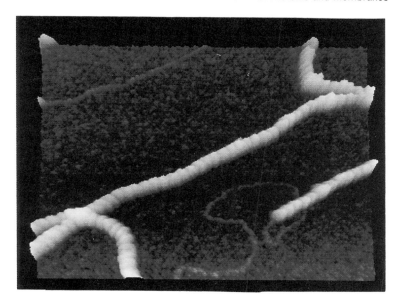

FIGURE 2.18 Scanning tunneling microscopic image in surface view representation of freeze-dried recA–DNA complexes coated with Pt-Ir-C. The area corresponds to 236 × 192 nm. Note the uncomplexed DNA segment that extends from a complexed region at the bottom right of the image.

STM images of coated and uncoated recA–DNA complexes are compared in Figure 2.19. The characteristic features appear in good agreement between the two methods. Each striation consists of three to four parts that presumably are the recA protomers.

2.6.1.2 POLYHEADS

Polyheads, morphologically aberrant capsids of mutated bacteriophage T even and lambda, have been the subject of extended structural analyses by TEM since, via their structure, they have provided an insight into the viral self-assembly processes (for reviews, see Kellenberger, 1978; Steven et al., 1976. They are composed of capsomeres hexameric gp 23, which are arranged on a near-hexagonal lattice that is folded into a cylinder. This cylinder is spread flat upon adsorption. The type III polyheads that we have used have quasi-p6 symmetry and a lattice constant of 13 nm (Steven et al., 1976). The helical geometry of the tubular polyheads, described by their circumference and by the angular pitch, varies within a certain range for the individual capsids (Yanagida et al., 1970).

In TEM projection the homogeneous, coherent Pt-Ir-C coat appears almost grainless and provides hardly any contrast on the rather smooth polyhead surface when rotary evaporated from a high-elevation angle (65°). The absence of a

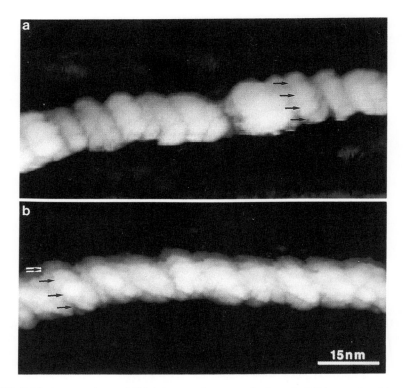

FIGURE 2.19 A comparison of STM images of an unstained recA–DNA complex (a) and a metal-coated complex (b). The characteristic features appear in good agreement for the two methods. In both case the striations are composed of three to four parts, corresponding to recA protomers (arrows).

FIGURE 2.20 Transmission electron microscopic micrograph of a freeze-dried polyhead, rotary shadowed with about 1.5 nm Pt-Ir-C at an elevation of 65° (a). The metal coat, used as conducting film for the STM experiments, is so uniform that hardly anything can be seen in the TEM besides a faint contour demarcating the polyhead border. The absence of a significant periodic signal in the optical diffraction pattern (b) indicates that the coating is uniformly thick and that no significant decoration (perferential nucleation) has occurred.

significant periodic structure indicates that the coating is uniformly thick and that decoration effects (i.e., preferential nucleation) are not dominating (Fig. 2.20). However, these films provide three-dimensional surface information regarding the structure underneath when imaged by STM (Amrein *et al.*, 1989). Figure 2.21 displays an STM topographical image of three polyheads in which the brighter parts are lying on top of capsids directly adsorbed to the support. The morphological units, the capsomeres, can easily be distinguished. The height difference between the support film and the top of the relief of the polyhead amounts to about 7 nm. This value has been reproduced within a range of 0.3 nm for all polyheads imaged.

2.6.1.3 HPI Layer

Figure 2.22 shows raw data from STM of the HPI layer freeze-dried and coated with Pt-Ir-C (Amrein *et al.*, 1991). The inner (i) and the outer (o) surfaces show clearly distinct reliefs. Cracks in the adsorbed layers (denoted with an arrow) are

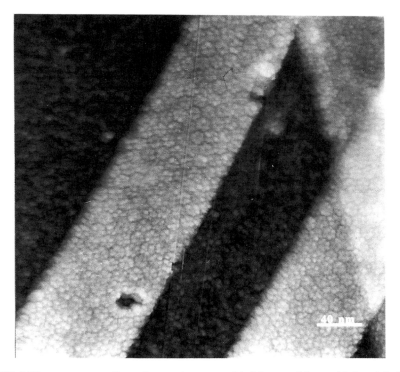

FIGURE 2.21 Scanning tunneling microscopic topographical image of freeze-dried and Pt-Ir-C–coated polyheads shown from a top view. The brighter parts of the polyheads lie on top of those directly adsorbed to the support.

FIGURE 2.22 Scanning tunneling microscopic raw data of the HPI layer, freeze-dried and coated with Pt-Ir-C. The surfaces originally directed toward the inner (i) side of the bacterium and the outer (o) surface show clearly distinct reliefs. Cracks in the adsorbed layers (denoted with an arrow) are often observed.

often observed. They are most likely due to adsorption of the spherical layer to a flat support. The edges of the layer in the upper right corner are smeared out due to the displacement of the tunneling current along the shank of the tip when the needle is moved over the steep edge of the HPI layer.

Figure 2.23a shows a high-magnification image of the outer surface of the layer. Most of the protein mass of the hexameric complexes is concentrated around the sixfold axis in a core region that contains a central cavity. The cores are interconnected via extensions, the spokes, across the twofold axis, leaving open large holes around the threefold axis. In the STM raw data the complexes appear rather heterogeneous, especially in the region of the spokes. This filigree structure is most probably prone to disruption during sample preparation. The central cavity of the inner side of the HPI layer (Fig. 2.23b) is much less

FIGURE 2.23 a: The outer surface of the layer at higher magnification. Most of the protein mass of the hexameric complexes is concentrated in a core region that contains a central cavity. The cores are interconnected via extensions, the spokes, across the twofold axis, leaving open large holes around the threefold axis. b: The inner side of the HPI layer. The central cavity is much less pronounced and the spokes lie much deeper than on the outer surface.

pronounced, and the spokes lie much deeper than on the outer surface. It proved to be rather difficult to obtain tunneling tips sharp enough truly to probe the spokes. Such tips were not very stable and quite often changed their shape during the acquisition of a single frame. However, they could usually be reshaped by pulsing.

2.6.2 Metal Replicas

To make STM of fluid biomaterials such as many membrane structures possible, Zasadzinski *et al.* (1988) used conventional freeze-fracture techniques (see Section 2.4). First the sample is quickly frozen to trap its fluid structure kinetically. It is then fractured under cryogenic temperatures to expose internal structure and subsequently replicated by platinum and carbon.

Zasadzinski and coworkers demonstrated, using the example of a naturally occurring phospholipid liquid crystal, DMPC, that such replicas may provide a better resolution when imaged with STM (Fig. 2.24b–d) than with TEM of the same replicas (Fig. 2.24a). They ascribe this gain in resolution to the fact that STM senses only the surface layer that was in direct contact with the original fracture surface, whereas in TEM the entire shape of the Pt crystallites determines the image.

Surface features may appear with an amplified height due the effect of attractive forces between the tip and the flexible replica. This effect may be strongly reduced by increasing the thickness of the replica and bonding them more firmly to a conductive support (Woodward *et al.*, 1991).

2.7 IMAGE PROCESSING AND QUANTITATIVE ANALYSIS OF STM DATA

2.7.1 Distortion Correction and Appropriate Display of STM Images

The following section deals with aspects of image display that are designed to enhance the visibility of the structures of interest and to correct distortions usually encountered in raw STM data.

Since the z-direction of the scanner is usually not perfectly perpendicular to the support, a linear ramp superimposes on the structural data of the specimens and causes a decrease in the contrast of features of interest and may interfere with height measurements. This ramp can be removed by subtracting a least-squares fitted plane from the data or by applying a high pass filter that suppresses the very low frequencies in the images. The latter, however, can influence the data of the investigated structures and makes further quantitative analysis questionable. Unequal sampling in x and y-directions (along and across scan lines) and linear angular distortions can occur due to inaccurate calibration of the microscope, thermal drift, or electrical creeping of the scanner. These distortions can be corrected by resampling the data points on new unit vectors that are determined by using the known lattice constants of a regular test specimen (e.g.,

2 STM of Proteins and Membranes 161

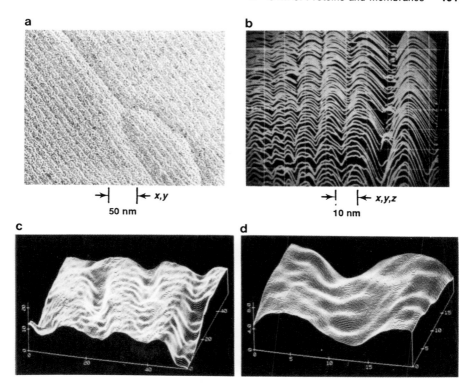

FIGURE 2.24 Scanning tunneling microscopic imaging of the freeze-fracture replica of DMPC. **a:** Transmission electron micrograph of a freeze-fracture replica of DMPC. Ripple periodicity is about 12 nm. Ripples are often interrupted by defects, such as the screw dislocation line that passes through this image. **b:** Analog STM image of the replica with no y-scan. Scan speed is increased from 15 Hz (bottom) to 30 Hz (top). Although detail is washed out in the fastest scans, the essential features are clear: (1) ripple periodicity is 13 nm and average amplitude is 4.5 nm and (2) the ripples are asymmetric, rising more steeply to the left than to the right. **c:** Digital STM image of the replica. The ripple amplitude and configuration are well defined as in b, although variations occur along the ripple. Note the fine structure that crosses the ripples roughly orthogonal to the ripple direction. The distance scale is in nanometers. **d:** Computer zoom of the right-central portion of c. The bands crossing the ripples are readily apparent. At present, it is difficult to say if the banding reflects an underlying molecular structure of the ripple phase, a structure inherent to the replica itself, or an unknown artifact. [From Zasadzinsky *et al.* (1988, Fig. 1, p. 1014).]

polyhead, HPI layer). The lattice constants are determined easily and accurately using the power spectrum of the Fourier transform computed from the image (see later). In addition, in Fourier space electrical or mechanical vibrations that have biased the data are detected much more sensitively than in the image itself.

Processing of STM images for displaying is documented in Figure 2.25 with the example of metal-coated polyheads. The data are displayed in gray tone

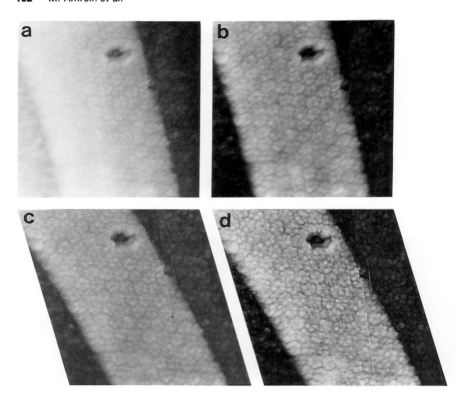

FIGURE 2.25 Digital image processing of STM data. **a:** Scanning tunneling microscopic STM raw data of a freeze-dried and Pt-Ir-C–coated polyhead. Usually the structural data in STM raw data are superimposed by a linear ramp. **b:** Scanning tunneling microscopic raw data after the removal of a least squares fitted plane. **c:** Unequal sampling in x- and y-directions and linear angular distortions may occur due to inaccurate calibration of the microscope, thermal drift or electrical creeping of the scanner. These distortions may be corrected by resampling the data points or new unit vectors that are determined by using the known lattice constants of a regular test specimen. **d:** The overall height variations in STM images of the polyheads are large compared with the corrugations of the capsomere structure. Here the capsomere contrast has been enhanced for better visibility by adding a high-pass-filtered version of the image to the data.

representation in top view. The overall height variations in STM images of the polyheads are large compared with the corrugations of the capsomere structure. Therefore the capsomere contrast was enhanced for better visibility in Figure 2.25d by adding a high pass filtered version of the image to the data.

2.7.2 Digital Image Averaging

The resolution of molecular details of protein surfaces is restricted by both the shape of the tunneling tip and the noise obscuring the structural information (signal). Sources of noise are intrinsic disorder in the specimen structure and artifacts induced by specimen preparation or metal coating. Noise stemming from the microscope itself hardly interferes, since it is typically smaller by one order of magnitude than the structural features that are resolved.

Resolution limitation that is due to an imperfect tip geometry can only be overcome by manufacturing sharper tips. Uniformly distributed noise, on the other hand, can be suppressed in the recorded images if they contain repeated images of structures such as single, isolated biological molecules, adsorbed in the same orientation, or biological specimens exhibiting helical or two-dimensional periodic order. It is assumed that the noisy images can be aligned and then summed to yield an average image. The signal-to-noise ratio of the resulting image improves by the square root of the number of averaged units. There is a number of established techniques developed for nose filtration of TEM micrographs that can be applied to a certain extent to STM topographical images without any further adaptations. They rely either on Fourier filtering (for introductionary review, see Aebi *et al.*, 1984) or on alignment and spatial averaging of the motifs.

Fourier filtration (Klug and DeRosier, 1966) allows the reduction of noise in images of two-dimensional lattices or helical structures. In the Fourier transform calculated from an image of a periodic array, all the structural information consistent with its crystal symmetry is confined to a set of discrete spots periodically sampled on the "reciprocal lattice." On the other hand, random noise contained in the image is more or less evenly distributed. This spatial separation of noise and signal allows filtration of the image in Fourier space by calculating the backtransform (Fourier synthesis) with only the periodic reflexes. The quality of the resulting average, however, depends very much on the perfection of the crystallinity. On the other hand, STM images of crystalline surfaces in our hands always showed long-range imperfections that were probably due to thermal drift, electrical creeping, or nonlinear behavior of the piezoscanner. Fourier filtration is therefore not very well suited to STM applications. The use of the Fourier transform for distortion corrections and for the detection of regular distortions due to electrical or mechanical vibrations during data acquisition is mentioned above.

Alignment and spatial averaging of repeating motifs in an image is usually referred to as *correlation averaging* (Saxton and Frank, 1977). The cross-correlation function of two image partitions that contain a common motif at different locations displays a distinct peak at an offset position, with the offset vector corresponding to the vector by which the motif appears to be shifted. By using this "self-detection" approach, motifs are aligned and summed to yield an average image. It is designed for single particle averaging and, in an adapted form (Saxton and Baumeister, 1982), for almost fully automatic processing of macromolecular structures in which a basically crystalline structure is degraded by imperfections of various kinds. This makes it very well suited to the slightly distorted STM images of regular arrays. Besides the average unit cell, the variance in the raw data may be calculated. This yields an estimate of the accuracy with which the heights of the structural features have been measured. The ratio of the periodic (signal) to the nonperiodic (noise) portion of the averaged images can be determined from a pair of independent averages. To obtain the single-to-noise ratio of the raw data, this value is then divided by the square root of the number of averaged motifs. In the following section correlation averaging is presented using the example of freeze-dried and metal-coated polyheads (see Section 2.6.1.2).

In Figure 2.26 the determination of the averaged capsomere morphology of a polyhead is documented. The polyhead used as input data for the correlation averaging is displayed in Figure 2.26a. The capsomere morphology obtained by correlation averaging (Fig. 2.26b) reveals a hexamer, with the peaks of its protomers lying on a 6 nm diameter circle. The sixfold symmetry of the averaged capsomeres indicates isotropic STM imaging. The maximum height of the relief measures about 1 nm (Fig. 2.26c). This may represent an underestimate of the total corrugation depth, since the tip geometry may have prevented the tip from truly probing the bottom of the depressions between capsomeres.

The heterogeneity of the motifs allows the evaluation of the mean values within the unit cells to at least ±0.2 nm at a confidence level of 95% (t-test). The averaged capsomere morphology proved to be reproducible in different experiments using different tips. The two data sets on the right in Figure 2.27 appear

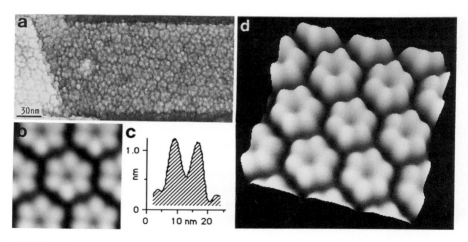

FIGURE 2.26 The capsomere morphology revealed after correlation averaging. a: The polyhead used as input data. Note that the capsomeres can be distinguished as hexamers already in the input data. The averaged capsomere morphology is shown in b from a top view, in c as a section along a line through the center of the capsomere and the tops of two protomers, and in d as a surface-view representation. The capsomere is composed of six protomers whose tops lie on a 6 nm diameter circle.

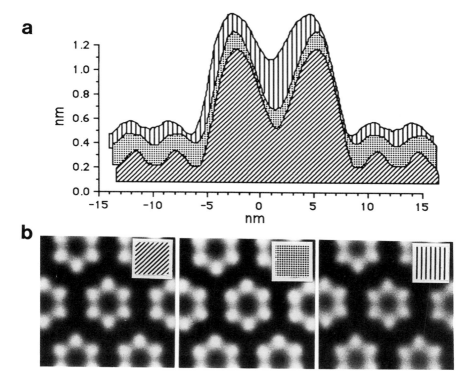

FIGURE 2.27 Comparison of data from three different experiments acquired with three different tunneling tips. a: The cross sections of the capsomeres whose top views are displayed below. b: The two data sets denoted with dots and tilted lines respectively are almost identical, whereas the capsomere shaded with "vertical lines" reveals less-well-resolved protomers. The latter may be due to a "flatter" tip.

practically identical, while in the data set on the left the protomers are less well resolved.

2.7.3 Height Measurement

Scanning tunneling microscopy principally provides very accurate data in the z-direction (i.e., the direction perpendicular to the support). It may therefore be used as a morphometric tool with which to study the absolute height of adsorbed membranes and protein structures, data that are difficult to obtain as accurately by other means (Amrein *et al.*, 1989; Fisher *et al.*, 1990; Wang *et al.*, 1990). First, however, the scanner must be calibrated very carefully in the z-direction, for example, by using the known height on monoatomic steps of the surface of crystalline gold. For reasons discussed earlier (see Section 2.5), metal-coated samples are best suited for absolute height measurements. Because the average thickness of the coating film can be assumed to be the same on both the specimen surface and the adjacent support, the difference in height can be taken as a measure of the thickness of the object. Before height measurements can be carried out, any overall height ramp in the pictures must be removed. The influence of the granularity of the coating film can be suppressed by averaging. By this means the height measurements can become very accurate. For example, the thickness of metal-coated, air-dried purple membrane has been determined to be 4.3 nm, with only 0.1 nm radial mean square (ms) (Wang et al., 1990).

High-precision height measurements not only represent important structural information. The method has been used, for example, to study how different preparation techniques affect the overall height of the metal-coated HPI layer (Wang et al., 1990). In a study by Fisher et al. (1990), thickness measurements of enzymatically modified purple membrane are evidence of conformational changes.

2.7.4 Comparison of TEM and STM Data

Considering that STM is a new technique in molecular structure research, the comparison of STM reliefs with structural data obtained by other means is obviously very important. In this discussion reliefs obtained from TEM micrographs of freeze-dried and metal-coated HPI layers are compared with STM data from the same specimen, prepared in a similar way. This allows the accuracy and reliability of the STM imaging to be judged in a straightforward way.

When a unidirectionally shadowed relief is observed in the TEM in a direction differing from that of evaporation, contrast modulation produced by locally varying coat thickness is related to the relief in a quantitative way. There are several published computational techniques that take advantage of this relationship in order to deduce the relief of biological macromolecular structures (Chalcroft, 1985; Guckenberger, 1985; Smith and Kistler, 1977). Direct scaling in the z-direction of the relief reconstructions has proven to be very difficult. Therefore, we have used the STM height measurements to scale the relief reconstruction of the HPI layer.

In Figure 2.28 STM raw data, an average STM image, and a TEM relief reconstruction of the outer surface of the HPI layer are compared. The major structural features are in good agreement for TEM and STM data even though the relief measured by STM appears to be broader. On the outer surface the peaks of the six protomers that form the core are lying on a 9.2 nm diameter circle in STM raw data (Fig. 2.28a) as well as in the relief reconstruction (Fig. 2.28c). They are located on an axis across the threefold and sixfold axes. The protomers in the core region are less well separated from each other in the STM average (Fig. 2.28b) than in the relief reconstruction and in the STM raw data. This indicates that their position in the STM raw data is not very well defined,

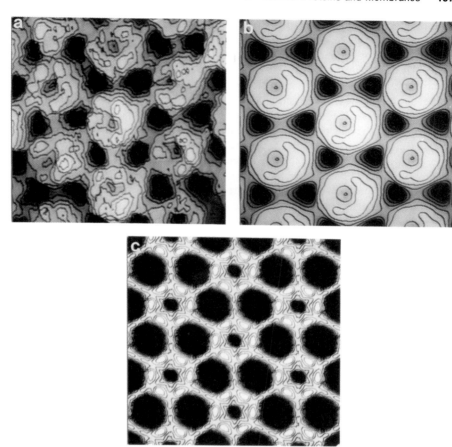

FIGURE 2.28 The outer surface of the HPI layer. **a**: Scanning tunneling microscopy raw data. **b**: average STM image as revealed by correlation averaging. **c**: Transmission electron microscopy relief reconstructions. Contour lines are spaced by 0.4 nm. The major structural features are in good agreement for TEM and STM data even though the relief measured by STM appears to be broader. The peaks of the six protomers that form the core are lying on a 9.2 nm diameter circle, about 1.2 nm above the height of the spokes. They are located on an axis across the threefold and the sixfold axies. The central cavity of the cores appear shallow in STM data when compared with TEM data.

and they are therefore blurred in the average structural unit. However, the height of these peaks with respect to the spokes is in good agreement for TEM and STM. On the other hand, the central cavities of the cores appear shallow in the STM data compared with the TEM data. The tip geometry has prevented the tip from probing the support film at the bottom of the depression around the threefold axis. These depressions are true holes since they show no evidence of any protein mass in TEM.

As a rule the sharper the tip is the less broadened the molecules appear and the better their structures are resolved. Actually the reliefs revealed by STM appear to be broader than the TEM relief reconstructions. In addition, narrow cavities and deep holes are usually not probed to the bottom. On the other hand, the heights of peaks seem to be measured with high accuracy as long as they are not too close together.

The signal-to-noise ratio of the raw data obtained from STM of the HPI layer is in the range of 1.5 to 2. The values are at least twice as high as for the raw data obtained from TEM of the same biological structure, freeze-dried and metal coated, and they are an order of magnitude higher than the signal-to-noise ratio of data from TEM of an unstained sample. However, care must be taken when using the signal-to-noise ratio to judge the quality of STM data. The tunneling tip in a way acts as a low pass filter that depends on the tip radius. Thus high spatial frequencies that usually contain most of the noise are suppressed and as a consequence the apparent signal-to-noise ratio is increased. STM images acquired with a broad tip may actually show a higher signal-to-noise ratio than those in which the structures are much better resolved. To compare the signal-to-noise ratios of STM and TEM data in a realistic way, a low pass filter has been applied to the raw data obtained from TEM of the metal-coated HPI layer.

Besides the influence of the tip geometry, artifacts induced by sample preparation can affect the quality of the STM topographical images. Noise stemming from the microscope itself is largely insignificant, since it is typically one order of magnitude less than the size of the structural features observed. The heterogeneity observed within the structural units of the HPI layer could be due to natural irregularities, structural alterations resulting from adsorption and dehydration, or the random distribution of the metal grains of the coating material. Results obtained on the polyheads suggest that the metal coat does not significantly account for the observed disorder of the HPI layer. Error introduced by the tip geometry could not cause additional heterogeneity, since it affects all structural units in a similar way.

Image averaging in the case of STM data from a freeze-dried and metal-coated HPI layer has not revealed any details other than those already gathered from almost every single unit cell. Some of the structural features that are directly observed with STM are obscured by noise in TEM data obtained from a metal-coated HPI layer. However, correlation averaging in the case of TEM data reveals these structures with a considerably better resolution than with STM. Nevertheless, the three-dimensional surface relief has yet to be reconstructed from averaged TEM images by an additional computational step.

2.8 CONCLUSIONS AND PROSPECTS

2.8.1 STM Imaging of Biological Macromolecular Structures

One cannot help but admit that STM imaging of uncoated membranes and proteins under conventional tunneling conditions in most cases suffers from the inconclusiveness and poor reproducibility of the results obtained. This obviously hampers any systematic investigation intended to relate the structure of a biological macromolecule or macromolecular complex to its biological function. Quite often STM data are distinct in the strongly variable appearance of the imaged structures. Nevertheless, a few very promising results justify the great expectations in imaging biological macromolecular structures with high resolution in a natural environment, notwithstanding the difficulties encountered thus far.

The specimen support, the immobilization of the biological objects, and their state of hydration may all turn out to be crucial parameters for successful STM imaging. It seems to be most important to characterize a sample, prepared as for STM, by additional means. TEM or even high-resolution SEM of a sample usually reveals immediately how densely a specimen is distributed on the surface and whether it is well spread or occurs in clusters. It also answers the question as to whether an observed structural variance in the STM data of some specimen is real or due to the STM imaging itself. The current state of the art is certainly not sufficient to judge the quality of STM data by merely comparing them with structural data acquired under completely different conditions (e.g., X-ray data).

To direct the sample preparation and the imaging conditions toward a more reliable STM data acquisition, a better understanding of the contrast mechanisms may be crucial. When both the sample and the support are good conductors, it is mainly the tip geometry that influences the quality of the topographical image (see Section 2.3). Thereby the absolute height of some object appears correct. In the case of uncoated proteins and membranes, however, the apparent height is usually much smaller than the expected height of the object. This is true even when the general aspect of the specimen is represented in a reasonable way. In some cases protein structures have been measured with almost no apparent height (Stemmer *et al.*, 1989) or even with reversed contrast (Mantovani *et al.*, 1990). This cannot be explained solely by a decrease in the tip-sample distance due to a change in the electronic structure, because the lack of height may be much larger than any possible tunneling gap. On the other hand, tunneling could always be between the substrate and the tip, and the contrast would arise from a local change of the work function of the substrate due to the organic adsorbates. However, this seems very unlikely for membranes and most proteins, since they have an extended three-dimensional structure. It is also generally believed that ionic conductivity in a hydration shell cannot account for the "tunneling" current and thus for the contrast formation, although the water and the ions may play important roles.

When the tunneling tip is moved from a support region to over a biological sample, it most likely must approach the surface considerably in order to establish the preset tunneling current. This will result in a strong repulsive force between the tip and the sample. Possibly any combination of a deformation of the tip, the specimen, and the substrate contributes to the observed height of the biological specimen (Mantovani et al., 1990; see also Figs. 2.8, 2.9). The contrast in the STM images may therefore depend on the electronic structure of the sample, as well as on the response to the force that arises during scanning.

The force not only contributes to the contrast in a way not fully understood, but it also causes the biological structures to be moved by the tip. Hence it may be the main reason for the poor reproducibility often encountered. It seems crucial to stabilize the biological structures mechanically so that they can withstand the mechanical stress (e.g., by allowing them to self-assemble in a regular array on a suitable support) and to lower the force between the tunneling tip and the sample (e.g., by choosing appropriate tunneling conditions). When STM imaging is carried out in air, a meniscus of a thin water film between the sample and the tip contributes considerably to force. This force can completely be avoided by imaging the samples either in a fluid or under vacuum conditions.

STM imaging at a tunneling voltage above 5 V is very reproducible. The contrast mechanism seems to be understood to a great extend, although the conduction mechanism within the biomaterials is not yet clear. Therefore the method may become a generally applicable way to acquire data from uncoated proteins and membranes. The large tip-sample separation allows very fast scanning without the risk of crashing the tip, and the forces that arise during scanning are probably negligible. The drawback to the large gap is the reduced lateral resolution. It cannot be better than roughly the tip-sample separation. Tunneling voltages no larger than absolutely necessary for this type of imaging and ever sharper tips may help to reduce the gap considerably and by this increase the resolution.

Interestingly the purple membrane could be imaged with both a high tunneling voltage and a voltage below 1 V. It seems that in the first case the conductivity within the biomaterial has been induced by the high electrical field and a voltage drop across the membrane of about 3.5 V has been deduced. In the second case the close vicinity of tip and sample must have induced a conducting mechanism at a much lower electrical field.

Scanning tunneling microscopy of metal-coated specimens has become a reliable technique when the uncoated samples cannot be imaged reproducibly. Furthermore, one can estimate the contribution of the elastic and electronic properties to the contrast on an uncoated specimen by comparison to STM data of a metal-coated sample of a similar specimen.

The quality of the structural data obtained from a coated object depends very much on how the sample has been prepared. Careful freeze-drying keeps the macromolecular structures in a near-native conformation that can be preserved by a three-dimensionally stable coating film. For optimal resolution the conductive film must be fine granular and cover the specimen homogeneously at a very small film thickness. Under optimal conditions the method permits direct three-dimensional visualization of structural details at a level at which individual subunits in protein complexes or even single domains of proteins can be resolved. In the case of a strongly corrugated object it may even be the shape of the tip that limits the resolution.

Scanning tunneling microscopy of metal replicas can reveal the molecular arrangements in the case of fluid biological structures that do not withstand isolation and adsorption in the native state. The method is certainly best suited for rather smooth structures such as membrane fracture surfaces.

2.8.2 STM in Molecular Structure Research

When investigating the structures of large biological macromolecules or macromolecular assemblies, STM is only an alternative to TEM or X-ray diffraction analysis. In many cases X-ray diffraction allows the determination of a structure down to atomic resolution. However, it requires perfectly ordered three-dimensional crystals that are quite often not attainable. TEM, on the other hand, is a very powerful technique with which to determine the sizes and shapes of biological macromolecules that cannot be crystallized into three dimensions. Any TEM micrograph, however, is basically a two-dimensional projection, and the three-dimensional structure of the object has to be reconstructed. Three-dimensional reconstructions are obtained by tomographic techniques (i.e., by combining the information from a series of projections with the specimen tilted at different angles with respect to the electron beam) or by relief reconstruction using micrographs of unidirectionally shadowed specimens. However, the resolution that is attained in the z-direction (i.e., perpendicular to the support) is usually considerably lower than the lateral resolution. In contrast, STM provides direct information about the specimen surface profile, and hence thickness, with high accuracy. Therefore high-precision height measurements by STM represent valuable information complementary to TEM data. Apart from the direct usefulness of such data, STM relief data could also be used to constrain three-dimensional reconstructions from TEM data, thus improving their z-resolution.

The quality of three-dimensional reconstructions from TEM data strongly depends on the degree to which the structural information is obscured by noise (i.e., signal-to-noise ratio). To attain resolution at a macromolecular level, the signal-to-noise ratio usually has to be increased by averaging over redundant structural information. Consequently subnanometer resolution has only been attained with highly ordered extended two-dimensional crystals. In contrast, the signal-to-noise ratio in raw STM data is usually much higher than in TEM micrographs of the same structure, and the z-dimension need not be reconstructed. Therefore, in contrast to TEM, STM may reveal the architecture of biological structures when the signal-to-noise ratio cannot be increased considerably by averaging.

ACKNOWLEDGMENTS

We are grateful to T. Hillebrand (MPI für Biochemie, Martinsried, BRD) for providing us with the unpublished STM images of HOPG presented in Figure 2.4 and to R. Wepf (Institute for Cell Biology, ETH Zurich, CH) for the sample preparations of native actin filamints and polyheads. We thank K. Fuchs (Institute for Cell Biology, ETH Zürich, CH) for assistance with digital image processing and J. Taylor and Reichelt for corrections.

REFERENCES

Aebi, U., Fowler, W. E., Buhle, E. L., Jr., and Smith, P. R. (1984). Electron microscopy and image processing applied to the study of protein structure and protein-protein interactions. *J. Ultrastruct. Res.* 88, 143-176.

Amrein, M., Stasiak, A., Gross, H., Stoll, E., and Travaglini, G. (1988). Scanning tunneling microscopy of recA-DNA complexes coated with a conducting film. *Science* 240, 514-516.

Amrein, M., Dürr, R., Stasiak, A., Gross, H., and Travaglini, G. (1989). Scanning tunneling microscopy of uncoated recA-DNA complexes. *Science* 243, 1708-1711.

Amrein, M., Dürr, R., Winkler, H., Travaglini, G., Wepf, R., and Gross, H. (1989). STM of freeze-dried and Pt-Ir-C coated bacteriophage T4 polyheads. *J. Ultrastruct. Res.* 102, 170-177.

Amrein, M., Wang, Z., and Guckenberger, R. (1991). Comparative study of a regular protein layer by scanning tunneling microscopy and transmission electron microscopy. *J. Vac. Sci. Technol.* B 9, 1276-1281.

Anders, M., Mück, M., and Heiden, C. (1988). SEM/STM combination for STM tip guidance. *Ultramicroscopy* 25, 123-128.

Bachmann, L., and Hildebrand, H. (1965). Sinterung von Aufdampfschichten aus Silber und Gold unter dem Einfluss von Adsorptionsschichten. In *Grundprobleme der Physik Dünner Schichten*. (Vandenhock and Rupprecht, Göttingen, eds.) pp. 77-82

Baró, A. M., Miranda, R., Alaman, J., García, N., Binnig, G., Rohrer, H., Gerber C., and Carrascosa, J. L. (1985). Determination of surface topography of biological specimens at high resolution by scanning tunnelling microscopy. *Nature* 315, 253-254.

Baumeister, W., Bart, M., Hegerl, R., Guckenberger, R., Hahn, M. and Saxton, W. O. (1986). Three-dimensional structure of the regular surface layer (HPI layer) of *Deinococcus radiodurans*. *J. Mol. Biol.* 187, 241-253.

Blackford B, and Jericho M. (1991). A metallic replica/anchoring technique for scanning tunneling microscope or atomic force microscope imaging of large biological structures. *J. Vac. Sci. Technol.* B 9, 1253-1258.

Bryant, P. J., Kim, H. S., Zeng, Y. C., and Yang, R. (1987). Technique for shaping scanning tunneling microscope tips. *Rev. Sci. Instrum.* 58, 1115.

Chalcroft, J. P. (1985). Considerations for the quantitative analysis of coated reliefs. *Ultramicroscopy* 16, 371-386.

Chicon, R., Ortuno, M., and Abellan, J. (1987). An algorithm for surface reconstruction in scanning tunneling microscopy. *Surf. Sci.* 181, 107-111.

Clemmer, C., and Beebe, T., P. (1991). Graphite: a mimic for DNA and other biomolecules in scanning tunneling microscope studies. *Science* 251, 640-642.

Coombs, J. H., and Pethica, J. B. (1986). Properties of vacuum tunnel currents: anomalous barrier heights. *IBM J. Res. Dev.* 30, 455–459.

Edstrom, R. D., Meinke, M. H., Yang, X., Yang, R., and Evans, D. F. (1990). Scanning tunneling microscopy of the enzymes of muscle clycogenolysis. *Ultramicroscopy* 33, 99–106.

Egelman, E. H., and Stasiak, A. (1986). Structure of helical RecA-DNA Complexes. *J. Mol. Biol.* 191, 677–697.

Emch, R., Descouts, P., and Niedermann, P. (1988). A small scanning tunneling microscope with large scan range for biological studies. *J. Micros.* 152, 85–92.

Fisher, K. A., Whitefield, S. L., Thomson, R. E., Yanagimoto, K. C., Gustafsson, M. G., and Clark, J. (1989). Measuring changes in membrane thickness by scanning tunneling microscopy. *Biochim. Biophys. Acta* 1023, 325–334.

Garcia, R., Keller, D., Panitz, J. Bear, D. and Bustamante, C. (1989). Imaging of metal-coated samples by scanning tunelling microscopy. *Ultramicroscopy* 27, 367–374.

Gerber, C., Binnig, G., Fuchs, H., Marti, O., and Rohrer, H. (1986). Scanning tunneling microscope combined with scanning electron microscope. *Rev. Sci. Instrum.* 57, 221–226.

Gross, H., Mueller, T., Wildhaber, I., and Winkler, H. (1985). High resolution metal replication, quantified by image processing of periodic test specimens. *Ultramicroscopy* 16, 287–304.

Gross, H. (1987). High resolution metal replication of freeze-dried specimens. In *Cryotechniques in Biological Electron Microscopy*. (R. A. Steinbrecht and K. Zierold, eds.) Springer-Verlag, Berlin, pp. 205–215.

Guckenberger, R. (1985). Surface reliefs derived from heavy-metal–shadowed specimens — Fourier space techniques applied to periodic objects. *Ultramicroscopy* 16, 357–370.

Guckenberger, R., Koesslinger, C., Gatz, R., Breu, H., Levai, N, and Baumeister, W. (1988). A scanning tunneling microscope (STM) for biological applications: design and performance. *Ultramicroscopy* 25, 111–122.

Guckenberger, R., Wiegräbe, W., Hillebrand, A., Hartmann, T., Wang, Z., and Baumeister, W. (1989). Scanning tunneling microscopy of a hydrated bacterial surface protein. *Ultramicroscopy* 31, 327–332.

Guckenberger, R., Hacker, B., Hartmann, T., Scheybani, T., Wang, Z. Wiegräbe, W., and Baumeister, W. (1991). Imaging of uncoated purple membrane by scanning tunneling microscopy. *J. Vac. Sci. Technol. B* 9, 1227–1230.

Haggerty, L., Watson, B. A., Barteau, M. A., and Lenhoff, A. M. (1991). Ordered arrays of proteins on graphite observed by scanning tunneling microscopy. *J. Vac. Sci. Technol. B* 9, 1219–1222.

Häussling, L., Michel, B., Ringsdorf, H., and Rohrer, H. (1991). Direct observation of streptavidin specifically adsorbed on biotin functionalized self-assembled monolayers with the scanning tunneling microscope. *Angewandte Chem. Int. English Edition* 30, 569–572.

Henderson, R. Baldwin, J. M., Ceska, T. A., Zemlin, F., Beckmann, E., and Downing K. H. (1990). Model for the structure of bacteriorhodopsin based on high-resolution electron cryo-microscopy. *J. Mol. Biol.* 213, 899–929.

Ichinokawa, T., Miyazaki, Y., and Koga, Y. (1987). Scanning tunneling microscope combined with scanning electron microscope. *Ultramicroscopy* 23, 115–118.

Hörber, J. K. H., Schuler, V., Witzemann, F. M., Schröter, K. H., Müller, H., and Ruppersberg, J. P. (1991). Imaging of cell membrane proteins with a scanning tunneling microscope. *J. Vac. Sci. Technol. B* 9, 1214–1217.

Howard-Flanders, P., West, S. C., and Stasiak, A. (1984). Role of RecA protein spiral filaments in genetic recombination. *Nature* (London) 309, 215–220.

Kellenberger, E. (1978). Possibility of detecting reproducibly some 5-10A Conformational differences by conventional techniques: physiologically defined lattice transformations in bacteriophage T4. Electron Microscopy 1978, Vol. III (J. M. Sturgess, ed) pp. 441–449.

Kellenberger, E. (1987). The response of biological macromolecules and supramolecular structures to the physics of specimen cryopreparation. In *Cryotechniques in Biological Electron Microscopy*. (R. A. Steinbrecht and K. Zierold, eds.) Springer-Verlag, Berlin, pp. 35–60.

Klug, A., and DeRosier, D. J. (1966). Optical filtering of electron micrographs: reconstruction of one sided images. Nature (London) 212, 29–32.

Lindsay, S. M., Thundat, T., Nagahara, L., Knipping, U., and Rill, R. L. (1989). Images of the DNA double helix in water. *Science* 244, 1017–1116.

Mantovani, J. G., Allison, D. P., Warmack, R. J., Ferrell, T. L., Ford, J. R., Manos, R. E., Thompson, J. R., Reddick, B. B., and Jacobson, K. B. (1990). Scanning tunneling microscopy of tobacco mosaic virus on evaporated and sputter-coated palladium/gold substrates. *J. Microsc.* 158, 109–116.

Michel, B., and Travaglini, G. (1988). An STM for biological applications: bioscope. *J. Microsc.* 152, 681–685.

Miles, M., McMaster, T., Carr, J., Tatham, A., Shewry, R., Field, J., Benton, P., Jeenes, D., Hanley, B., Whittam, M., Cairns, P., Morris, V., and Lambert, N. (1990). Scanning tunneling microscopy of biomolecules. *J. Vac. Sci. Technol.* A 8, 698–702.

Olk, C., Heremans, J., Lee, P., Dziedzic, D., and Sargent, N. (1991). IgG antibody and antibody-antigen complex imaging by scanning tunneling microscopy. *J. Vac. Sci. Technol.* B 9, 1268–1271.

Panitz, J. A. (1987). A TEM study of electron tunneling in biological macromolecules. In *Proc. Annual EMSA Meeting*. Baltimore, MD. (G. W. Bailey, ed.) San Francisco Press, San Francisco, p. 140.

Robards, A. W., and Sleyter, U. B. (1985). Low temperature methods in biological electron microscopy. In *Pratical Methods in Electron Microscopy* (A. M. Glauert, ed.) Elsevier, Amsterdam, Vol. 10.

Saxton, W. O., and Frank, J. (1977). Motive detection in quantum noise-limited micrographs by cross-correlation. *Ultramicroscopy* 2, 219–227.

Saxton, W. O., and Baumeister, W. (1982). The correlation averaging of a regularly arranged bacterial cell envelope protein. *J. Microsc.* 127, 127–138.

Smith, P. R., and Kistler, J. (1977). Surface reliefs computed from micrographs of heavy metal-shadowed specimens. *J. Ultrastruct. Res.* 61, 124–133.

Sogo, J., Stasiak, A., De Bernardin, W., Losa, R., and Koller, T. (1987). Binding of protein to nucleic acids. In "Electron microscopy in molecular biology" (J. Sommerville and U. Scheer U, eds), pp. 61–79. IRL Press, Oxford, Washington DC.

Spence, J. C. H. (1988). A scanning tunneling microscope in a side-entry holder for reflection electron microscopy in the Phillips EM400. *Ultramicroscopy* 25, 165–170.

Stemmer, A., Engel, A., Häring, R., Reichelt, R., and Aebi, U. (1988). Miniature-size scanning tunneling microscope with integrated 2-axes heterodyne interferometer and light microscope. *Ultramicroscopy* 25, 171–181.

Stemmer, A., Hefti, A., Aebi, U., and Engel, A. (1989). Scanning tunneling microscopy and transmission electron microscopy on identical areas of biological specimens. *Ultramicroscopy* 30, 263–280.

Steven, A. C., Couture, E., Aebi, U., and Showe, M. K. (1976). Structure of T4 polyheads. *J. Mol. Biol.* 106, 187–221.

Studer, D., Moor, H., and Gross., H. (1981). Single bacteriorhodopsin molecules revealed on both surfaces of freeze-dried and heavy metal-decorated purple membranes. *J. Cell Biol.* 90, 153–159.

Travaglini, G., Rohrer, E., Amrein, M., and Gross, H. (1987). Scanning tunneling microscopy on biological matter. *Surf. Sci.* 181, 380–390.

Travaglini, G., Rohrer, H., Stoll, E., Amrein, M., Stasiak, A., Sogo, J., and Gross, H. (1988). Scanning tunneling microscopy of recA–DNA complexes. *Physica Scribta* 38, 309–314.

Venables, J. A., Spiller, G. D. T., and Hansbücken, M. (1984). Nucleation and growth of thin films. *Rep. Prog. Phys.* 47, 399–459.

Voelker, M. A., Hameroff, S. R., Jackson, H. D., Dereniak, E. L., McCuskey, R. S., Schneiker, C. W., Chvapil, T. A., Bell, L. S., and Weiss, L. B. (1988). STM imaging of molecular collagen and phospholipid membranes. *J. Microsc.* 152, 557–566.

Wang, Z., Hartmann, T., Baumeister, W., and Guckenberger, R. (1990). Thickness determination of biological samples with a z-calibrated scanning tunneling microscope. *Proc. Natl. Acad. Sci. U.S.A.* 87, 9343–9347.
Welland, M. E., Miles, M. J., Lambert, N., Morris, V. J., Coombs, J. H., and Pethica J. B. (1989). Structure of the globular protein vicilin revealed by scanning tunneling microscopy. *Int. J. Biol. Macromol.* 11, 29–32.
Wepf, R., Amrein, M., Bürckli, U., and Gross, H. (1991). Pt-Ir-C, a high resolution shadowing material for TEM, STM and SEM of biological macromolecular structures. *J. Microsc.* 163, 51–64.
Wildhaber, I., Gross, H., Engel, A., and Baumeister, W. (1985). The effects of air-drying and freeze-drying on the structure of a regular protein layer. *Ultramicroscopy* 16, 411–422.
Woodward, J., Zasadzinski, J., and Hansma, P. (1991). Precision height measurements of freeze fracture replicas using the scanning tunneling microscope. *J. Vac. Sci. Technol. B* 9, 1231–1235.
Yanagida, M., Boy de la Tour, E., Alff-Steinberger, C., and Kellenberger, E. (1970). Studies on the morphopoeisis of the head of bacteriophage T-even. *J. Mol. Biol.* 50, 35–58.
Zasadzinski, J. A. N., Schneir, J., Gurley, J., Elings, V., and Hansma, P. K. (1988). Scanning tunneling microscopy of freeze-fracture replicas of biomembranes. *Science* 239, 1013–1015.

CHAPTER 3
Protein Assemblies and Single Molecules Imaged by STM

M. J. Miles and T. J. McMaster
H. H. Wills Physics Laboratory
University of Bristol
Bristol, United Kingdom

3.1 Introduction
3.2 Elongated Proteins and Polypeptides
    3.2.1 Ordered Arrays
    3.2.2 Wheat Seed Storage Protein
    3.2.3 Collagen
    3.2.4 Elastin
    3.2.5 Model Polypeptides
3.3 Globular Proteins
    3.3.1 Introduction
    3.3.2 Vicilin
    3.3.3 RNA Polymerase Holoenzyme
    3.3.4 Cell Membrane Protein
    3.3.5 Other Globular Proteins
3.4 Conclusions
References

3.1 INTRODUCTION

Scanning tunneling microscopy (STM) has several potential advantages for the study of proteins. The unparalleled resolution of STM under favorable conditions has led many experimental groups to apply the technique to obtain structural and conformational data regarding proteins. Many proteins of interest are not crystallizable and thus are not amenable to high-resolution X-ray diffraction, and nuclear magnetic resonance (NMR) techniques are limited in the size of proteins that can be studied. Scanning tunneling microscopy image contrast depends on local electrical conductivity of the specimen, not on local electron density, as is the case for transmission electron microscopy (TEM). This means that specimens consisting of low-atomic-number elements such as those commonly comprising biological molecules can be studied only if electrical conductivity is sufficiently high. In TEM, this type of specimen must usually be coated or stained with a compound containing a high-atomic-number metal. As well as introducing possible disruption of the biomolecular structure, such sample preparation methods prevent the study of kinetic processes. Some studies have, however, shown the possibility of imaging unstained macromolecular biological structures in a frozen-hydrated state with TEM. A further major advantage of STM is that is can operate in air under high humidity and even in liquid, so that water molecules involved in maintaining the protein structure are not removed as they may be in the vacuum of the TEM or the metal-coating apparatus. The protein structure should therefore be less disrupted and more closely resemble the native state. Finally, radiation damage is not a problem in STM, as the highest electron energies used are about 10 eV compared with 100 keV in the TEM.

There are various problems that limit the routine collection of high-resolution, reproducible results for protein specimens. The action of the STM tip on scanning the surface more often than not causes movement of the molecule. This is compounded by weak adsorption of isolated proteins to commonly used STM substrates. A further related problem is the low conductivity of most proteins, leading to a smaller tip-sample working distance than is desired, exacerbating the tip–sample interaction problem. Counterbalancing this is the fact that too strong of an adsorption to the substrate can cause serious distortion of the molecule. Ideally, the strength of the adsorption should be controlled by matching the substrate to the particular molecule to be studied.

The image-formation mechanism remains poorly understood. However, with the increasing number of molecules being studied, some trends are emerging. Sometimes water seems to play an important role in image contrast, but it is not yet clear if it can be applied more generally.

In this chapter, we concentrate on sample preparation techniques for STM imaging of protein molecules. Instrumentation considerations and tip preparation methods are comprehensively reviewed in Chapter 2, this volume. We shall relate successful imaging of protein molecules to mechanisms of immobilizing the molecules on the substrate surface. The principal methods considered are (i) the formation of ordered stabilized arrays of molecules, resulting from the formation of liquid-crystal or two-dimensional crystal structures; and (ii) increasing the attractive force between the molecule and the substrate.

3.2 ELONGATED PROTEINS AND POLYPEPTIDES

3.2.1 Ordered Arrays

The immobilization of proteins on the surface of the substrate can be achieved by the formation of an ordered monolayer. The successful imaging of alkane monolayers and of low molecular-weight organic liquid crystal molecules (see Chapter 4, this volume) illustrate both the effect of physical adsorption forces and the stability of an ordered array of extended molecules. The advantages in forming a liquid-crystal structure in the bulk as a precursor to obtaining an ordered monolayer on the substrate surface have been recognized for other molecules. Flory (1956) calculated the liquid-crystal phase diagram for solutions of rod-like molecules. Liquid-crystal formation is favored at high-volume fractions of rods and poorer solvents. In choosing systems of molecules and solvents and their concentration ranges, note must be taken of the possibility that crystallization may be a competing process. To encourage molecular-ordered monolayer adsorption, it is necessary to match the system to the substrate.

3.2.2 Wheat Seed Storage Protein

3.2.2.1 HIGH MOLECULAR WEIGHT SUBUNIT OF WHEAT GLUTEN

The high molecular weight (HMW) subunit protein from wheat gluten dissolved in trifluoroethanol (TFE) appeared to be a potentially successful candidate for the liquid-crystal approach to stable monolayer formation. The protein consists of a central repetitive domain composed of hexapeptide and nonapeptide motifs flanked by nonrepetitive terminal domains that are thought to be predominantly α-helical when dissolved in a solvent such as TFE. The central repetitive domain, containing over 80% of the amino acid sequence, is predicted (Tatham *et al.*, 1985) to form regularly repeated β-reverse turns, resulting in a β-spiral conformation. The presence of β-reverse turns is supported by circular dichroism and by Fourier transform infrared spectroscopy (Tatham *et al.*, 1990a). The overall structure is, therefore, expected to be extended and rodlike. Hydrodynamic studies of a single purified HMW from durum wheat suggest that the molecule adopts an extended rodlike conformation with dimensions of 55×1.8 nm in 0.05 M acetic acid and 50% aqueous 1-propanol and 65×1.55 nm in TFE (Field *et al.*, 1987). This aspect ratio of greater than 30:1 is well above the minimum value of about 3 required for liquid-crystal formation according to the Flory analysis. However, the molecule is not so long that entanglements might pose a kinetic obstacle to liquid-crystal formation. The protein does not readily crystallize, and thus this is not a competing process to liquid-crystal formation. Therefore, under appropriate solvent conditions and concentrations, liquid-crystal ordering of HMW molecules should be possible. In addition, the hydrophobic character of the protein and of the graphite surface should, in the presence of water, promote the adsorption of an ordered monolayer of HMW protein.

A single HMW subunit protein from durum wheat was used (Miles et al., 1991). To prevent aggregation as a result of disulfide-bond formation, disulfide bonds were reduced and pyridethylated. The protein was purified by reversed-phase high-performance liquid chromatography (HPLC) with the water/acetonitrile/trifluoroacetic acid solvent system. A solution containing 200 mg/ml of HMW protein in TFE was prepared, and a 10 ml drop was deposited on each freshly cleaved highly oriented pyrolytic graphite (HOPG) substrate used. The solvent was allowed to evaporate slowly so that the concentration required for liquid-crystal formation existed for a time sufficient to allow ordering of the protein molecules. The initial concentration was chosen so as to give on average considerably more than a monolayer of protein on the surface of the substrate, as experience had shown that after evaporation of the solvent the surface distribution of protein was highly nonuniform. An initial concentration is therefore required that optimizes the production of monolayer areas.

The specimens were examined in an STM operated in air in the constant-current mode. An electrochemically etched tungsten tip was used. Typical tunneling conditions were a +700 mV sample bias and a 70 pA tunneling current. Regions of ordered monolayers were found, although, as expected, many regions of thick aggregates were also found. The STM image in Figure 3.1a has two domains of rodlike features that are apparently monolayers. The Fourier transform (Fig. 3.1b) of the lower domain contains sharp maxima consistent with (two-dimensional) crystalline order rather than a less-ordered liquid-crystalline structure. This may be a result of epitaxial adsorption. In the original image (Fig. 3.1a) it is possible to discern a periodicity along the molecules. From the Fourier transform the repeat distance of 1.5 nm can be measured, while the intermolecular periodicity is found to be 3.1 nm. By partial subtraction of the nonperiodic component of the Fourier transform and performing the inverse transform of all the sharp maxima of Figure 3.1b, the Fourier-filtered images in Figure 3.1c,d were obtained. The helical molecular structures in these images have a diameter of 1.95 nm, which is consistent with the hydrodynamic data. The adsorption forces between any molecule and substrate will inevitably cause some distortion of the molecule and therefore its image in STM and in conventional microscopy. However, in the present case, it is not considered that this would cause a major

---

FIGURE 3.1 a: Unprocessed STM image, 60 × 58.5 nm, of two domains of rodlike HMW molecules on HOPG. The aligned molecules meet at what may be a step in the graphite surface. b: Fourier transform pattern of the lower domain of a. The degree of order of the pattern indicates that the molecules lie on a two-dimensional lattice. c: The Fourier-filtered version of the lower domain of a. 54.5 × 37.4 nm. d: A three-dimensional representation of part (16 × 11 nm) of c. (From Miles et al., 1991, with permission.)

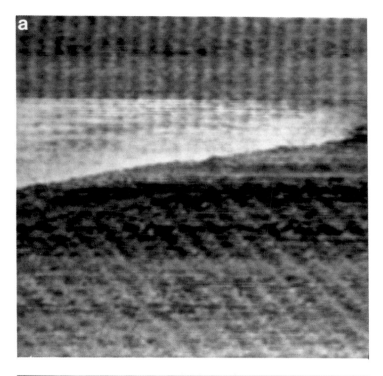

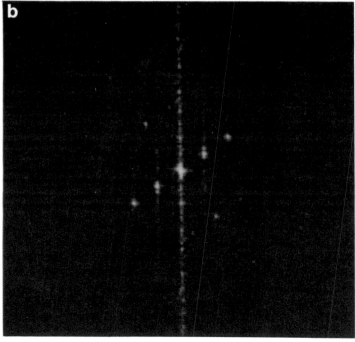

FIGURE 3.1 *(Continues)*

FIGURE 3.1 *(Continued)*

problem, because the HMW structure is known to be stable in a range of solvents and temperatures and resistant to denaturation; thus adsorption forces would not be expected to cause a serious distortion of the conformation.

3.2.2.2 GLIADINS FROM WHEAT

The gliadins, like vicilin (see Section 3.3.2) from the pea and the HMW subunit protein from wheat, are seed-storage proteins. The γ-gliadins are an important group of the sulfur-rich gliadin proteins found in wheat, other cereals, and meadow grasses. They consist of a repetitive (N-terminal) domain and a nonrepetitive (C-terminal) domain. Circular-dichroic studies and secondary-structure prediction indicate a reverse-turn rich conformation for the repetitive domain, as was the case for the HMW subunit protein, and an α-helical-rich, globular conformation for the nonrepetitive domain (Tatham et al., 1990a,b). The molecule has been shown to be asymmetric with the possibility that the repetitive domain is extended.

For STM studies (Thomson et al., 1992), gliadin was deposited on HOPG from solutions in either TFE or acetic acid over a wide range of concentrations up to 200 mg/ml. Regions of molecular aggregates were frequently found. However, monolayer regions were also common. This distribution of material on the surface was confirmed over larger areas by TEM of replicas taken from the STM specimens. The nature of the monolayer is different from those observed with STM of HMW subunit proteins, cyanobiphenyls, or alkanes. The long axis of the gliadin molecule is approximately perpendicular to the substrate surface (Fig. 3.2) instead of parallel to it, as in the other cases. The surface pressures and potentials of gliadin films have been studied at an air–water interface, and two-dimensional micelles have been detected. It is expected that the repetitive domain, which contains few charged animo acid residues, will be more hydrophobic and therefore more likely to adsorb to the hydrophobic HOPG surface, leaving the nonrepetitive domain on the upper surface of the monolayer. Such a structure is very similar to that of a lipid monolayer. The packing of molecules in the monolayer again provides a self-stabilizing structure, but there was evidence that the scanning tip was causing some rearrangement in the layer.

To image individual gliadin molecules, it was necessary to use concentrations lower by up to a factor of 1000. Many areas were found to be devoid of protein, but protein molecules could be found (Fig. 3.3). They indicate a rodlike region with a globular head, probably corresponding to the repetitive and nonrepetitive domains, respectively. Dimerization is often seen.

3.2.2.3 COLLAGEN

Collagen makes up 25% of human protein. It is the principal protein of connective tissue and basement membranes. There is a hierarchy of structure from the

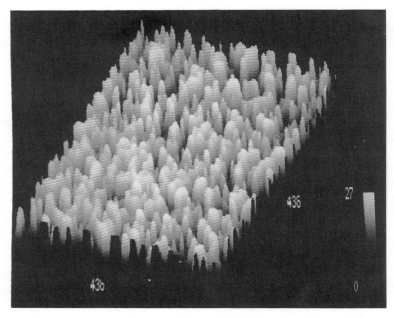

FIGURE 3.2 Three-dimensional rendering of a monolayer of γ-gliadin deposited from TFE at a concentration of 200 mg/ml. The molecules are observed with their long axes perpendicular to the substrate. Sizes are in angstroms, and tunneling conditions were 10 pA and 1 V. (From Thomson *et al.*, 1992, with permission.)

individual polypeptide molecules up to collagen fibers that have diameters of several micrometers. The individual molecules form left-handed helices (α-chains), three of which are wound into a right-handed superhelix. This stiff, triple-stranded rope is about 300 nm long and 1.5 nm in diameter. These are grouped together in fives in a staggered, overlapping arrangement to form microfibrils (~ 8 nm cross section), and these aggregate into fibrils (diameter ~ 300 nm) that assemble into collagen fibers). This structure is self-assembling and thus can be made to occur on the STM substrate surface.

There are seven genetically distinct collagen polypeptides (α-chains). About a dozen types of collagen have been found. The major types are known as I, II, III, IV, and V. About 90% of collagen in the body is type I. After secretion into extracellular space, types I, II, and III assemble into collagen fibrils. However, type IV, which is the main collagen in basal laminae, does not form fibrils. Guckenberger *et al.* (1988) have compared STM and TEM images of type IV collagen molecules aerosol deposited from 50% glycerol solution on glass and carbon, respectively. For both microscopies, the molecules were shadowed at low angle with tantalum–tungsten. Both sets of images show the molecule as a

3 Protein Assemblies and Single Molecules Imaged by STM 185

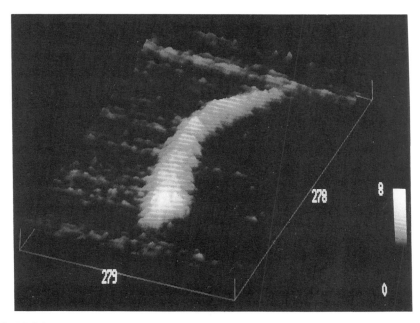

FIGURE 3.3 γ-Gliadin molecules deposited from 1% acetic acid at a concentration of 0.2 mg/ml. The distance along the molecular object is consistent with two molecules arranged head to tail. The bright patches at the bottom and the middle of the image correspond to the globular hydrophobic heads of the two molecules. (From Thomson *et al.*, 1992, with permission.)

thread on the surface. No intramolecular detail is resolved owing to the granular nature (~2 nm) of the metallic shadowing.

Gathercole and Miles (1993) have used STM to image uncoated type I skin collagen in air. The specimen was prepared by depositing acid-soluble calf skin collagen onto HOPG from a dilute solution in acetic acid. Sheaves of cylindrical fibrils with a minimum diameter of 8 nm were observed (Fig. 3.4a). They showed the native D-period molecular stagger pattern. These images are directly comparable with SEM images of coated material. In regions where the specimen forms a sufficiently thin and uniform layer, a repeated pattern of light and dark banding contrast along the microfibrils is imaged (Fig. 3.4b). This pattern corresponds to the well-known a–e TEM banding in acid-stained type I collagen specimens (Doyle *et al.*, 1974).

This provides important information for the general understanding of STM image contrast in biomolecules. For thin layers of organic molecules, such as the alkyl cyanobiphenyls, higher tunnel currents have been obtained over aromatic groups, probably as a result of electron delocalization. Using the collagen type I α_1- and α_2-chain sequence data in order to determine the distribution of aro-

matic amino acids along the microfibril, no correlation was found with the light and dark banding in the STM images. However, the bright regions correlated with the occurrence of charged amino acids massed at equivalent positions across the microfibril (Fig. 3.5), as has been shown by Meek *et al.* (1979) for the stained collagen case. The bright regions of the banding in the STM image correlate with the regions having a high density of charged amino acids. Because these regions consist of both positive and negative charges, the contrast is unlikely to arise from changes in the local barrier height due to local variations in potential. Being hydrophilic, these charged regions will have a greater number of water molecules associated with them than the hydrophobic regions. It is likely that this will increase the local surface conductivity, resulting in a lighter region in the STM image than was observed. The result suggests that water molecules adsorbed to the surface of a protein molecule can play an important role in image formation.

3.2.4 Elastin

The other principal protein found in skin is elastin. Elastic fibers are vital to the functioning of many mammalian organs such as the aorta, lung, skin, and spinal ligaments. Elastin is an insoluble polymeric protein providing the elastomeric force in these structures. Elastin fibers are formed by extensive crosslinking of the single chain precursor tropoelastin into a macroscopic aggregate. Therefore, α-elastin, a water-soluble 70 kDa acid-hydrolyzed derivative that contains virtually all of the elastin molecule, should form a more suitable specimen for study by STM. As in the case with collagen molecules, α-elastin retains the ability to self-assemble into larger elastinlike aggregates. Transmission electron microscopy studies of elastin have not been successful as the solvents used in fixation and embedding procedures cause conformational changes in this protein, rendering the elastin component amorphous in an ultramicrotomed section of elastic tissue. X-ray diffraction has also failed to give much information because of the low degree of order in the molecular structure.

α-Elastin was extracted by the method of Partridge from a sample prepared, according to the procedures of Lansing *et al.* (1952), from bovine ligamentum nuchae. The purified α-elastin was dissolved in water to give a solution with a

---

FIGURE 3.4 a: 400×388 nm scan of polymeric collagen I arranged in overlapping sheaves. The maximum height of the image is 12.5 nm, although this may not correspond to the thickness of the aggregate. The minimum diameter of the filaments is around 8 nm, and there is some evidence of the D-period banding along the fibers. b: Magnified view of another deposit of collagen I. In this thinner specimen the D-period banding is more clearly seen, as is the kinking of the molecules which is probably occurring at the molecular hole-overlap interface (see Fig. 3.5). The image size is 185×173 nm. Conditions for both figures were 10 pA and 1.5 V sample bias. (From Gathercole *et al.*, 1993, with permission.)

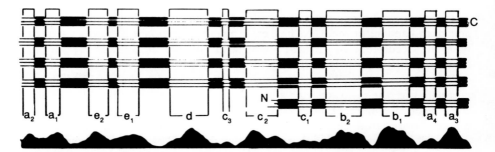

FIGURE 3.5 The top portion of the figure is a schematic representation of the collagen triple helices grouped together in a microfibril. The boxed sections represent the occurrence of charged amino acid residues as they are grouped across and along the collagen microfibril. The bottom portion is a smoothed densitometer trace of the TEM bands along the microfibril axis. The position of bright bands in the STM image (see Fig. 4b), correlates with the spacing of the TEM banding in acid-stained collagen. (After Meek *et al.*, 1979, with permission.)

concentration of 200 mg/ml. A 10 ml drop was placed on HOPG and allowed to evaporate.

Using a sample bias of 800 mV and a tunnel current of 0.1 nA, regions could be found sufficiently thin to be imaged without serious dragging of the sample by the tip. Some regions showed extensive layers of nonbranching filaments (Fig. 3.6). Each filament was composed of spherical beadlike units with a diameter of about 5 nm.

Two models have been proposed to explain the elastomeric nature of elastin fibers. The first model describes elastin as a random entropic elastomeric network. In the second model, elastin fibers are proposed to have a structure resembling a string of beads. In this model, deformation results in conformational changes in the spherical protein molecules such that internal hydrophobic regions become exposed to the aqueous environment, thus providing the restoring force. The STM images of these specimens clearly show beadlike structures consistent with the second model.

3.2.5 Model Polypeptides

Synthetic polypeptides may provide valuable information for understanding STM images of natural proteins. They have known chemical and physical structures, and, because these conformations exist in natural proteins, they act as models for secondary structures in proteins. By imaging these model polypeptides in the STM, it may be possible to gain information on the origins of image contrast and study the effect of adsorption forces on the conformation of the molecule.

FIGURE 3.6 178 × 178 nm scan of beaded filaments of elastin. The filaments are seen partly end-on, giving a "candlewick" appearance.

3.2.5.1 POLYTYROSINE

The repeat unit of polytyrosine is shown in Figure 3.7. The aromatic group in the tyrosine residue is expected to result in a higher local conductivity, as was observed in STM images of cyanobiphenyl molecules. This aromatic group is intimately associated with the α-helical structure and thus might facilitate the imaging of the helix. Polytyrosine was deposited from solutions in dimethylformamide (DMF) onto HOPG. Isolated polytyrosine molecules or monolayers were

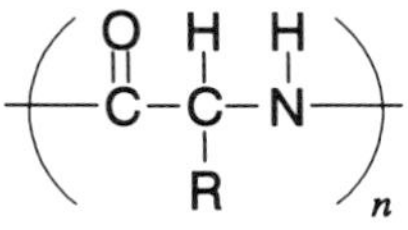

$$R = CH_2C_6H_4OH$$

FIGURE 3.7 Repeat unit of polytyrosine.

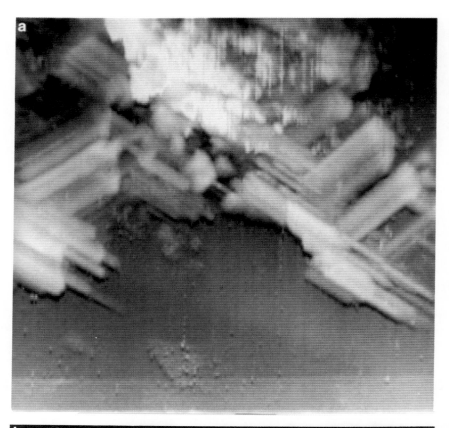

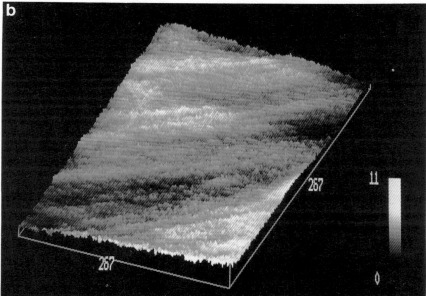

not reproducibly observed in the STM. However, crystallike structures such as those shown in Figure 3.8a were frequently found. Polytyrosine appears more readily to crystallize than to form ordered monolayers on the substrate surface. The surface of the crystals showed regular striations on a molecular scale (Fig. 3.8b).

3.2.5.2 POLY(γ-BENZYL-L-GLUTAMATE)

The synthetic polypeptide poly(γ-benzyl-L-glutamate) (PBLG; Fig. 3.9) forms lyotropic liquid-crystal structures in DMF and other solvents at volume fractions greater than 0.15 (Parry and Elliott, 1967; Squire and Elliott 1969, Watanabe and Uematsu 1984). Although PBLG can be crystallized (Wee and Miller, 1971; Russo and Miller, 1988), it forms crystals less readily than does polytyrosine. Specimens for STM imaging were prepared by depositing on HOPG a 5 ml drop of solution consisting of 1 mg/ml of PBLG, molecular mass 70 kDa, in DMF. The graphite and solution were covered so that solvent evaporation was slow, taking about 24 hr. This procedure provided the conditions for liquid-crystalline ordering of the PBLG molecules and adsorption of a monolayer to at least some regions of the graphite surface.

Although thick deposits unsuitable for STM imaging were common, regions sufficiently thin were also found (McMaster *et al.*, 1990). These images, such as those in Figure 3.10a, show helical structures of pitch 1.61 nm, width 2.9 nm, and lateral spacing 4.12 nm. The Fourier transform (Fig. 3.10b) shows this monolayer structure to be a two-dimensional crystal rather than a liquid-crystalline structure. Figure 3.10c shows a Fourier-filtered image of Figure 3.10a. In other regions, helices of similar pitch and width were found but with a smaller lateral spacing of 3.3 nm (McMaster *et al.*, 1991). This difference may be accounted for by either a different set of planes in the same crystal structure or one of the other crystalline forms of PBLG known to exist (Watanabe and Uematsu, 1984; Watanabe *et al.*, 1986).

The benzyl groups on the end of the side chains are expected to be imaged strongly with STM. The observed width of the helices is consistent with fully extended side chains. The characteristic conformation of PBLG is the 18_5 α-helix, consisting of 18 residues in 5 turns over the repeat distance of 2.7 nm. This helical structure shows a near repeat at 10.8 residues after 3 turns corresponding to a repeat of 1.62 nm. This is close to the observed repeat of the STM images. It appears that adsorption to the HOPG surface favors a distortion of the helical structure to produce an 11_3 helix. Watanabe and Uematsu (1984) have

---

FIGURE 3.8 a: A 256 × 256 nm scan of crystals of polytyrosine deposited semirandomly on the STM substrate and imaged at 10 pA and 1 V bias. The white patch at the top of the micrograph appears to be more amorphous material being moved by the tip. b: The surface of polytyrosine crystals on a 26.7 × 26.7 nm scale. The surface displays striations on a molecular scale.

R = CH$_2$CH$_2$COCH$_2$C$_6$H$_5$

FIGURE 3.9 Repeat unit of PBLG.

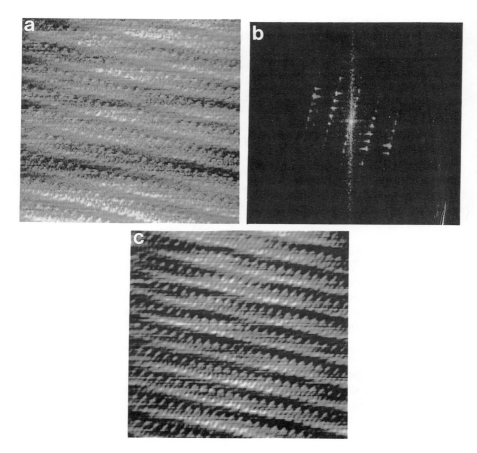

FIGURE 3.10 a: A 340 × 340 nm scan of PBLG on HOPG. The data are unprocessed, and tunneling conditions were 100 pA and 500 mV. b: Fourier transform pattern of a. The degree of order may be compared with that in Figure 3.1b. c: Fourier-filtered version of a. The periodic features that are apparent in the raw data have been emphasized. (From McMaster *et al.*, 1990, with permission.)

found the closely related 7_2 helices in PBLG films cast from chloroform solutions. They attributed the metastability of this structure to interhelical stacking of the benzyl rings. In this STM case, it is likely that this helical structure is stabilized by the interaction of the benzyl rings with the graphite surface.

3.3 GLOBULAR PROTEINS

3.3.1 Introduction

STM offers unique opportunities for three-dimensional structural studies of globular proteins, and there have been several notable achievements in this field. The focus of this section is on those proteins that do not form liquid crystallike arrays, as do some of the extended proteins and polypeptides described above, and on those that do not naturally occur as ordered two-dimensional membrane layers (e.g., the purple membrane and the hexagonally packed intermediate layer protein; see Chapter 2, this volume).

3.3.2 Vicilin

This early work (Welland et al., 1989) is notable as it shows the first high-resolution STM image of a single protein molecule. It was imaged in air and without a conductive coating, and the experiment also demonstrated the use of I-V spectroscopy to provide extra information on sample deposition.

Vicilin is a legume seed storage protein that exists as a trimer of three subunits, each of 50 kDa molecular mass. The arrangement of the subunits into a flattened pseudohexagonal structure, with each subunit divided into two spherical domains with a volume ratio of 2:1, has been shown by small-angle X-ray scattering (SAXS) (l'Anson et al., 1988). Drops of protein solutions in the concentration range 10–100 μg/ml were deposited onto amorphous carbon substrates. After 2 min, the drops were removed and the residue allowed to dry in air.

Figure 3.11a shows a typical image of a deposited molecule. The authors observe that the size of the structure is 10 nm, consistent with the diameter of the pseudohexagonal model obtained from synchrotron X-ray scattering. The substrate topography is shown in Figure 3.11b and the root-mean-square (rms) roughness was measured at 0.25 nm.

The substrate appears rougher than HOPG, the conventional substrate for globular proteins. However, the more hydrophilic nature of the amorphous carbon surface of the X-ray mirror probably ensures stronger physisorption of the vicilin molecules and therefore more reliable imaging.

The values of bias voltage and tunnelling current in this study were 4 V (sample positive) and 20–30 pA, respectively. This combination of high bias voltage and relatively low tunnel current is unusual in STM of proteins, as most studies utilize biases on the order of 1 V or less. The work of Guckenberger et al.

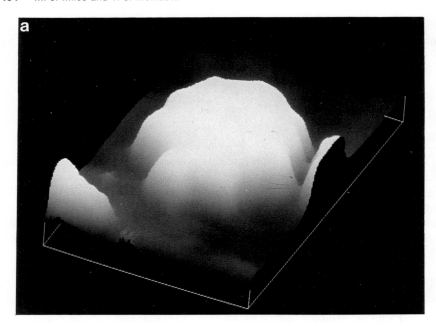

FIGURE 3.11 a: Three-dimensional rendering of a single vicilin molecule. The objects to left and right are the edges of other molecules. The molecule width is 10 nm. b: Typical 100 × 100 nm scan of the amorphous carbon substrate used in this study. RMS roughness is of the order of 0.25 nm. c: Topographical image of vicilin molecules *in vacuo*. The outline is shrunken with respect to that in a. It is imaged on the edge of an aggregate. d: Conductivity-voltage plots measured with the STM tip over the amorphous carbon substrate (top view), and over a vicilin molecule (bottom view). (From Welland et al., 1989, with permission.) *(Continues)*

(1991) on purple membrane (see Chapter 2, this volume) shows images obtained with biases greater than 5 V and subpicoamp currents. They also reported that humidity and positive contrast were directly related. This has implications for the understanding of the mechanism of image contrast formation, discussed in this chapter with regard to STM imaging of collagen fibers.

Removal of bound water by vacuum was observed to have a marked effect on the image topography, as shown in Figure 3.11c, and results in a shrunken structure. The authors note that the shape approximates that observed with TEM. The change in shape observed on vacuum drying is an indication that the STM images in air are not dehydrated and that the original native protein structure is substantially preserved.

The use of current-voltage spectroscopy (Fig. 3.11d) to check that observed images are not substrate features has since been applied to DNA on HOPG (Allison et al., 1990). HOPG is a very convenient substrate for biomolecular experiments because of its extreme flatness and the ease with which it can be cleaved. However, as recent studies have shown (Clemmer and Beebe, 1991;

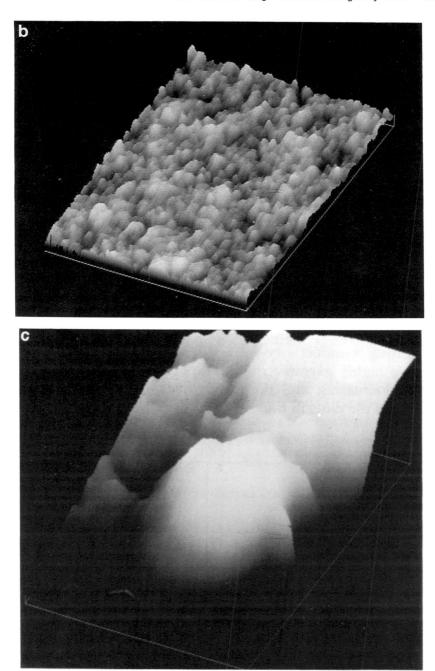

FIGURE 3.11 *(Continued)*

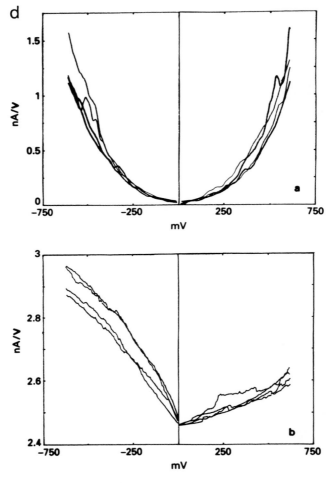

FIGURE 3.11 *(Continued)*

Chang and Bard, 1991), a great variety of artifactual structures may be observed on HOPG, occurring principally at step edges and defects in the crystal plane of the graphite. The use of dI/dV spectroscopy to indicate the presence of biomolecular species may not be helpful on HOPG, as step edges may show local barrier height variations leading to artifactual dI/dV changes. In this respect the use of an amorphous carbon substrate in preference to HOPG has advantages in addition to the improved adsorption properties.

3.3.3 RNA Polymerase Holoenzyme

Keller *et al.* (1991) have obtained images of the DNA-dependent enzyme RNA polymerase by using a combination of electrodeposition onto a gold surface and imaging in a high-humidity environment. The enzyme has a molecular mass of 459 kDa, and the core enzyme consists of up to eight subunits. The gold surface, Au(111), was prepared by the method of flame annealing gold wire into small spheres (Hsu and Cowley, 1983; Schneir et al., 1989). The protein was dissolved in buffer solution containing 50% v/v glycerol, and the gold balls (~1–2 mm diameter) were suspended in this solution. Positive voltages were applied to the gold, and after electrodeposition the balls were removed from the solution and imaging was carried out with a small thin film of the solution remaining on the surface. This provided the high humidity environment for the biomolecule and necessitated the use of insulated tips.

The authors report differences in deposited structures with crystalline arrangements only formed at voltages of ~5 V for deposition times of greater than 5 hr and random deposition observed at higher voltages (Fig. 3.12a).

The heterogeneity of sizes in the randomly deposited molecules is explained by subunit dissociation, and analysis of molecular weights estimated from the sizes of the STM images gives a distribution similar to that observed for various subunit assembly intermediates. The electrochemical method seems to be useful in binding the protein more strongly to the surface, but it may be a cause of molecular dissociation. The charge on the Au(111) surface in this work would appear to have an analogous effect to the positively charged lipid layers used to obtain two-dimensional crystals of the same molecule by Darst et al., (1988, 1989).

The presence of a buffer solution layer means that the molecule is fully hydrated in an environment close to that of its native state. This overcomes the dehydration and resultant molecular distortion that occurs to varying extents when air drying and vacuum methods are used for sample preparation. The tip-sample forces that often cause sample movement when STM imaging is performed in air are substantially reduced when both tip and sample are immersed in a fluid layer. However, the authors do report sample movement by the tip even in solution, when an electrodeposition time of 10 sec was used. Additionally in this case holes were observed in the surface on removal of the protein, which indicates the presence of an embedding layer of oxide or organic material or buffer salts formed during electrodeposition. The presence of this layer may explain the two- to three-fold reduction in height measured for RNA polymerase molecules in the more stable deposited structures.

Varying the buffer solution to provide amphoteric conditions was observed to prevent dissociation, and a more homogeneous distribution was imaged (Fig. 3.12b). The authors draw attention to the presence of some molecules with the characteristic jaw shape that has been observed by TEM and STEM, although the electrodeposition method used is unlikely to produce a unique surface orientation.

The combination of imaging in solution and electrodeposition appears to lead to a high success rate in observing molecules. This is in contrast to the experience of many workers in this field where weak adsorption and tip-sample interaction make routine imaging difficult. The mechanism of image formation

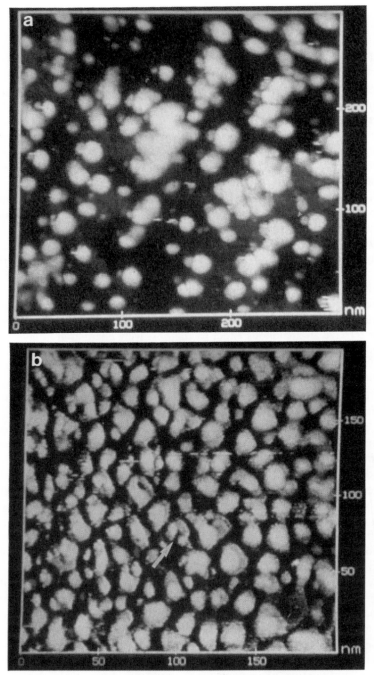

FIGURE 3.12 a: RNA polymerase randomly deposited from storage buffer onto gold by the application of 1 V bias for 1 hr. There is evidence of fragmentation in the wide range of sizes observed. b: Molecules deposited from amphoteric buffer under the same conditions. In this case, some of the molecules showed the characteristic jaw shape obtained in electron microscopic results (arrow). (From Keller *et al.*, 1991, with permission.)

by STM is not settled, though the authors speculate on the role of mobile ions in a protein surface conduction mode. This observation of the importance of the role of water is supported by other experimental results on collagen fibers (see Section 3.2.2.3).

3.3.4 Cell Membrane Protein

The importance of comparing structures observed by STM with data from more established techniques has also been stressed by Hörber et al. (1991, 1992) in their work on cell membrane proteins. They argue that highly resolved pictures of metal-shadowed or -coated material may pose difficulties in interpretation because of the grain size of the shadowing material. Tackling the interpretative problems associated with STM of uncoated proteins favors the development of experiments using molecules with well-defined topographical structures.

The authors chose to study the nicotinic acetylcholine receptor channel protein complex (AChR). This is responsible for the triggered change of the electric potential between the inside and the outside of the cell in the electric organ of the *Torpedo marmorata* organism. The protein does not form a two-dimensional crystal, but is highly enriched in the native membranes of the organism. The protein-rich membrane fragments were prepared using the electron microscopic method of Elliot et al. (1980) and adsorbed onto HOPG without coating. The protein was imaged in solution with insulating tips and in air.

The problem of sample movement, which is a particular problem for STM imaging of isolated protein molecules, is overcome in this case because the protein is naturally located within a larger, more stable membrane fragment. Some of the deposited structures may not, however, be suitable for STM investigation because the membrane fragments may form vesicles or other aggregated structures that are too thick for resolution of individual protein molecules.

Figure 3.13a shows an AChR dimer imaged in air. Tunneling currents of 0.5 – 5 pA were used, which is low compared with many biological STM experiments, illustrating the requirement to maximize the tip-sample distance. In general, this may mitigate against high-resolution studies, although in this case the central pore is observed and the topography agrees well with electron microscopic data (Fig. 3.13b). In their earlier work, the authors noted that the pore was not as well-resolved in STM as in electron microscopy. This was attributed to the shape of the tip and to the inability of the tip to follow steep slopes, with consequent broadening of molecule structure. This phenomenon is well known, and attempts have been made to quantify it (Engel, 1991; Keller, 1991). The authors assert that the surface topography and ring asymmetry are better resolved by STM, although there is evidence of the tip interfering with the sample from the streakiness of the STM image.

The simultaneous acquisition of spectroscopic information while collecting topographical data may aid the interpretation of complex structures. Hörber *et*

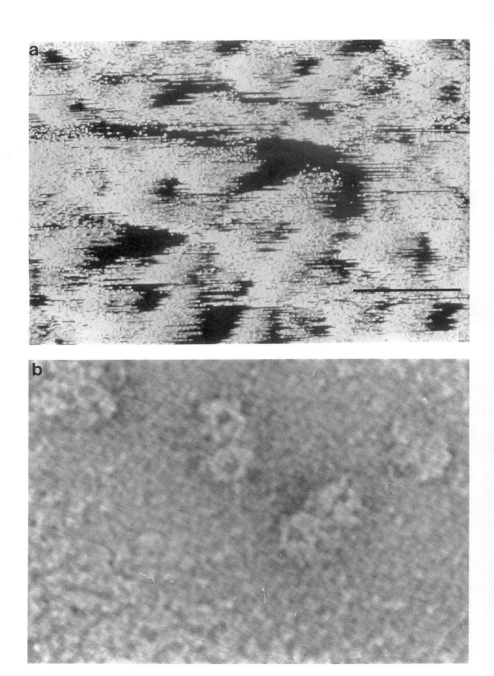

FIGURE 3.13 a: Scanning tunneling microscopic image of an AChR dimer imaged in air. The bar marker represents 7 nm, the approximate measured diameter of a monomer ring unit. b: An electron microscopic image of an AChR dimer. (From Hörber *et al.*, 1992, with permission.)

al. (1991) have developed a method of incorporating the mean curvature of the I-V relationship (related to local barrier height variation) with the topographical data. They have applied this to imaging of dimers of the cyclic decapeptide gramicidin and tentatively identify aliphatic and aromatic parts of the dimer. They state, however, that application of this method requires improvement of preparation techniques, particularly for buffer solutions containing surface active molecules.

3.3.5 Other Globular Proteins

The work of Leatherbarrow *et al.* (1991) on immunoglobulin G (IgG) molecules deposited from buffer onto HOPG shows images with a trilobed structure, which the authors assign to the two Fab arms and the Fc region of the IgG molecule. The IgG molecule is a challenging system for STM investigation because the Fab arms and Fc region may be crystallized separately, but no whole intact molecule has been crystallized. The measured dimensions of the Fab portions are approximately twice that obtained from the X-ray structure, while the measured height of 3 nm is found to be consistent with the X-ray data. The effect of tip geometry causing an increase in the lateral dimensions of molecules has been mentioned above, although the magnitude of the effect is usually not of this size. Values of molecule height have generally been found to be substantially smaller than expected from X-ray data. Lee *et al.* (1991) and Edstrom *et al.* (1991), with their work on the glycogenolytic enzymes, phosphorylase kinase, and phosphorylase b, have made helpful comparisons between STM measurements both in air and in solution and X-ray data. They found that the lateral dimensions of phosphorylase kinase as measured at the solid–liquid interface were 74–78% of the values they obtained at the solid–air interface. For phosphorylase b oligomers the air measurements show an increase of about 25% over the X-ray data. They also calculate that the measured height of the enzyme in solution STM is very close to that measured in air, but this value is only 12% of that measured by scanning force microscopy (SFM) and predicted from the molecular weight of the molecule. This result emphasizes the need to compare STM results with structural and topographical data from other sources when possible.

Various efforts have been made to image isolated globular proteins on HOPG using air-drying alone. With only the graphite-molecule interaction and no other immobilizing effect, the molecules are very susceptible to movement by the tip. This has been observed by Jericho *et al.* (1990) in their attempt to image native uncoated pepsin molecules on HOPG. Even the hopping mode of STM operation did not ensure successful imaging of the relatively small molecules. Feng *et al.* (1989) deposited amino acids, lysozyme, albumin, and fibrinogen onto HOPG. They encountered the phenomena of sample movement and of aggregation of weakly bound structures, which limited the attainment of high-resolution images in some cases.

Häussling et al. (1991) have sought a novel way of anchoring protein to a substrate for STM investigation. Strongly bound chemisorbed monolayers of a biotin-substituted thiol, disulfide, or lipid are formed on a flat gold surface. Two of the four binding sites on the biotin molecule are still available for binding with the bacterial protein streptavidin. The authors observed protein docked onto the monolayer and obtained their best results using the biotinylated lipid.

3.4 CONCLUSIONS

In spite of the generally low conductivity of biological molecules, images of a wide variety of protein molecules have been obtained. A principle factor in obtaining interpretable STM images is the immobilization of the protein molecules. This has been illustrated with reference to systems in which the molecules order in two-dimensional surface layers or form semiordered or self-assembled aggregates. The use of a specially prepared substrate, electrodeposition, and observing molecules in their native environment has also been successful. The ideal of examining molecules in their native environment using STM is likely to remain difficult to achieve because of the constraint of using a conducting substrate. This obstacle is overcome with SFM (see the "SFM in Biology" part), and this technique has been increasingly used for biomolecular studies, although at present the achievable resolution is inferior to that of the STM.

REFERENCES

Allison, D. P., Thompson, J. R., Jacobson, K. B., Warmack, R. J., and Ferrell T. L. (1990). Scanning tunneling microscopy and spectroscopy of plasmid DNA. *Scann. Microsc.* 4, 517–522.
Chang, H. P., and Bard, A. J. (1991). Observation and characterization by STM of structures generated by cleaving highly-oriented-pyrolytic-graphite. *Langmuir* 7, 1143–1153.
Clemmer, C., and Beebe, T. (1991). Graphite: a mimic for DNA and other biomolecules in STM studies. *Science* 251, 640–642.
Darst, S. A., Ribi, H. O., Pierce, D. W., and Kornberg, R. D. (1988). 2-D crystals of *E. coli* RNA polymerase holoenzyme on positively charged lipid layers. *J. Mol. Biol.* 203, 269–273.
Darst, S. A., Kubalek, E. W., and Kornberg, R. D. (1989). Three-dimensional structure of *E. coli* RNA polymerase holoenzyme determined by electron crystallography. *Nature (London)* 340, 730–732.
Doyle, B. B., Hulmes, D. J. S., Miller, A., Parry, D. A. D., Piez, K. A., and Woodhead-Galloway, J. (1974). Axially projected collagen structures. *Proc. R. Soc. Lond. Ser. B* 187, 37–46.
Edstrom, R. D., Miller, M. A., Eilings, V., Yang, X., Yang, R., Lee, G., and Evans, D. F. (1991). STM and AFM visualization of the components of the skeletal muscle glycogenolytic complex. *J. Vac. Sci. Technol. B* 9, 1248–1252.
Elliott, J., Blanchard, S. G., Wu, W., Miller, J., Strader, C. D., Hartig, P., Moore, H. -P., Racs, J., and Raftery, M. A. (1980). Purification of *Torpedo californica* post synaptic membranes and fractionation of their constituent proteins. *Biochem. J.* 185, 667–677.
Engel, A. (1991). Biological applications of scanning sensor microscopes. *Annu. Rev. Biophys. Biophys. Chem.* 20, 79–108.
Feng, L., Andrade, J. D., and Hu, C. Z. (1989). STM of proteins on graphite surfaces. *Scann. Microsc.*, 3, 399–410.

Field, J. M., Tatham, A. S., and Shewry, P. R. (1987). The structure of a high-M_r subunit of Durum-wheat (*Triticum durum*) gluten. *Biochem. J.* 247, 215–221.

Flory, P. J. (1956). Phase equilibria in solutions of rod-like particles. *Proc. R. Soc. Lond. Ser. A* 234, 73–89.

Gathercole, L., Miles, M. J., McMaster, T. J., and Holmes, D. F. (1993). Scanning probe microscopy of collagen I and pN collagen I and the relevance to STM contrast generation in proteins. *Chemical Society Faraday Transactions*. Accepted for publication.

Guckenberger, R., Wiegräbe, W., and Baumeister, W. (1988). STM of biomacromolecules. *J. Microsc.* 152, 795–802.

Guckenberger, R., Hacker, B., Hartmann, T., Scheybani, T., Wang, Z., Wiegräbe, W., and Baumeister, W. (1991). Imaging of uncoated purple membrane by STM. *J. Vac. Sci. Technol. B* 9, 1227–1230.

Häussling, L., Michel, B., Ringsdorf, H., and Rohrer, H. (1991). Direct observation of Streptavidin specifically adsorbed on biotin-functionalized self-assembled monolayers with the STM. *Angewandte Chem. Int. English Edition* 30, 569–572.

Hörber, J. K. H., Schuler, F. M., Witzemann, V., Schröter, K. H., Müller, H., and Ruppersberg, J. P. (1991). Imaging of cell membrane proteins with an STM. *J. Vac. Sci. Technol. B* 9, 1214–1218.

Hörber, J. K. H., Schuler, F. M., Witzemann, V., Müller, H., and Ruppersberg, J. P. (1992). Imaging biological membrane structures with an STM. In *AIP Conference Proceedings 241, Scanned Probe Microscopy*. (H. K. Wickramasinghe, ed.) Santa Barbara.

Hsu, T., and Cowley, J. M. (1983). Reflection electron microscopy of fcc metals. *Ultramicroscopy* 11, 239–250.

I'Anson, K. J., Miles, M. J., Bacon, J. R., Carr, H. J., Lambert, N., Morris, V. J., and Wright, D. J. (1988). Structure of the 7S globulin (vicilin) from pea (*Pisum sativum*). *Int. J. Biol. Macromol.* 10, 311–317.

Jericho, M. H., Blackford, B. L., Dahn, D. C., Frame, C., and Maclean, D. (1990). STM imaging of uncoated biological material. *J. Vac. Sci. Technol. A* 8, 661–666.

Keller, D. (1991). Reconstruction of STM and AFM images distorted by finite-size tips. *Surf. Sci.* 253, 353–364.

Keller, R. W., Bear, D., and Bustamante, C.(1991). A method for imaging *E. coli* RNA polymerase holoenzyme with the STM in an aqueous environment. *J. Vac. Sci. Technol. B* 9, 1291–1297.

Lansing, A. I., Rosenthal, T. B., Alex, M., and Dempsey, W. (1952). The structure and chemical characterization of elastic fibres as revealed by elastase and by EM. *Anat. Rec.* 114, 555–575.

Leatherbarrow, R. J., Stedman, M., and Wells, T. N. C. (1991). Structure of immunoglobulin G by STM. *J. Mol. Biol.* 221, 361–365.

Lee, G., Evans, D. F., Eilings, V., and Edstrom, R. D. (1991). Observation of phosphorylase kinase and phosphorylase b at solid-liquid interfaces by STM. *J. Vac. Sci. Technol. B* 9, 1236–1241.

McMaster, T. J., Carr, H. J., Miles, M. J., and Morris, V. J. (1990). Polypeptide structures imaged by the STM. *J. Vac. Sci. Technol. A* 8, 648–651.

McMaster, T. J., Carr, H. J., Miles, M. J., Cairns, P., and Morris, V. J. (1991). STM of PBLG. *Macromolecules* 24, 1428–1430.

Meek, K. M., Chapman, J. A., and Hardcastle, R. A. (1979). The staining pattern of collagen fibrils. *J. Biol. Chem.* 254, 10710–10714.

Miles, M. J., Carr, H. J., McMaster, T. J. I'Anson, K. J., Belton, P. S., Morris, V. J., Field, J. M., Shewry, P. R., and Tatham, A. S. (1991). STM of a wheat seed storage protein reveals details of an unusual supersecondary structure. *Proc. Natl. Acad. Sci. U.S.A.* 88, 68–71.

Parry, D. A. D., and Elliott, A. (1967). The structure of a paracrystalline phase of PBLG in DMF. *J. Mol. Biol.* 25, 1–13.

Russo, P., and Miller, W. G. (1988). Coexistence of liquid crystalline phases in PBLG-DMF. *Macromolecules* 16, 1690–1693.

Schneir, J., Harary, H. H., Dagata, J. A., Hansma, P. K., and Sonnenfeld, R. (1989). STM and

fabrication of nanometer scale structures at the liquid–gold interface. *Scann. Microsc.* 3, 719–724.
Squire, I. M., and Elliott, A. (1969). Liquid crystalline phases of PBLG in solution. *Mol. Cryst. Liquid Cryst.* 7, 457–468.
Tatham, A. S., Miflin, B. J., and Shewry, P. R. (1985). The β-turn conformation in wheat gluten proteins—relationship to gluten elasticity. *Cereal Chem.* 62, 405–412.
Tatham, A. S., Drake, A. F., and Shewry, P. R. (1990a). Conformational studies of synthetic peptides corresponding to the repetitive regions of the high-molecular-weight (HMW) glutenin subunits of wheat. *J. Cereal Sci.* 11, 189–200.
Tatham, A. S., Masson, P., and Popineau, Y. (1990b). Conformational studies of peptides derived by the enzymic hydrolysis of a γ-type gliadin. *J. Cereal Sci.* 11, 1–13.
Thomson, N. H., Miles, M. J., Tatham, A. S., and Shewry, P. R. (1992). Molecular images of cereal proteins by STM. *Ultramicroscopy* 42–44, 1204–1213.
Watanabe, J., and Uematsu, I. (1984). Anomalous properties of PBLG film composed of unusual 7/2 helices. *Polymer* 25, 1711–1717.
Watanabe, J., Imai, K., and Uematsu, I. (1986). On novel crystalline forms of PBLG. *Macromolecules* 19, 1489–1491.
Wee, E. L., and Miller, W. G. (1971). Liquid crystal–isotropic phase equilibria in the system PBLG–DMF. *J. Phys. Chem.* 75, 1446–1452.
Welland, M. E., Miles, M. J., Lambert, N., Morris, V. J., Coombs, J. H., and Pethica, J. B. (1989). Structure of the globular protein vicilin revealed by STM. 11, 29–32.

CHAPTER 4
Ordered Organic Monolayers Studied by Tunneling Microscopy

D. P. E. Smith
IBM Physics Group Munich,
Munich, Germany

J. E. Frommer
IBM Almaden Research Center,
San Jose, California
and Institut für Physik der Universität,
Basel, Switzerland

4.1 Introduction
4.2 Adsorbate Mobility
4.3 Chemisorption
4.4 Self-Assembly on van der Waals Surfaces
    4.4.1 Liquid Crystals
    4.4.2 *n*-Alkanes
    4.4.3 Alkyl Derivatives
    4.4.4 Other Molecules
4.5 Langmuir–Blodgett Films
4.6 Conducting Solids
4.7 Conclusions
Appendix
References

4.1 INTRODUCTION

In this chapter, scanning tunneling microscopy (STM) studies of monolayers of organic molecules adsorbed onto substrates are presented. The molecules we discuss are simple compared with complex biomolecules. Their simplicity means that sample preparation and interpretation of the STM images are relatively easy and well defined. The highest degree of success has been obtained with highly ordered adsorbates. In particular, it is possible to prepare and observe two-dimensional organic crystalline films that are homogeneous over many micrometers. Samples with this kind of long-range order allow for less ambiguous interpretation of STM images, since a repetitive molecular structure can be easily recognized even when instrumentational distortions are present. Organic samples that suffer from a lack of order often yield uninterpretable STM images. Atomic resolution imaging has been achieved on certain monolayer systems. The experience gained in the study of these relatively uncomplicated cases may be applicable to the more complex "real" biological systems. The results discussed in this chapter might suggest to the optimistic biologist the ultimate resolution obtainable with the STM.

Scanning tunneling microscopy depends on a current of tunneling electrons flowing between a conductive tip and a conductive surface. To study most organic layers with the STM, it is therefore necessary that the layers be sufficiently thin to allow the electrons to tunnel through to the conducting substrate. In this chapter we focus on molecular films that are believed to be only one monolayer thick (see Sections 4.3, 4.4). Alternatively, if the organic layer itself is conductive, as in the case of charge transfer salts (see Section 4.4), normal conduction can occur, and even very thick films can be imaged. A puzzling third case has presented itself: STM of relatively thick layers of organics that traditionally have been considered electronically insulating (Section 4.5; Michel et al., 1989; and Chapter 2, this volume). Alternate electron pathways between tip and conductive support have been proposed to rationalize the tunneling currents detected through these "insulating" samples. Among the proposals are surface and defect conductivity (García and García, 1990) and conduction through a thin water film (Yuan et al., 1991). The mechanism of conductivity in these samples remains enigmatic.

The use of the STM and atomic force microscopes (AFM) in the domain of organic molecules is gaining rapid popularity and is the subject of several reviews (see Frommer, 1992; Heckl, 1992; Fuchs, 1993).

4.2 ADSORBATE MOBILITY

A critical feature that can render adsorbates elusive to the STM is their mobility. The problem is keeping the organic material to be studied in one place while it is being probed by the intrusive STM "tip." Several approaches have been taken to immobilize molecules on STM supports. The two most well-defined cases are chemical reaction with the substrate and self-assembly into vast planes. These two cases are discussed in Sections 4.3 and 4.4, respectively.

One requirement in all circumstances is to keep the tip sufficiently far away from the material being investigated to avoid injuring the sample. For STM this is equivalent to maintaining a high tunnel resistance between the tip and the sample. A typically high tunnel–gap resistance in the case of physisorbed organic molecules would correspond to a gap voltage of 1 V and a tunnel current of 10–100 pA. This represents a tunnel resistance of 10^{11} to 10^{10} Ω. Sano and Kunitake (1991) have examined this problem systematically and found for the graphite/(carboxy)azobenzene system that the minimum tunneling resistance is 2×10^9 Ω. Below this value the graphite substrate became visible, and the molecular structure would not return if the resistance was increased to the original value. Molecules were irreversibly removed from the scanned area. In samples in which a drop of liquid crystal is deposited on the substrate, molecules can apparently be swept away by low gap resistances, but the molecular structure will often reappear once the resistance is again increased (Smith et al., 1989, 1990). The fluid drop apparently acts as a reservoir to fill in the missing molecules in the lattice.

In air, the maximum bias voltage with which one can operate is approximately 1.5 V. (This refers to the voltage drop across the tunnel gap; with a highly resistive substrate some voltage drop occurs across the substrate itself.) Above this voltage the tunnel current often becomes unstable, perhaps as the result of ionization phenomena. [In ultrahigh vacuum (UHV) and at liquid helium temperatures these instabilities do not occur, and biases of several volts can indeed be used.] Because the maximum voltage is limited, high gap resistances or large gap spacings are achieved by using low tunnel currents. In principle, one wishes to operate at the lowest current possible which still gives good feedback response (see Instrumentation, the Appendix). Molecules that are chemisorbed to a support can be imaged with gap resistances as low as 10^7 Ω, but a threshold is nevertheless observed below which the image is disrupted.

4.3 CHEMISORPTION

Three experiments that have been performed in ultrahigh vacuum have demonstrated the use of reactive substrates to immobilize molecules. The first involves exposure of the (111) surface of a rhodium crystal to measured doses of benzene and carbon monoxide gases (Ohtani et al., 1988). In Figure 4.1 is a high-resolution STM image of this system, clearly showing the benzene rings. The rings' threefold symmetry is probably due to the registry of the benzene with the rhodium lattice. The small bumps in the dark areas at the corners of the benzene rings are assigned to CO. The fact that the molecules were in a closely packed array also likely helped to prevent excessive molecular motion and led to stable STM imaging.

FIGURE 4.1 C_6H_6 and CO coadsorbed on the rhodium (111) surface imaged with a scanning tunneling microscope. The benzene appears as threefold rings and the CO as the darker spots at the corners of the rings. (Courtesy of R. J. Wilson, IBM Almaden Research Center.)

In the second example, copper phthalocyanine was imaged after being sublimated onto a copper (100) surface (Lippel et al., 1989). The STM images detail the internal structures of even isolated molecules. In contrast, attempts to image copper phthalocyanine on gold (Gimzewski et al., 1987) and silver (Lippel et al., 1989) surfaces were unsuccessful. This could be attributed to insufficient or inappropriate bonding of the molecules to the substrate.

The third UHV study involved the adsorption of naphthalene ($C_{10}H_8$) on a platinum (111) surface (Hallmark et al., 1991a,b). In this study, the cleaned platinum crystal was dosed with naphthalene gas and then heated to between 100 and 200°C to create an ordered overlayer. The long axis of the molecule aligned with the three equivalent crystallographic directions of the metal surface. The bilobed character of the molecules is evident in the STM images, but internal features of the molecule were not discerned. Attempts at higher resolution imaging by using lower bias voltages or slower scan rates were unsuccessful, apparently due to strong tip-molecule interactions. The naphthalene molecules were typically observed to lie flat on the substrate, but some defects were observed that may correspond to the rings being tilted on edge. At submonolayer coverages the molecules often underwent discrete rotations among the three equivalent orientations and translations by single platinum lattice spacings. At high molecular coverages molecular motion was observed rarely and only at grain boundaries.

On the reactive surfaces used in the benzene/CO and copper-phthalocyanine studies, the tunnel gap resistances used were as low as 5×10^7 and 5×10^6 Ω, respectively. On the noble metal surface used in the naphthalene study, the lower limit was found to be about 5×10^8 Ω. These gap resistances are significantly lower than the limit of 5×10^9 Ω possible with the physisorbed system of alkylcyanobiphenyls on graphite (Smith *et al.*, 1989).

It seems clear that for successful STM imaging it is preferable to have molecules either chemisorbed or strongly physisorbed to a substrate. This situation is unfortunately difficult to realize outside of the carefully controlled UHV environment. A reactive surface exposed to air or liquid is unlikely to remain reactive. Electrochemical approaches to the deposition of organic material may be able to overcome this (Nagahara *et al.*, 1990; De Rose *et al.*, 1991; see also Chapter 5, this volume).

Another approach is to prepare samples under vacuum and then image them under ambient conditions. This approach has proven successful for obtaining well-resolved images of perylene dianhydride physisorbed onto graphite (Ludwig *et al.*, 1992). In this study, as in the benzene/CO one discussed above, the images are dominated by the hydrocarbon moieties, with little STM evidence for the carbonyl groups.

Finally, an increasingly popular method for preparing well-ordered organic monolayers involves the reaction of ω-thiols with a gold surface. In this method, the sulfur reacts with the gold to form a covalent link between the substrate and the adsorbate. The alkyl chains then orient perpendicular to the interface with the substrate and self-assemble into a fairly continuous film, providing a smooth, organic surface composed of chain termini. These surfaces have been studied by STM with molecular resolution (Widrig *et al.*, 1991; Häussling *et al*, 1991), results that are consistent with STM studies on Langmuir-Blodgett films (see Section 4.5).

4.4 SELF-ASSEMBLY ON VAN DER WAALS SURFACES

Another approach toward adsorbate immobilization is to take advantage of the formation of an ordered layer of molecules in registry with a relatively inert surface such as the basal plane of graphite. [Scanning tunneling microscopy experiments done on highly oriented pyrolytic graphite (HOPG) have given the same results as on single crystal or natural graphite.] On these surfaces, presumably only van der Waals forces anchor the molecules to the substrate. Most of the remainder of this chapter is devoted to studies in this category. Sample preparation involves the self-assembly of organic molecules lying on the substrate into vast planes (typically micrometers in dimension). The masses of these planes and their interactions with the underlying surface immobilize, or lock in, the individual molecules. The driving force for assembly comes from both intermolecular and molecule-surface interactions. The most widely studied system in this category is liquid crystals.

4.4.1 Liquid Crystals

A growing number of STM results are emerging from organic adsorbates that have liquid crystalline phases. Briefly, liquid crystals are molecules that are more highly ordered than a liquid but more disordered than a crystal (de Gennes, 1983). They are rodlike molecules that align along a common axis but are otherwise disordered (nematics); or they order in layers, with each layer behaving like a distinct two-dimensional liquid or solid (smectics). Liquid crystals are favorable molecules for STM study, since they naturally self-organize due to intermolecular interactions. They, therefore, form a stable molecular lattice even when deposited on an inert surface. Moreover, most liquid crystal molecules are rod shaped because they contain alkyl groups, and these groups, as discussed in Section 4.4.2, provide good anchoring to the graphite basal plane. The STM results show that a large number of liquid crystal molecules from highly ordered two-dimensional molecular "crystals" at the surface of graphite. This observation has been extended to include the substrate MoS_2, a transition-metal dichalcogenide. Very likely other materials will also be shown to be effective as substrates.

4.4.1.1 ALKYLCYANOBIPHENYLS

Foster and Frommer (1988) were the first to use STM to study the ordering of smectic and nematic liquid crystals on graphite. The first material they tried, 4-n-octyl-4'-cyanobiphenyl (8CB), has proven to be especially popular and has since been studied by several other groups (Foster et al., 1989; Smith et al., 1989, 1990, 1991; McMaster et al., 1990b; Mizutani et al., 1990a,b; Hara et al., 1990). An STM image of this smectic A material on the graphite basal plane is shown in Figure 4.2. The generic class of 4-n-alkyl-4'-cyanobiphenyls (mCB) is commercially available in various alkyl chain lengths, in pure form and in a wide variety of mixtures having low-temperature smectic and nematic phases (BDH Ltd., Poole, England). The mCB molecules studied order somewhat similarly on graphite. The long axes of the molecules lie parallel to the substrate; the phenyl groups orient either tilted or parallel with respect to the graphite plane, depending on intermolecular spacing. In Figure 4.2, the biphenyl groups appear as pairs of bright rings and the aliphatic tails as the darker zigzag chains of spots, each spot associated with an alternating methylene group. The position of the nitrogen atom in the cyano group can also be seen as a bright spot near the biphenyl rings, *para* to the biphenyl linkage. The appearance of the phenyl groups is similar to that of pure benzene (see Fig. 4.1) and that of the aliphatic tails to the pure n-alkanes (see Fig. 4.4).

4 Ordered Organic Monolayers 211

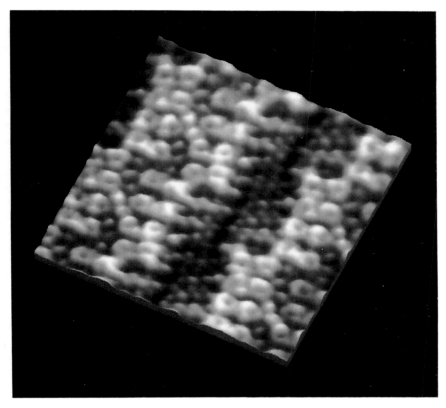

FIGURE 4.2 4'-n-Octyl-4-cyanobiphenyl (8CB) adsorbed on graphite. The size of the image is 55 × 55 Å². The biphenyl groups appear as bright pairs of rings and the alkyl groups as somewhat dimmer zigzag chains.

Several different preparation schemes for depositing and successfully imaging liquid crystal molecular films have evolved. Three are presented here: (1) A droplet, typically several micrograms, is applied to the freshly cleaved graphite (or other van der Waals) surface. The liquid crystal is then heated into the isotropic phase so that it spreads over the surface; from observation by optical microscopy we believe that the molecules are wetting the surface. The resulting film is allowed to cool slowly in order to form a well-ordered structure. (2) The liquid crystals are heated to around 100°C. At this temperature the vapor pressure of many liquid crystals is fairly high, typically in the range of 1 mTorr. The molecules sublimate and deposit on the substrate held several centimeters above the liquid. (3) The molecules are dissolved in a polar solvent such as an alcohol. The dilute solution is applied to the substrate, and the solvent is allowed to evaporate.

That all these deposition methods provide viable samples for STM imaging illustrates the thermodynamic stability of the ordered molecules on the surface. The STM, being sensitive only to those molecules very close to the interface between the substrate and adsorbate, serves as a powerful probe to confirm this. The total thickness of the deposited organic film does not appear to affect the STM results, a feature also of STM studies of paraffin on graphite (see Section 4.4.2) and of alkyl derivatives (see Section 4.4.3). Since the organic adsorbates are generally considered to be electrically insulating, it is assumed that the STM tip is no further than the distance of one or two monolayers from the substrate, but this remains unproven.

The best STM results were achieved when the sample temperature was held near the bulk crystalline-to-smectic (or crystalline-to-nematic) transition temperature during imaging. At this temperature the molecules in the vicinity of the graphite surface are presumably crystallized and thereby immobilized. The bulk of the molecules must remain in the fluid state in order for the STM tip to be able to penetrate the film and image the interface. If the samples are too warm, order is difficult or impossible to discern, most likely because the molecules near the surface are too fluid. At the other extreme, if the films are cooled too far below their bulk solidification temperature, imaging also becomes difficult, possibly because the molecules no longer wet the substrate.

The STM images in Figure 4.3 show the different ordering of 6CB, 8CB, 10CB, and 12CB molecules on HOPG and on MoS_2. MoS_2—like graphite, a layered material whose basal plane is very inert—was first shown to cause ordering of liquid crystal molecules by Hara *et al.* (1990). As shown in Figure 4.3, the same molecules order differently on graphite and on MoS_2 due to the differences in lattice constant and surface interactions. For instance, 8CB on graphite forms a unit cell composed of eight molecules in which the cyanobiphenyl groups lie next to each other. In contrast, 8CB on MoS_2 forms a unit cell of four molecules in which a cyanobiphenyl group lies between two alkyl groups. The molecule 12CB forms the same eight-molecule-unit cell as 8CB on graphite but has a unit cell with 18 molecules on MoS_2. The details of the ordering of the alkylbiphenyls on graphite versus MoS_2 have been described in a series of articles (Smith and Heckl, 1990; Iwakabe *et al.*, 1990, 1991; Smith *et al.*, 1992).

On both graphite and MoS_2 the alykl chains are observed to align with the substrate lattice (Smith *et al.*, 1990; Smith *et al.*, 1992). This is determined by decreasing the tunnel gap resistance below about 10^7 Ω so that the STM tip penetrates the molecular layer. The substrate surface then becomes visible. The alignment of the alkyl groups suggests that these groups are registered with the substrate, similar to how *n*-alkanes order on graphite (see Section 4.4.2). A reasonable conclusion is that it is the alkyl groups in the alkylcyanobiphenyl molecules that lead to the observed mode of adsorption and ordering on the studied surfaces. The biphenyl groups do not appear to play as active a role in the ordering process. The importance of the alkyl tail is borne out by the observation that alkylcyanobiphenyl molecules with an even number of carbons in the

FIGURE 4.3 Top row: alkylcyanobiphenyl molecules adsorbed onto MoS₂. From left to right: 6CB, 8CB, 10CB, 12CB. Bottom row: the same alkylcyanobiphenyl molecules on graphite. 8CB forms completely different structures on MoS₂ and graphite; 12CB forms a slightly different structure. The 6CB and 10CB structures are essentially the same on both substrates.

aliphatic moiety all form rather similar structures on graphite (see Fig. 4.3), whereas those with odd numbers of carbons form very different structures (for 7CB, see Mizutani et al., 1990a; 9CB and 11CB, Smith et al., unpublished data).

4.4.1.2 OTHER TYPES OF LIQUID CRYSTALS

The number of liquid crystal molecules whose ordered monolayers have been observed by STM is increasing rapidly. We are aware of the following experiments done with graphite as a substrate: 4-(*trans*-4-*n*-pentylcyclohexyl)benzonitrile (Foster and Frommer, 1988); 5-nonyl-2-*n*-nonoxlphenylpyrimidine, 4-cyano-4'-*n*-butoxybiphenyl, and 4-*n*-propyl-4'-cyanobicyclohexane (Spong et al., 1989a,b); 4-(4'-pentyl-*trans*-cyclohexl)cyanobenzene (Mizutani et al., 1991); and 4-(4'-*n*-heptylcyclohexyl)-cyanophenyl (Mizutani et al., 1990a, 1990b). The ordering of phenyl benzoate ferroelectric chiral smectic C materials has been studied on MoS_2 (Hara et al., 1991). A mixture of alkoxyphenylbenzoate and alkoxybiphenylbenzoate on graphite has shown a variety of smectic A and smectic C structures (Sautière et al., 1992). The sample in these various experiments were prepared simply by melting the organic crystal on the cleaved graphite surface. All of the molecules contain alkyl groups, pointing out again the probable importance of these groups in the adsorption process.

In the experiments reported thus far the alkyl groups always appear darker than the aromatic, cycloalkane, and cyano groups. This effect may be related to the lowest unoccupied molecular orbitals (as suggested by Smith et al., 1990) and how these molecular orbitals interact with the substrate to affect the potential barrier in the tunnel gap (Spong et al., 1989a). It was predicted by Smith et al. (1990) that AFM images of liquid crystals would probably not show the same contrast as STM, since AFM senses the total charge density whereas STM senses the molecular orbitals between the Fermi level of the substrate and the applied bias across the tunnel gap. The study by Yamada et al. (1992) appears to support this prediction. Their AFM images of 8CB on graphite showed the same lattice as seen with STM, but the contrast between the alkyl and aryl moieties was reversed: the alkyl group was brighter than the aryl group. This apparent greater "height" of the alkyl group could be explained on the basis of van der Waals radii if the aromatic rings are lying down flat on the graphite in which case the height of the alkyl chain would be ~4.0 angstroms, and the height of the aromatic ring would be ~3.5 angstroms. A more definitive explanation awaits our deeper understanding of the contrast mechanisms of the scanning probe microscopes (SXM).

Of particular interest to the biological community are the results on proteins and polypeptides in extended helical conformation (McMaster et al., 1990a; Miles et al., 1990; see Chapter 3, this volume). These rodlike molecules show liquid crystalline behavior and form ordered structures on graphite. Poly (g-benzyl-L-glutamate) (PBLG), a synthetic polypeptide, was dissolved in dimethylformadine at a concentration of 1 mg/ml. A 5 μl drop was deposited on graphite and allowed to dry. As the solvent evaporated the PBLG passed though a liquid-crystal phase that may have contributed to ordering on the substrate. The STM images of PBLG clearly show ordered rows of helical structures. In a related study, high molecular weight (HMW) wheat protein was thought to form a rare conformation based on β-reverse turns. The HMW wheat protein, extracted from durum wheat, was dissolved in trifluoroethanol at a concentration of 200 μg/ml (Miles *et al.*, 1990), deposited on graphite, and left to dry. It was believed that in the solvent the protein had a rodlike shape and thus might again form a liquid-crystalline phase. The resulting periodic structure visible with STM is consistent with the predicted β-spiral conformation.

4.4.2 n-Alkanes

Long-chain n-alkanes have provided one of the most striking examples of STM imaging of organic monolayers (Fig. 4.4) (McGonigal *et al.*, 1990, 1991; Rabe and Buchholtz, 1991a). The relatively simple chemical structures of these molecules yields clear-cut images of hydrocarbon order and registry on a graphite surface. The alkanes used in this study belong to the class of higher molecular weight hydrocarbons known as *paraffins*. Paraffins have long been known to form close-packed monolayers of horizontally oriented molecules on the graphite basal plane (Everett and Findenegg, 1969; Groszek, 1970). Little further adsorption occurs after the monolayer is complete. Using calorimetry, Groszek (1970)

FIGURE 4.4 n-$C_{32}H_{66}$ molecules adsorbed in close-packed rows on graphite. The field of view is 100 $\times$ 100 Å^2. (Courtesy of D. J. Thomson, University of Manitoba.)

attributes the strong adsorption of paraffins to the decrease in potential energy when the molecules adhere to the surface and orient in such a way that the hydrogens in the methylene groups are positioned over the centers of the carbon hexagons in the graphite basal plane. The heat of adsorption increases uniformly with the chain length and reaches very high values (209 kJ/mol) for dotriacontane (n-$C_{32}H_{66}$).

McGonigal et al. (1990, 1991) have obtained atomic resolution STM images of n-alkanes adsorbed at the liquid–graphite interface. As shown in Figure 4.4, an image of n-$C_{32}H_{66}$, the layers possess a high degree of two-dimensional ordering. The long axes of the molecules are parallel to each other, forming 40 Å wide stripes of aligned molecules. The stripes are separated from each other by 5 Å wide, dark troughs. Across these dark troughs the molecules are offset by one-half of their parallel spacing. The structure is consistent with a commensurate ordering of paraffins on graphite. The closest commensurate packing of molecules on the graphite surface gives a parallel spacing of $\sqrt{3}a_0 = 4.26$ Å for the molecules ($a_0 = 2.46$ Å, the lattice spacing of graphite). The spacing observed by STM is 4.5 ± 0.5 Å. This is the first direct evidence that paraffins form commensurate layers on graphite, confirming the earlier calorimetric results (Groszek, 1970). Remarkably similar images of n-$C_{44}H_{90}$ have been obtained with high-resolution electron microscopy (Dorset and Zemlin, 1990). The 2.5 Å resolution images of paraffin crystals, epitaxially grown on benzoic acid and held at a temperature near 4 K, reveal lamellae 47.5 Å in width.

To prepare the sample of n-$C_{32}H_{66}$ shown in Figure 4.4, a solution of n-$C_{32}H_{66}$ in iso-octane was applied to the freshly cleaved graphite surface, followed by application of a solution of n-$C_{32}H_{66}$ in decane (McGonigal et al., 1990, 1991). The heat of adsorption of n-$C_{32}H_{66}$ from iso-octane is very high (120 kJ/mol). Unfortunately, the iso-octane is very volatile, so the second solution was necessary to maintain a stable solid–liquid interface. The preference of graphite to adsorb longer length n-alkanes resulted in the n-$C_{32}H_{66}$ being preferentially adsorbed from solution, producing the ordered layer observed in Figure 4.4. It is known from earlier studies that adsorbed layers of liquids can take on solidlike properties (Everett and Findenegg, 1969). It is likely that the hydrocarbon planes imaged with the STM have this solidlike property, in this study as well as in many of the other STM studies of adsorbed organic molecules.

McGonigal et al. (1991) point out how calorimetric and surface mass measurements can aid STM investigators in identifying those molecules that are likely to form physisorbed monolayers at the liquid–substrate interface. This kind of analysis also gives the temperature range in which ordering of stable monolayers is expected.

4.4.3 Alkyl Derivatives

In Sections 4.4.1 and 4.4.2 the important role played by the alkyl groups in stabilizing molecular monolayers on van der Waals surfaces was stressed. It is

4 Ordered Organic Monolayers 217

therefore not surprising that alcohol (McGonigal *et al.*, 1991; Rabe and Buchholtz, 1991b; Yackoboski *et al.*, 1992) and fatty acid alkane derivatives (Kuroda *et al.*, 1991; Rabe and Buchholtz, 1991b; Kishi *et al.*, 1992), as well as dialkylarene (Spong *et al.*, 1989a,b; Foster *et al.*, 1989; Buchholz and Rabe, 1991), are also observed to form ordered lamellar structures. In the experiments of Rabe and Buchholtz, the molecules were typically dissolved in phenyloctane. Other solvents such as diphenylsulfide and 1,3,5-trimethyl-1,1,3,5,5-pentaphenyltrisiloxane gave the same results. Direct application of decanol to graphite was also an effective preparation method (McGonigal *et al.*, 1991; Yackoboski *et al.*, 1992). Alkoxy-, bromoalkoxy-, and hydroxy-(carboxy)azobenzene derivatives could be directly melted onto graphic under N_2 (Sano and Kunitake, 1991).

A different approach to preparing STM samples is to deposit fatty acids with the Langmuir-Blodgett (LB) technique (Kuroda *et al.*, 1991). The LB technique (see Section 4.5) creates ordered films whose molecules are aligned perpendicular to the substrate. Kuroda *et al.* (1991) found that the LB method was an effective way of bringing monolayer quantities of behenic acid onto graphite. He then observed by STM that they appeared to rearrange to form an oriented monolayer parallel to the substrate. The images do not look significantly different from those prepared by placing a solution of molecules on the graphite, but the quantity of material deposited is better regulated with the LB technique. Another advantage of working with an LB film is that there is no solvent. The absence of a macroscopically thick fluid overlayer facilities scattering or diffraction experiments being performed on the same samples. An STM image of LB-deposited behenic acid is shown in Figure 4.5a (Kishi *et al.*, 1992). The structure is similar to that observed by Rabe and Buchholtz (1991b) for stearic acid adsorbed from a solution of phenyloctane. Figure 4.5b,c shows the underlying graphite substrate and an overlay showing how the alkyl chains align with the graphite lattice. The carboxyl groups are noticeably brighter than the alkyl groups, and there is a periodic variation in this brightness that is attributed to only every fifth molecule being fully registered with the substrate (Kishi *et al.*, 1992). Simultaneous work-function measurements suggest that differences in the barrier height do not fully account for the observed contrast between the carboxyl and alkyl groups, as has been suggested by Spong *et al.* (1989a).

Contrast changes within molecules have been observed on varying the bias voltage applied while imaging decanol on graphite (Yackoboski *et al.*, 1992). Thus, the hydroxy group could be differentiated from the alkyl group. Furthermore, a threshold voltage of between 0.1 and 0.3 V was identified below which the graphite substrate was imaged and above which the alcohol adsorbate was imaged. This allowed accurate correlation of adsorbate-substrate superposition. This work joins a growing number of reports of near simultaneous imaging of adsorbate and substrate (Mizutani *et al.*, 1990a; Smith *et al.*, 1990; Heckl *et al.*, 1991; Ludwig *et al.*, 1992).

The possibility of studying molecular dynamics at the molecular scale has been demonstrated by the motion of defects at grain boundaries in didodecyl-

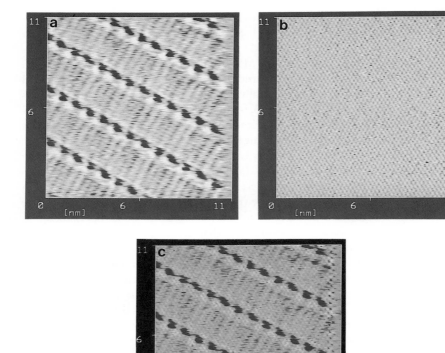

FIGURE 4.5 a: Behenic acid on graphite prepared with the Langmuir-Blodgett technique. The carboxyl groups are seen as bright spots at the ends of the alkyl chains. The field of view is 100×110 Å^2. The tunneling current was 3 pA, and the sample bias was -1.0 V. b: Graphite substrate below the behenic acid lemellae observed with 100 pA tunnel current and $+100$ mV bias. c: Superimposed image of a and b showing the alignment of the molecules with the substrate. (Courtesy of E. Kishi, Canon Research Center.)

benzene and octadeconol (Rabe and Buchholtz, 1991a,b) adsorbed on graphite. If the STM mechanics and electronics are very high speed (Besenbacher et al., 1991) or if the variable-current mode of STM operation is used (Bryant et al., 1986), then dynamics on the time scale of milliseconds can be followed.

The general property of alkyl chains to become immobilized on van der Waals surfaces means that one can pin a variety of chemically active groups in a monolayer matrix if a suitable alkyl derivative can be made. For example, 4-n-hexadecyl-4'-purine has been found to form an ordered monolayer on graphite (Smith and Heckl, unpublished data). Purine is a fluorescent dye molecule; an ordered two-dimensional assembly of these molecules may possess novel optical properties.

Two forces microscope experiments have confirmed results on alkyl derivatives first obtained by STM. Yamada *et al.*, (1992), using films provided by Kuroda, observed lamellar structures in behenic acid under water. Eng *et al.* (1992) were able to resolve the molecular rows of didodecylbenzene on graphite, but only when using a metal-coated cantilever biased at $+4$ V. The applied field in the STM tunnel gap may thus play an important role in the imaging process. While it is possible that the electric field is inducing order in the films (see voltage Pulses, Appendix A), Eng *et al.* (1992) propose that it is the image force that is the important factor. The additional attractive force between cantilever and substrate acts to push the tip through the organic film down to the film–substrate interface. Since Yamada *et al.* (1992) did not need an applied bias to image either behenic acid or 8CB with their force microscope, an applied field does not appear to be a prerequisite.

4.4.4 Other Molecules

The STM has also shown that the simple polar molecules acetone and dimethylsulfoxide form ordered molecular organic layers on graphite (Hubacek *et al.*, 1989). Acetone was allowed to condense as a vapor onto the graphite and was observed to form a commensurate 2×2 structure. Dimethylsufoxide was applied as a drop of liquid and formed a more closely packed incommensurate structure. The adsorption and desorption processes could in both cases be observed as a function of temperature.

The decoding of DNA via STM requires being able to recognize the four bases of the genetic code. Allen *et al.* (1991) prepared monolayer films of adenine on graphite by applying an aqueous solution of adenine to heated graphite. Heckl *et al.* (1991), using a similar technique, grew monolayer films of guanine on graphite and on MoS_2. Both groups resolved the individual purine molecules and some internal molecular structure. On graphite, the double ring structures associated with the two purine bases could be distinguished from each other, as well as from the single ring structures of the two pyrimidine bases cytosine and thymine (W. Heckl, personal communication). On MoS_2, guanine formed a well-defined molecular lattice with four molecules per unit cell (Heckl *et al.*, 1991). The stability of the array was attributed to both substrate registry and intermolecular hydrogen bonding. However, as was found in the case of alkylcyanobiphenyls (Smith *et al.*, 1992), STM images of guanine on MoS_2 did not show the internal detail possible when graphite was used as a substrate. This presumably has to do with differences in the electronic structures of the two substrates and with how the molecular orbitals interact with substrate states to affect the local density of states a few angstroms above the surface.

Polyimide, a technologically important material composed of chains of phenyl and imide groups, was deposited by the LB technique onto graphite by Fujiwara et al. (1991). The highly aromatic polymer chains could be seen to align parallel to the deposition direction of the LB film. The zigzag structure seen in the STM images agreed with molecular dynamics simulations carried out on graphite.

4.5 LANGMUIR-BLODGETT FILMS

Langmuir-Blodgett films were among the first organic systems to be studied by STM. These films are assembled in a water trough and transferred to a substrate. In contrast, the systems described in Section 4.4 self-assemble on a substrate. In the LB technique, molecules are dissolved in a solvent such as chloroform, spread on the surface of pure water, and compressed to form a monolayer. Monolayers are transferred to the substrate by iterative dipping into the water trough. Cadmium arachidate (CdA) is the most well-studied system in the STM field. The first images of a 54 Å thick CdA bilayer on graphite contain a periodic array of ellipsoidal features, 3 Å in diameter and spaced, center-to-center, 5.84 Å along the a-axis and ≈ 4 Å along the b-axis (Smith et al., 1987). The mean molecular density observed was one molecule per 19.4 Å^2, in good agreement with the calculated densities of molecules on the water surface. These results were reproduced by others (Lang et al., 1988; Hörber et al., 1988; Mizutani et al., 1988; Fuchs, 1988). Fuchs et al. (1990) have observed similar images of 22-tricosenoic acid monolayers (thickness 32 Å) deposited on WSe$_2$. In none of these experiments has atomic resolution been obtained; rather, the 3 Å diameter of the elliptical features probably represents a radius of gyration of the hydrocarbon terminus of the lipid molecule. This conclusion applies equally well to the STM images recorded from gold-supported alkane thiol self-assembled monolayers (see Section 4.4).

At the intersection of two previously discussed classes of molecules, liquid crystals and LB films, lies an azobenzene derivative that has been studied with STM (Loo et al., 1990). This molecule contains moieties that allow it to assemble as both a liquid crystal and an LB film. STM samples were prepared on graphite by the methods of both disciplines: dipping into an ordered monolayer to form a bilayer and self-assembly on the substrate. Both sets of samples yielded STM images in which the molecule appears to orient perpendicular to the graphite plane. The molecular density observed was one molecule per 20.5 Å^2 for the LB method and one per 26.5 Å^2 for the self-assembly technique. The LB density is quite close to the density at which the film was fabricated, with the self-assembled density corresponding well with the extrapolated zero pressure density.

In addressing the dilemma of explaining tunneling probability through a hydrocarbon layer several nanometers thick, critics have assigned the STM-observed periodicities to distorted images of the underlying graphite or WSe$_2$ substrate. This criticism has been countered by several recent experiments. In the first study, remarkably similar STM images of CdA were obtained on indium-tin-oxide glass slides, an amorphous substrate (Göbel et al., 1992). Another corroboration is a force microscope experiment that recorded molecular resolution from CdA on the amorphous surface of a silicon wafer (Meyer et al., 1991). In another AFM experiment, arachidic acid bilayers were imaged on silanated silicon wafers, and the 54 Å thickness of the bilayer was observed at a discontinuity of the LB film (Bourdieu et al., 1991). These results and many subsequent AFM studies on LB films (Frommer, 1992) suggest that one must take seriously the earlier STM results showing molecularly resolved lattices on thick insulating films. It has been pointed out by Yuan et al. (1991) that in principle a water bridge can be formed between the tip and sample of radius 10 to 20 Å. Conduction could then be enhanced by ionic transport through the films or over their surfaces.

4.6 CONDUCTING SOLIDS

The chemisorbed and self-assembled organic films described in Sections 4.3 and 4.4 could be imaged with the STM because the imaged layer was only a few angstroms from the conducting substrate. Scanning tunneling microscopy imaging of thicker organic films is possible if the material itself is electrically conductive, as shown by the work on TTF–TCNQ, a conducting organic crystal composed of stacks of tetrathiofulvalene and tetracyanoquinodimethane (Sleater and Tycko, 1988). The TTF and TCNQ moieties could be resolved in the STM images, and there was qualitative agreement between the images and a simple molecular orbital calculation. More recent STM studies on charge transfer complexes of TCNQ with 4-ethyl pyridine and triethylammonium reveal stacked rows of TCNQ moieties in which the benzene rings and triply bonded nitrogen of the cyano groups are distinguishable (Magonov and Cantow, 1990). These are similar to the orbitals resolved most clearly in the images of the alkylcyanobiphenyls (Fig. 4.2).

Several groups (Yang et al., 1990; Lacaze et al., 1992; Magonov and Cantow, 1990) have performed preliminary STM investigations with conductive polymers such as polypyrrole, polyarenes, and polythiophene. Doping organic layers to render them electrically conductive is a technique that has not yet been extensively explored as an STM sample preparation technique. It certainly deserves attention if thicker organic films are to be studied with the STM.

4.7 CONCLUSIONS

In this chapter, we presented many adsorbate–substrate systems as seen through the circuits of the STM, often on the molecular and sometimes on the atomic scale. At times we are surprised by the arrangement of molecules at the interface when viewed so closely. At other times we recognize the packing as being consistent with the results of other analytic methods. We have seen that the ordering of molecules on surfaces is directed by many forces, including intermolecular, surface, and interfacial forces. Many questions still remain. For example, in recording these surface arrangements, how does the STM affect the distribution of molecules? What is the capability to manipulate willfully the imaged surface? What is the mechanism of imaging organic layers, especially thick layers that have traditionally been considered insulating? Finally, can we learn to distinguish different functional groups and thereby decode unknown molecules?

APPENDIX

Voltage Pulses

It has been observed in a number of studies that the application of a brief surge in tunnel gap voltage results in an increased resolution of organic molecules on certain surfaces (Foster *et al.*, 1988; 1989; Smith *et al.*, 1989; Bernhardt *et al.*, 1990; McGonigal *et al.*, 1990; Rabe and Buchholtz, 1991b). The pulses are typically in the range of 3–4 V above the applied DC bias and 100 nsec to 10 μsec in duration. They often dramatically change the nature of the image, causing an ordered molecular lattice to appear where an atomically flat and featureless surface had been observed before. It may simply be that the voltage pulses remove molecules from the tunnel gap, thereby cleaning or reshaping the tip. Another possibility is that when molecules are deposited onto a substrate they are sometimes in an amorphous state. The pulses could provide the energy necessary to crystallize the film fully or provide pinning sites to anchor the molecules. It has been reported that voltage pulses are unnecessary if liquid-crystal molecules are carefully prepared to ensure good crystallization by slow cooling through the phase-transition temperature (Hara *et al.*, 1990). In that study Pt-Ir tips were used, known to be somewhat cleaner than the tungsten tips used by the other investigators.

Instrumentation

Working with very low tunnel currents demands some care with the STM instrumentation. For instance, when operating with such low currents one must keep track of any voltage offsets and noise sources in the current to voltage (I-V) converter and the feedback loop. We have found that an I-V converter with a gain of 10^8 V/A works well for the applications discussed here. The bandwidth for a simple op-amp–based amplifier is typically around 20 kHz, a good range in which to avoid ambient noise problems and thermal drift. The resolution is in principle determined by the Johnson (thermal) noise of the feedback resister that determines the gain of the amplifier. The Johnson noise is $i_n = \sqrt{(4K_B T/R_f)} = (128 \text{pA}/\sqrt{\text{Hz}})/\sqrt{R_f}$, where R_f is the feedback resistor. For $R_f = 10^8 \, \Omega$ the gain is 10^8 V/A and the thermal noise is 1.8 pA for a bandwith of

20 kHz. If the cable between the amplifier and the sample is very long, the source capacitance reacts with the input noise voltage of the amplifier to give an additional broadband noise. It is worth keeping this cable length as short as just a few centimeters.

Another thing to keep in mind when using low tunnel currents is the response of a logarithmic amplifier, if any, in the feedback loop. All log amps have rapidly decreasing bandwidths as the input current (or voltage) becomes smaller. For example, the Analog Devices AD759 log amp has a bandwidth of 200 kHz at an input voltage of 10 mV (corresponding to 100 pA tunnel current with 10^8 V/A gain) but only 1.8 kHz at a tunnel current of 1 pA. To avoid having an unusably small feedback bandwidth, one can add a DC current offset either at the current amplifier or at the logarithmic amplifer. For example, an offset corresponding to 10 pA improves the bandwidth of the log amp to 25 kHz. Finally, some feedback circuits have an electronic rectifier before the log amp. Since these switch as the current passes through zero, they should be adjusted so that they do not oscillate when the current is small.

A variety of tunnel tip materials have been used: tungsten, Pt-Ir, gold, carbon, and likely many others. Tungsten tips have the advantage that they are easily etched to form very sharp (radius of about 1000 Å) tips with very hard mechanical properties. Typical etching conditions are 1 M solution of KOH with 1 amp AC current and a Pt or carbon counterelectrode. Tungsten tips do not stay clean for very long and tend to be covered primarily with a layer of carbon. Removing this layer may necessitate the voltage pulses discussed in this appendix. Pt-Ir tips are also widely used, both chemically etched (1 g KCN + 5 ml conc. aq. sol. NH_4OH + 5 ml glycerine) or mechanically cut. They appear to stay clean longer, and at least one group reports that with Pt-Ir tips voltage pulses were not necessary to observe the ordering of liquid-crystal molecules (Hara *et al.*, 1990). Gold tips can be etched in HCl + HNO_3 (1:1). Although not very mechanically strong, they are very clean and give acceptable results. Carbon tips can be easily made with 0.5 mm mechanical pencil lead. The softness of the tips and the ease with which multiple tips are created make them difficult to use. However, they might be useful when trying to image very soft biological samples that might be damaged by a hard tip such as tungsten.

REFERENCES

Allen, M. J., Balooch, M., Subbiah, S., Tench, R. J., Siekhaus, W., and Balhorn, R. (1991). Scanning tunneling images of adenine and thymine at atomic resolution. *Scann. Microsc.* 5, 625–629.

Bernhardt, R. H., McGonigal, G. C., Scheider, R., and Thomson, D. J. (1990). Mechanisms for the deposition of nanometer-sized structures from organic fluids using scanning tunneling microscopy. *J. Vac. Sci. Technol.* A 8, 667–671.

Besenbacher, F., Jensen, F., Laegsgaard, E., Mortensen, K., and Stensgaard, I. (1991). Visualization of the dynamics in surface reconstructions. *J. Vac. Sci. Technol.* B 9, 854–878.

Bourdieu, L., Silberzan, P., and Chatenay, D. (1991). Langmuir-Blodgett films: from micron to angstrom. *Phys. Rev. Lett.* 76, 2029–2032.

Brandow, S. L., DiLella, D. P., and Colton, R. J. (1991). Evidence of pronounced positional order at the graphite–liquid crystal interface of a bulk nematic material. *J. Vac. Sci. Technol. B* 9, 1115–1118.

Bryant, A., Smith, D. P. E., and Quate, C. F. (1986). Imaging in real time with the tunneling microscope. *Appl. Phys. Lett.* 48, 832–834.

Buchholtz, S., and Rabe, J. P. (1991). Conformation, packing, defects and molecular dynamics in monolayers of dialkyl-substituted benzenes. *J. Vac. Sci. Technol. B* 9, 1126–1128.

deGennes, P. G. (1983). Liquid Crystals. In *Concise Encyclopedia of Solid State Physics* (R. G. Lerner and G. L. Trigg, eds.) Addison-Wesley, Reading, MA, pp. 143–148.

De Rose, J. A., Lidsay, S. M., Nagahara, L. A., Oden, P. I., and Thundat, T. (1991). Electrochemical deposition of nucleic acid polymers for scanning probe microscopy. *J. Vac. Sci. Technol. B* 9, 1166–1170.

Dorset, D. L., and Zemlin, F. (1990). Direct phase determination in electron crystallography: the crystal structure of an n-paraffin. *Ultramicroscopy* 33, 227–236.

Eng, L. M., Fuchs, H., Buchholz, S., and Rabe, J. P. (1992). Ordering of didodecylbenzene on graphite: a combined SFM/STM study. *Ultramicroscopy* 42–44, 1059–1066.

Everett, D. H., and Findenegg, G. H. (1969). Calorimetric evidence for the structure of films adsorbed at the solid/liquid interface: the heats of wetting of "Graphon" by some n-alkanes. *Nature (London)* 223, 52–53.

Foster, J. S., and Frommer, J. E. (1988). Imaging of liquid crystals using a tunnelling microscope. *Nature (London)* 333, 542–545.

Foster, J. S., Frommer, J. E., and Arnett, P. C. (1988). Molecular manipulation using a tunnelling microscope. *Nature (London)* 331, 324–326.

Foster, J. S., Frommer, J. E., and Spong, J. K. (1989). Imaging of liquid crystals using a tunneling microscope. *Proc. SPIE* 1080, 200–208.

Frommer, J. (1992). Scanning tunneling microscopy and atomic force microscopy in organic chemistry. *Angew. Chem. Int. Ed. Engl.* 31, 1298–1328.

Fuchs, H. (1988). High resolution STM—studies on graphite and Langmuir-Blodgett films. *Phys. Sci.* 38, 264–268.

Fuchs, H., Akari, S., and Dransfeld, K. (1990). Molecular resolution of Langmuir-Blodgett monolayers on tungsten diselenide by scanning tunneling microscopy. *Z. Phys. B* 80, 389–392.

Fuchs, H. (1993). Atomic force and scanning tunneling microscopies of organic surfaces. *J. Molec. Structure* 292(1–3), 29–48.

Fujiwara, I., Ishimoto, C., and Seto, J. (1991). Scanning tunneling microscopy study of a polyimide Langmuir-Blodgett film. *J. Vac. Sci. Technol. B* 9, 1148–1153.

Garcia, R., and García, N. (1990). Electron conductance in organic chains: why are STM experiments possible on bare biological samples? *Chem. Phys. Lett.* 173, 44–50.

Gimzewski, J. K., Stoll, E., and Schlittler, R. R. (1987). Scanning tunneling microscopy of individual molecules of copper phthalocyanine adsorbed on polycrystalline silver surfaces. *Surf. Sci.* 181, 267–277.

Göbel, H. D., Hörber, J. K. H., Gerber, C., Leitner, A., and Hänsch, T. W. (1992). Molecular structures on lipid monolayers on ITO-glass and on graphite imaged by an STM. *Ultramicroscopy*, 42–44, 1260–1268.

Groszek, A. J. (1970). Selective adsorption at graphite/hydrocarbon interfaces. *Proc. R. Soc. Lond. A* 314, 473–498.

Hallmark, V. M., Chiang, S., Brown, J. K., and Wöll, C. (1991a). Real-space imaging of the molecular organization of naphthalene on Pt (111). *Phys. Rev. Lett.* 66, 48–51.

Hallmark, V. M., Chiang, S., and Wöll, C. (1991b). Molecular imaging of ordered and disordered naphthalene on Pt (111). *J. Vac. Sci. Technol. B* 9, 1111–1114.

Hara M., Iwakabe, Y., Tochigi, K., Sasabe, H., Garito, A. F., and Yamada, A. (1990). Anchoring structure of smectic liquid-crystal layers on MoS_2 observed by scanning tunnelling microscopy. *Nature (London)* 344, 228–230.

Hara, M., Umemoto, T., Takezoe, H., Garito, A. F., and Sasabe, H. (1991). Novel quadruple zigzag structure of antiferroelectric liquid crystal molecules observed by scanning tunneling microscopy. *Jpn. J. Appl. Phys.* 30, L2052–L2054.

Häussling, L., Michel, B., Ringsdorf, H., and Rohrer, H. (1991). Direct observation of Streptavidin specifically adsorbed on biotin-functionalized self-assembled monolayers with the scanning tunneling microscope. *Angew. Chem. Int. Ed. Engl.* 30, 569–572.

Heckl, W. M., Smith, D. P. E., Binnig, G., Klagges, H., Hänsch, T. W., and Maddocks, J. (1991). Two-dimensional ordering of the DNA base guanine observed by scanning tunneling microscopy. *Proc. Natl. Acad. Sci. U.S.A.* 88, 8003–8005.

Heckl, W. (1992). Scanning tunneling microscopy and atomic force microscopy on organic and biomolecules. *Thin Solid Films* 210/211, 640–647.

Hörber, J. K. H., Lang, C. A., Hänsch, T. W., Heckl, W. M., and Möhwald, H. (1988). Scanning tunneling microscopy of lipid films and embedded biomolecules. *Chem. Phys. Lett.* 145, 151–158.

Hubacek, J. S., Brockenbrough, R. T., Gammie, G., Skala, S. L., and Lyding, J. W. (1989). Scanning tunnelling microscopy of graphite-adsorbed molecular species. *J. Microsc.* 152, 221–227.

Iwakabe, Y., Hara, M., Kondo, K., Tochigi, K., Mukoh, A., Garito, A. F., Sasabe, H., and Yamada, A. (1990). Two types of anchoring structure in smectic liquid crystal molecules. *Jpn. J. Appl. Phys.* 29, L2243–L2246.

Iwakabe, Y., Hara, M., Kondo, K., Tochigi, K., Mukoh, A., Yamada, A., Garito, A. F., and Sasabe, H. (1991). Correlation between bulk ordering and anchoring structures of liquid crystals studied by scanning tunneling microscopy. *Jpn. J. Appl. Phys.* 30, 2542–2546.

Kishi, E., Matsuda, H., Kuroda, R., Takimoto, K., Yamano, A., Eguchi, K., Hatanaka, K., and Nakagiri, T. (1992). Barrier height imaging of fatty acid Langmuir-Blodgett films. *Ultramicroscopy,* 42–44, 1067–1072.

Kuroda, R., Kishi, E., Yamano, A., Hatanaka, K., Matsuda, H., Eguchi, K., and Nakagiri, T. (1991). Scanning tunneling microscope images of fatty acid Langmuir-Blodgett bilayers. *J. Vac. Sci. Technol. B* 9, 1180–1183.

Lacaze, E., Garbarz, J., Quillet, V., Schott, M., Pham, M., Moslih, J., and Lacaze, P. (1992). STM of conducting polymer thin films formed by electropolymerization on graphite and gold. *Ultramicroscopy* 42–44, 1037–1043.

Lang, C. A., Hörber, J. K. H., Hänsch, T. W., Heckl, W. M., and Möhwald, H. (1988). Scanning tunneling microscopy of Langmuir-Blodgett films on graphite. *J. Vac. Sci. Technol. A* 6, 368–370.

Lippel, P. H., Wilson, R. J., Miller, M. D., Wöll, C., and Chiang, S. (1989). High-resolution imaging of copper-phthalocyanine by scanning-tunneling microscopy. *Phys. Rev. Lett.* 62, 171–174.

Loo, B. H., Liu, Z. F., and Fujishima, A. (1990). Scanning tunneling microscopic images of an azobenzene derivative differently deposited on highly oriented pyrolytic graphite surfaces. *Surf. Sci.* 227, 1–6.

Ludwig, C., Gompf, B., Glatz, W., Petersen, J., Eisenmenger, W. (1992). Video-STM, LEED, and X-ray diffraction investigations of PTCDA on graphite. *Z. Phys. B.* 86, 397–404.

Magonov, S. N., and Cantow, H.-J. (1990). Applications of scanning tunneling microscopy to layered materials, organic charge transfer complexes and conductive polymers. In *Scanning Tunneling Microscopy and Related Methods.* (R. J. Behm, N. Garcia, and H. Rohrer, eds.) Kluwer Academic Publishers, Dordrecht, pp. 367–376.

McGonigal, G. C., Bernhardt, R. H., and Thomson, D. J. (1990). Imaging alkane layers at the liquid/graphite interface with the scanning tunneling microscope. *Appl. Phys. Lett.* 57, 28–30.

McGonigal, G. C., Bernhardt, R. H., Yeo, Y. H., and Thomson, D. J. (1991). Observation of highly ordered, two-dimensional *n*-alkane and *n*-alkonol structures on graphite. *J. Vac. Sci. Technol. B* 9, 1107–1110.

McMaster, T. J., Carr, H., Miles, M. J., Cairna, P., and Morris, V. J. (1990a). Polypeptide structures imaged by the scanning tunneling microscope. *J. Vac. Sci. Technol. A* 8, 648–651.

McMaster, T. J., Carr, H., Miles, M. J., Cairns, P., and Morris, V. J. (1990b). Adsorption of liquid crystals imaged using scanning tunneling microscopy. *J. Vac. Sci. Technol. A* 8, 672–674.

Meyer, E., Howald, L., Overney, R. M., Heinzelmann, H., Frommer, J., Güntherodt, H.-J., Wagner, T., Scheir, H., and Roth, S. (1991). Molecular-resolution images of Langmuir-Blodgett films using atomic force microscopy. *Nature* 349, 398–400.

Michel, B., Travaglini, G., Rohrer, H., Joachim, C., and Amrein, M. (1989). Images of crystalline alkanes obtained with scanning tunneling microscopy. *Z. Physiol. Biol.* 76, 99–105.

Miles, M. J., McMaster, T., Carr, H. J., Tatham, A. S., Shewry, P. R., Field, J. M., Belton, P. S., Jeenes, D., Hanley, B., Whittam, M., Cairns, P., Morris, V. J., and Lambert, N. (1990). Scanning tunneling microscopy of biomolecules. *J. Vac. Sci. Technol. A* 8, 698–702.

Mizutani, W., Shigeno, M., Saito, K., Watanabe, K., Sugi, M., Ono, M., and Kajimura, K. (1988). Observation of Langmuir-Blodgett films by scanning tunneling microscropy. *Jpn. J. Appl. Physiol.* 27, 1803–1807.

Mizutani, W., Shigeno, M., Sakakibara, Y., Kajimura, K., Ono, M., Tanishima, S., Ohno, K., and Toshima, N. (1990a). Scanning tunneling spectroscopy study of adsorbed molecules. *J. Vac. Sci. Technol. A* 8, 675–678.

Mizutani, W., Shigeno, M., Ono, M., and Kajimura, K. (1990b). Voltage-dependent scanning tunneling microscopy images of liquid crystals on graphite. *Appl. Phys. Lett.* 56, 1974–1976.

Mizutani, W., Shigeno, M., Ohmi, M., Suginoya, M., Kajimura, K., and Ono, M. (1991). Observation and control of adsorbed molecules. *J. Vac. Sci. Technol. B* 9, 1102–1106.

Nagahara, L. A., Thundat, T., Oden, P. I., Lindsay, S. M., and Rill, R. L. (1990). Electrochemical deposition of molecular adsorbates for in situ scanning probe microscopy. *Ultramicroscopy* 33, 107–116.

Ohtani, H., Wilson, R. J., Chiang S., and Mate, C. M. (1988). Scanning tunneling microscopy of benzene molecules on the Rh(111)-(3 × 3) (C_6H_6 + 2CO) surface. *Phys. Rev. Lett.* 60, 2398–2401.

Rabe, J. P., and Buchholz, S. (1991a). Direct observation of molecular structure and dynamics at the interface between a solid wall and an organic solution by scanning tunneling microscopy. *Phys. Rev. Lett.* 66, 2096–2099.

Rabe, J. P., and Buchholz, S. (1991b). Commensurability and mobility in two-dimensional molecular patterns in graphite. *Science* 253, 424–427.

Sano, M., and Kunitake, T. (1991). Imaging of bromo- and hydrogen-terminated azobenzene derivatives by scanning tunneling microscopy. *J. Vac. Sci. Technol. B* 9, 1137–1140.

Sautière, L., Day, S., and Miles, M. J. (1992). Smectic structures in a two-component liquid crystal mixture. *Ultramicroscopy* 42–44, 1054–1058.

Sleator, T., and Tycko, R. (1988). Observation of individual organic molecules at a crystal surface with use of a scanning tunneling microscope. *Phys. Rev. Lett.* 60, 1418–1421.

Smith D. P. E. (1991). Defects in alkylcyanobiphenyl molecular crystals studied by STM. *J. Vac. Sci. Technol. B* 9, 1119–1125.

Smith, D. P. E., Bryant, A., Quate, C. F., Rabe, J. P., Gerber, C., and Swalen, J. D. (1987). Images of a lipid bilayer at molecular resolution by scanning tunneling microscopy. *Proc. Natl. Acad. Sci. U.S.A.* 84, 969–972.

Smith, D. P. E., and Heckl, W. M. (1990). Surface phases. *Nature (London)* 346, 616–617.

Smith, D. P. E., Hörber, H., Gerber, C., and Binnig, G. (1989). Smectic liquid crystal monolayers on graphite observed by scanning tunneling microscopy. *Science* 245, 43–45.

Smith, D. P. E., Hörber, J. K. H., Binnig, G., and Nejoh, H. (1990). Structure, registry and imaging mechanism of alkylcyanobiphenyl molecules by tunneling microscopy. *Nature (London)* 344, 641–644.

Smith, D. P. E., Klagges, H. A., and Heckl, W. M. (1992). Ordering of alkylcyanobiphenyl molecules

at MoS$_2$ and graphite surfaces studied by tunneling microscopy and molecular dynamics. *Surf. Sci.* 278, 166–174.
Spong, J. K., Mizes, H. A., LaComb, L. J., Dovek, M. M., Frommer, J. E., and Foster, J. S. (1989a). Contrast mechanism for resolving organic molecules with tunnelling microscopy. *Nature (London)* 338, 137–139.
Spong, J. K. LaComb, L. J., Jr., Dovek, M. M., Frommer, J. E., and Foster, J. S. (1989b). Imaging of liquid crystals using a tunneling microscope. *J. Phys.* 50, 2139–2146.
Widrig, C., Alves, C., Porter M. (1991). Scanning tunneling microscopy of ethanethiolate and *n*-octadecanethiolate monolayers spontaneously adsorbed at gold surfaces. *J. Amer. Chem. Soc.* 113, 2805–2810.
Yackoboski, K., Yeo, Y. H., McGonigal, G. C., and Thomson, D. J. (1992). Molecular position at the liquid/solid interface measured by voltage-dependent imaging with the STM. *Ultramicroscopy* 42–44, 963–967.
Yang, R., Evans, D., Christensen, L., Hendrickson, W. (1990). Scanning tunneling microscopy evidence of semicrystalline and helical conducting polymer structures. *J. Phys. Chem.* 94, 6117–6125.
Yamada, H., Akamine, S., and Quate, C. F. (1992). Imaging of organic molecular films with the atomic force microscope. *Ultramicroscopy*, 42–44, 1044–1048.
Yuan, J.-Y., Shao, Z., and Gao, C. (1991). Alternative method of imaging surface topologies of nonconducting bulk specimens by scanning tunneling microscopy. *Phys. Rev. Lett.* 67, 863–866.

CHAPTER 5

Potentiostatic Deposition of Molecules for SXM

S. M. Lindsay
Department of Physics
Arizona State University
Tempe, Arizona

N. J. Tao
Department of Physics
Florida International University
Miami, Florida

5.1 Introduction
5.2 Theoretical Background
5.3 Experimental Procedures
5.4 The Au(111) Surface under Aqueous Electrolytes
    5.4.1 Clean and Dirty Gold
    5.4.2 Electronic Contrast: Adenine on Au(111)
    5.4.3 Phosphate Solutions on Au(111)
5.5 Ligated DNA Oligomers on Au(111)
    5.5.1 Choice of DNA Sample
    5.5.2 Synthesis and Characterization of DNA
    5.5.3 STM Imaging of DNA Adsorbates
    5.5.4 What Holds the DNA in Place?
    5.5.5 Reversibility and the Role of the $23 \times \sqrt{3}$ Reconstruction
    5.5.6 Statistical Analysis: Preferential Adsorption of Small Molecules
5.6 Discussion and Conclusions
References

5.1 INTRODUCTION

One of the most remarkable aspects of the scanned probe microscopies is their ability to work under water (Sonnenfeld and Hansma, 1986; Drake et al., 1989). This opened a new area of study in structural biology. Some examples from our group include scanning tunneling microscopic (STM) images of the DNA double helix in water (Lindsay and Barris, 1988; Lindsay et al., 1989a,b) atomic force microscopic (AFM) images of long double helical DNA under water (Lyubchenko et al., 1992; Lindsay et al., 1992b), and high-resolution mapping of the elastic properties of a biological composite (Tao et al., 1992). In this chapter we describe another new application: the imaging of molecular adsorbates held onto an electrode maintained under potential control (Tao et al., 1991; Lindsay et al., 1992a). This method works for both the STM and AFM, but we have obtained higher resolution images with the STM. In addition, STM images are sensitive to local electronic states. This allows molecules to be identified, which suggests that the microscope might be used to sequence nucleic acids or proteins in certain conditions.

In this chapter, we describe our electrochemical methods for depositing and imaging molecular layers on the Au(111) surface. These layers are rich in novel electronic properties, often associated with the $23 \times \sqrt{3}$ reconstruction on Au(111). We now understand that many "artifacts" on this surface are a consequence of interaction between sites on the $23 \times \sqrt{3}$ reconstruction and adsorbate molecules. However, it is also possible to produce relatively undistorted images that permit direct measurement of the geometry of hydrated DNA (and other) molecules with unprecedented resolution.

The electrochemical method we have developed is a little tricky to use (a characteristic of any approach to *atomic-level* control of surface coverage). However, it has the following advantages.

1. The procedure we outline here will hold small molecules in place against the strong (and uncontrolled) interactions encountered in the STM. The resolution of the STM should be similar to that of the AFM operated in contact mode. The width of the point spread function of an STM is given approximately by $2\sqrt{0.7R/K}$, where R is the radius of the tunnel tip (often near-atomic) and K may have a value near 1 Å^{-1}. In an AFM image a molecule of radius r yields an image width of $4\sqrt{Rr}$ in contact mode. However, *in practice,* we find that AFM resolution is often relatively much poorer than these estimates would suggest, an effect we do not fully understand.
2. The method allows selective attachment of the biomolecules. Other procedures that dry the sample (or precipitate it in an uncontrolled manner) may cause salt to be deposited along with the polymer in ways that complicate the image (our earlier work relied on nonreversible electrochemical reactions of salt–DNA complexes and suffered from these complications (Lindsay et al., 1989a,b).
3. With adequate control of contamination, the method allows *quantitative* deposition of a *uniform* layer of adsorbate. *The substrate does not have to be*

surveyed for "recognizable" images. This stands in sharp contrast to most other methods for depositing biopolymers, with which control of the substrate at an atomic level is rather difficult.
4. Control of the interfacial charge is trivial, allowing one to explore the role of charge in both the image contrast and the conformation of the biopolymer by direct imaging.
5. The hydration of the adsorbate is maintained throughout the imaging process.

The major disadvantage of the method is that small molecules are preferentially adsorbed. Contamination poses problems, and the technique cannot rival such feats as the imaging of the entire λ phage genome by the AFM (Lyubchenko et al., 1992).

The underlying idea is very simple: DNA is negatively charged in solution and thus should be attracted to a positively charged surface. Until recently, various problems have prevented us from forming images of the adsorbate *while the positive surface charge is maintained*, so we had resorted to imaging complexes that had been reacted onto the surface (Lindsay and Barris, 1988; Barris *et al.*, 1988; Lindsay *et al.*, 1988a,b; Nagahara *et al.*, 1990).

We have now overcome these obstacles, and in this chapter we describe how DNA and other molecules can be adsorbed onto a gold electrode and imaged *in situ* even at a relatively high positive surface charge. By dint of good luck, we have avoided the great graphite fiasco (Clemmer and Beebe, 1991), but gold surfaces are not simple either, and we will illustrate some problems we have encountered in interpreting contrast on this substrate.

5.2 THEORETICAL BACKGROUND

Here, we introduce some relevant details of the liquid–solid interface. However, it should be noted at the outset that one of the primary theoretical problems lies in interpreting STM contrast. Some aspects of this problem have been discussed in detail elsewhere (Lindsay, 1992). It may be that many different mechanisms of conduction can give rise to what we call a *tunnel current*. Our interpretation is entirely phenomenological, and many of our conclusions are therefore somewhat speculative.

Figure 5.1 shows an electrochemistry cell schematically. When a charged surface is immersed in an electrolyte, its surface charge is neutralized by an accumulation of counter ions. Our substrate is the working electrode (Fig. 5.1, WE). If the ions bind specifically to the metal atoms at the surface, a new interface [the inner Helmholtz plane (IHP)] is formed. Any excess surface charge that remains is compensated by a diffuse cloud of counter ions. The electrostatic attraction (or repulsion for co-ions) is balanced by thermal fluctuations. The balance is described by the Poisson-Boltzmann equation. A solution of this equation for point ions at plane electrodes shows that most of the potential

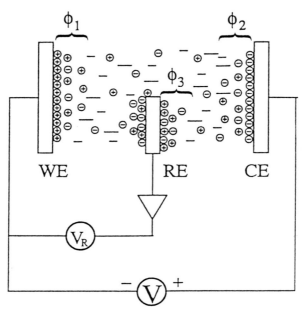

FIGURE 5.1 Schematic arrangement of the electrochemistry cell. The applied voltage (V) is dropped at the two double layers that form at the working electrode (WE) and counter electrode (CE). The potential drop at the WE can be monitored by a reference electrode (RE). (From Lindsay, 1992, Fig. 4.6. Reprinted with permission by VCH Publishers © 1992.)

change occurs over a distance given by

$$\kappa^{-1} = \frac{0.3}{z\sqrt{C}} \text{ nm} \qquad (5.1)$$

for dilute aqueous solutions at 25°C (Bard and Faulkner, 1980). C is the molarity of a $z:z$ electrolyte, and κ^{-1} is the Debye length, which is around 100 Å for the low ionic strengths used in this work. The potential that causes the excess charge at the IHP to be zero (so that the anion concentration equals the cation concentration at the interface) is called the *potential of zero charge* (PZC). For the gold electrodes and neutral solutions used here, it is found to be near 0 V (with respect to a silver quasireference electrode [AgQRE], P.I. Oden, unpublished data). Thus, at potentials more positive than 0 V (versus AgQRE), the interface acquires a net positive charge. The range of potentials around the PZC that give rise to polarization currents only (i.e., no reactions occur that involve electron transfer across the cell) is called the *double-layer* region of potential.

This description is little more than a cartoon. The possibilities for the formation of some quite intricate structures are almost limitless. Perhaps the scanned probe microscopes themselves constitute the best introduction to the electrochemical interface.

The change in charge density across the interface is equivalent to a change in potential (ϕ_1 in Fig. 5.1). This may be on the order of ~ 1 V, even for the large Debye lengths discussed above, and the local electric field at the interface may be quite high. Unfortunately, there is no way to characterize this field directly. Introducing another electrode into the cell (e.g., the counter electrode; CE in Fig. 5.1) introduces another unknown potential drop (ϕ_2). This problem is solved by introducing a third electrode that draws no current and has a surface potential (ϕ_3 in Fig. 5.1) pinned by equilibrium with dissolved ions. As long as the polarization at this third electrode *does not change during an experiment*, voltages can be referred to it in a way that allows the field (i.e., charge) at the working electrode to be reproduced.

We use a silver wire QRE. Experiments with several aqueous electrolytes show that this electrode gives results that are identical to those with the saturated calomel electrode (SCE) shifted by 240 mV (0 V versus AgQRE = 240 mV versus SCE; this is not the case when certain reactive anions — such as iodine — are present). The substrate potential specified in this manner is called V_s here.

Knowing the PZC in terms of the reference voltage, the surface charge density can be calculated from measurements of the interfacial capacitance. For small departures from the PZC with aqueous electrolytes in contact with Au(111), Oden (unpublished data) has estimated that the PZC occurs at ~ 0 V, with the charge reaching approximately $+e/30$ per surface atom at $+0.3$ V (versus AgQRE).

The foregoing discussion ignores the finite size of the hydrated ions. If they are packed densely enough, they cannot undergo thermal motion, and the potential varies linearly with distance as it would for a parallel plate capacitor. The Poisson-Boltzmann equation can only be applied outside this layer [the edge of which is called the *outer Helmholtz plane* (OHP)] (Bard and Faulkner, 1980). We use a phosphate-buffered solution as an electrolyte. Our experiments suggest that the PO_4^{3-} ion may bind specifically on the Au(111) surface, although DNA does not appear to do so. Nonetheless, DNA can become trapped within this layer, allowing packing forces to hold it in place as the scanning probe is dragged over the surface. This is a key point, because estimates of the interaction forces in an STM suggest that most attractive interactions are too weak to hold an isolated molecule in place and repulsive packing forces are probably required (Lindsay, 1992). This situation is illustrated schematically in Figure 5.2.

Control of surface coverage is a key problem. We use Au(111) that has been grown onto mica epitaxially. This surface is quite flat so that an electrode area can be estimated that facilitates quantitative control of the deposition. No current flows when the cell is operated in the double-layer region, so the transport of DNA into the Debye layer must be driven Fick's law of diffusion. The positively charged electrode will act as a sink for the negatively charged DNA, but only over a distance $\sim \kappa^{-1}$. Thus, the DNA must be allowed to adsorb

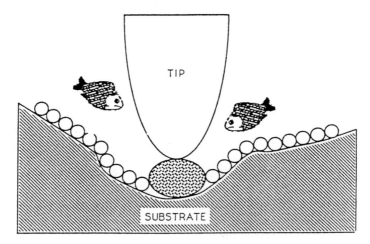

FIGURE 5.2 Retention of a molecular adsorbate by packing forces. The large biopolymer (shaded) is held in place by the surrounding small ions that are packed onto the substrate. Physical contact with the tip, adsorbate, and substrate isolate the tunnel junction from the surrounding liquid (illustrated here by the fish). (From Lindsay, 1992, Fig. 4.1. Reprinted with permission by VCH Publishers © 1992.)

for a time long enough to allow diffusion from the bulk of the electrolyte. This is a slow process compared with the electrophoresis that occurs in the high-field region. The time scale can be estimated using the Einstein–Smoluchowski relation. This gives the time to diffuse a root mean square (RMS) distance, l, as

$$t \sim \frac{l^2}{2D_t}, \tag{5.2}$$

where D_t, the translational diffusion constant, has a value for small DNA molecules of $\sim 10^{-8}$ cm^{2}s^{-1} (de la Torre and Bloomfield, 1981). For a cell depth, l, of a fraction of a centimeter, hours must be allowed for diffusion of DNA to the interface. Since the scanning probe obscures the substrate, it must be withdrawn for this period to allow diffusion. D_t depends on the molecular weight of the molecule, so this time will depend on sample size also.

If all the molecules in the cell were adsorbed, formation of a dense monolayer would require a concentration (g/ml)

$$c = \frac{\rho d}{l}, \tag{5.3}$$

where ρ is the density of DNA, and d is its diameter (20 Å). A cell depth, l, of a fraction of a centimeter would require $c \sim 1$ μg/ml for a monolayer. However, we will show that the layer that is imaged appears to stabilize after a period of time that is shorter than that required for diffusion, inhibiting further adsorption of DNA. Thus a somewhat greater concentration is required to achieve a full monolayer.

Finally, diffusion into the high-field region does not necessarily result in formation of a stable adsorbate. Indeed, at low surface charge densities, polymer adsorption is inhibited by the consequent loss of configurational entropy and other factors (Muthukumar, 1987; Åkeson et al., 1989). We see evidence of this in the selective adsorption of shorter fragments.

In outline, the procedure is as follows: A solution of electrolyte containing a few μg/ml of DNA is placed onto a clean gold electrode, and the electrode potential is set to give a positive surface charge. After a period of ~1 hr, the STM tip is lowered and the surface imaged while potential control is maintained. Our experimental procedures are described and some problems and important controls are outlined in the remainder of this chapter.

5.3 EXPERIMENTAL PROCEDURES

An electrochemistry cell for use in an STM is shown schematically in Figure 5.3. It consists of a small (~5 mm inner diameter) glass tube polished at one end so as to form an interference fit onto the substrate. The cell is pulled into place by the surrounding magnetic base with magnets inserted into the STM base. The

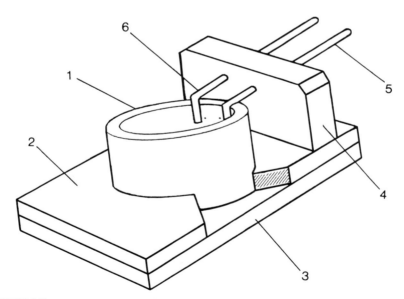

FIGURE 5.3 The electrochemistry of a cell for STM. The Au(111) on mica (3) is held onto the STM base by a magnetic stainless steel plate (2). The glass cell (1) forms an interference fit onto the substrate (shown by the cut-away). The insulating block (4) holds the counter (5) and reference (6) electrodes. (From Lindsay et al., 1992b, Fig. 1. Reproduced from the Biophysical Journal, 1992, 61, pp. 1570–1584, by copyright permission of the Biophysical Society.)

counter electrode is a 0.2 mm diameter Pt wire that dips about 2 mm into the solution. The QRE is a 0.5 mm diameter Ag wire.

Cleanliness is essential for reproducibility: The inside of the cell is cleaned with standard glass cleaning procedures, and fresh wire is degreased and cut for each electrode for each run. The substrates are Au(111) grown epitaxially onto mica, as described elsewhere (DeRose *et al.*, 1991).

Even gold becomes contaminated almost immediately on exposure to air (Smith, 1980), so our criterion for "cleanliness" is based on observation of the 23 × √3 reconstructed surface (Tao and Lindsay, 1991, 1992). We show examples in Figure 5.4A,B. The procedure outlined by DeRose *et al.* (1991) gives almost perfectly uniform substrates that consist of atomically flat gold steps of ~0.2 μm dimension. The following points deserve emphasis:

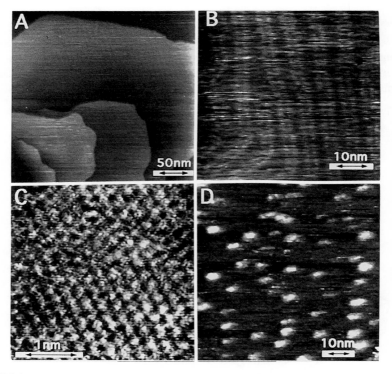

FIGURE 5.4 Characterizing the Au(111) substrates on several length scales (images A–C are obtained under $HClO_4$ at 0 V versus AgQRE). All images are raw data. A shows three flat terraces (the 23 × √3 reconstruction is just visible). B is a zoom-in showing the reconstruction (the corrugation is 0.1 Å). C shows the surface at atomic resolution. D is taken under an aqueous solution extracted from a polyacrylamide gel (0.1 V versus AgQRE) and illustrates some characteristic features seen when the surface is contaminated.

1. The substrates are prepared in ultrahigh vacuum (UHV; 10^{-9} torr).
2. Prolonged baking of the substrates at 420–460°C appears to be essential. Good thermal contact with the substrate heater is required, and the substrate holders should be designed so as to avoid strains due to thermal cycling.
3. Green mica gives better results than ruby mica.
4. Only rather limited exposure to ambient conditions is acceptable: After preparation of the substrates, the UHV chamber is backfilled with a positive pressure of nitrogen. The Conflat port containing the substrate heater is removed and placed in a laminar flow hood. The substrates are then loaded into clean glass containers that are stacked in a vacuum desiccator. Exposure to air is no more than ~20 minutes. The desiccator is evacuated and backfilled with argon. The substrates are transferred as needed to the STM stage in the laminar flow hood and flooded with clean electrolyte. With this procedure, substrates remain "good" for a week or more of storage under argon.

Very low Faradaic leakage is required for the STM tips in order to avoid electrochemical effects owing to the tip. We have described a procedure for coating the tips with wax (Nagahara et al., 1989). Improvements in our procedure for making these now result in a 90% yield of tips that give atomic resolution and a leakage current of less than 1 pA (with 0.14 V applied between the tip and a large CE in 1 M NaOH).

The tips are washed in clean water and introduced into the cell after deposition of the adsorbate. We have conducted tests in which the tip potential is controlled either with respect to the substrate or with respect to the reference electrode. As long as the tips remain well insulated, we see usually no changes caused by repeated scanning in either case.

Mixing of solutions also requires some care. Many sources of "clean" water cause modification of the gold surface. We have investigated various sources, and the following perform at about the same level (about 5% of the gold surface is modified due to what appear to be contaminants): Water from a new bioresearch grade Nanopure water system (Barnstead, Dubuque, Iowa) fed with the campus-distilled water; water from this system redistilled with $KMnO_4$; water for HPLC from Aldrich Chemical; and water from the campus-distilled water system triply distilled in a custom system in the laboratory of A. Beiber at Arizona State University. Many sources were worse than this, so the optimum procedure at a particular site needs to be established empirically. Other solutions are described in the appropriate sections.

We have also carried out extensive voltammetry on the various constituents of DNA, the results of which are reported elsewhere (Lindsay et al., 1992b). We find that

1. Components of the DNA backbone (sugar, diester phosphates) are not electrochemically active on Au(111) at positive charges.
2. Most of the common buffer salts are quite easily oxidized. Phosphate buffer

is an exception, not reacting until a potential of ~1.2 V (AgQRE). There is a small (and reversible) adsorption peak near +0.2 V.

Most images were obtained with a Nanoscope II STM/AFM from Digital Instruments (Santa Barbara, California). Imaging conditions are reported in the various experiments. The scans were calibrated with atomic resolution of the graphite surface; however, both drift and uncontrolled variations in tip length led to a run-to-run uncertainty in dimensions of as much as ±10%.

5.4 THE AU(111) SURFACE UNDER AQUEOUS ELECTROLYTES

A thorough characterization of the substrate surface is a prerequisite to imaging adsorbates. Here we describe "good" and "bad" surfaces. The procedure we have outlined for preparing substrates is elaborate, but we have found no other approach that gives a *reproducible* surface. By reproducible, we mean that

1. The surface consists of atomically flat terraces (like those shown in Fig. 5.4A) *everywhere* unless the tip has been crashed into it.
2. Most (≥50%) of the terraces show the $23 \times \sqrt{3}$ reconstruction.

In our earlier work with these substrates we often saw sharp steps that met at 60° (DeRose et al., 1991). We believe that these were slip planes caused by tip crashes. The Nanoscope servo is superior to that of the STM we used originally, and we rarely see these features now (they can be artificially produced by bending the substrate).

5.4.1 Clean Gold and Dirty Gold

We have recently described a study of phase transitions at the Au(111)–HClO$_4$ interface (Tao and Lindsay, 1992b). It is remarkable that the Au(111) surface maintains the $23 \times \sqrt{3}$ structure (the structure seen on the *clean* Au[111] surface in *UHV*) under a covering electrolyte. Moreover, this structure is stable up to potentials in excess of 350 mV (versus AgQRE). We show some images from this work (similar images have been obtained under pure water; Lindsay et al., 1992) in Figure 5.4A–C (data were acquired at a tunnel current, I_t, of 1 nA, with a tip bias with respect to the substrate, V_t, of −100 mV). The image in Figure 5.4A has been leveled to improve contrast on the terraces, but the data are otherwise unprocessed. The steps in Figure 5.4A are 3 Å in height and correspond to single-atom height steps on the gold surface. Careful inspection reveals the stripes to be associated with the $23 \times \sqrt{3}$ reconstruction. Figure 5.4B is a zoom-in showing the reconstruction (there is a step at the extreme left of the image). The height of the stripes is ~0.1 Å. The image in Figure 5.4C was taken at yet higher magnification to show the underlying gold atoms. The corrugation is about 0.1 Å, and the lattice constant is 2.5 Å (within 10% of the 2.8 Å expected for the Au[111] surface). These images change little over the range $0 < V_s < 0.3$ V (but see Tao and Lindsay, 1992).

There are *many* ways of ruining this beautiful surface. However, the images obtained in the presence of contamination often have a rather characteristic appearance. Figure 5.4D shows a scan obtained under water that we believe was contaminated by the constituents of a polyacrylamide gel. It was taken at about the same magnification as shown in Figure 5.4D and with $V_s = 0.1$ V (versus AgQRE). The reconstruction is gone, and "blobs" of about 3 Å height cover the surface. In this case the extra atoms associated with the reconstruction may have formed islands (the blobs) because their height is close to that of a step on the (111) surface. We have observed this phenomenon as the clean interface is transformed rapidly at high potentials (Tao and Lindsay, 1992). The gold islands tend to align along the path of the high point of the $23 \times \sqrt{3}$ reconstruction so that the surface becomes covered with stripes composed of blobs. This may account for the 70 Å periods we have seen in some experiments in which DNA was deposited onto the surface (Lindsay and Sankey, 1991).

However, similar patterns can also be seen when no such islands have formed. The contrast is still associated with the $23 \times \sqrt{3}$ reconstruction, but is mainly electronic in origin.

5.4.2 Electronic Contrast: Adenine on Au(111)

Many "dirty" experiments produce images of "blobs" lying along the path of the $23 \times \sqrt{3}$ reconstruction, but the height of the blobs is much less than a single atomic step on Au(111). Here we present evidence for electronic contrast associated with adsorbates that interact with the reconstruction (our experiments on the $23 \times \sqrt{3}$ reconstruction under $HClO_4$ suggest that localized electronic states contribute to the contrast even in this simple case; Tao and Lindsay, 1992).

Figure 5.5 shows images of an Au(111) held at 0 V (versus AgQRE) under an aqueous solution of adenine. Prior to the introduction of adenine, the surface was clean, displaying the $23 \times \sqrt{3}$ reconstruction over most of the terraces. After deposition of adenine, the surface appears to be unaltered (viewed at low magnification), showing the same patterns of the $23 \times \sqrt{3}$ reconstruction as observed before adsorption of the adenine. However, the corrugation associated with the "reconstruction" is increased by *more than 1 order of magnitude*. Figure 5.5A shows a typical region of double stripes (compare with Fig. 5.4B), with a peak corrugation of 1.5 Å. Zooming in on one of the stripes (Fig. 5.5B) shows that it is composed of pairs of bright spots surrounded by rather well-ordered wormlike structures. A scan at yet higher magnification shows that the "worms" have a periodic structure of ~3 Å and a width of ~5 Å. We believe that they consist of stacked adenines because the period seen along the "worms" is similar to the base-stacking distance seen in DNA.

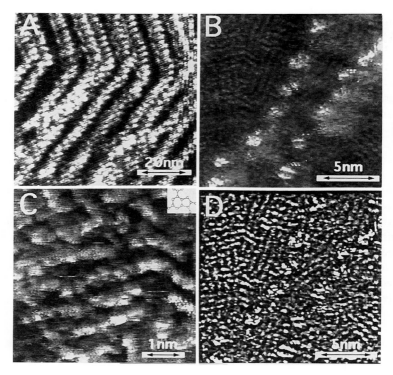

FIGURE 5.5 Adenine adsorbate (under aqueous solution at 0 V versus AgQRE). A shows the decorated $23 \times \sqrt{3}$ reconstruction (the contrast is 1.5 Å). B is a zoom-in showing that the "reconstruction" pattern is composed of sets of bright dots. Wormlike features cover the background. A further zoom-in C shows that the "worms" are composed of stacked adenines (the inset shows the adenine structure approximately to scale). A–C are raw data. D is a high-pass filtered version of C showing that the "high spots" in C are not high at all (the "worms" pass right through them), so the contrast is electronic in origin. (Here $V_t = -100$ mV and $I_t = 0.05$ nA.)

This result is of interest because studies of the DNA bases on graphite indicate that they lie flat (Heckl *et al.*, 1991; Srinivasan *et al.*, 1991), although images obtained on graphite may be difficult to interpret (Garbarz *et al.*, 1991). Clean gold is known to be hydrophilic (Smith, 1980). If this is true of our electrode surface also, it may account for the observed base stacking (similar results were obtained with guanine; Tao and Lindsay, unpublished data).

In Figure 5.5D we show a version of Figure 5.5B that has been high-pass filtered (Fig. 5.5A–C show raw data) that demonstrates quite clearly that the worms pass right "through" the bright blobs. This implies that the blobs are in fact not changes in substrate height but are local regions of enhanced electron density at the Fermi level or enhanced electron transmission caused by some interaction of the adenines with sites on the $23 \times \sqrt{3}$ reconstruction. Similar false contrast has been observed for dislocation networks on graphite (Garbarz et al., 1991).

5.4.3 Phosphate Solutions on Au(111)

Salt is needed to maintain the conformation of double helical DNA (and also, as we shall see, to help trap DNA on the substrate). Voltammetry indicated that sodium phosphate was a suitable choice, so we have examined the Au(111) surface under various concentrations of an NaH_2PO_4/Na_2HPO_4 buffer (ph 6.5). At the rest potential (CE disconnected) the surface displays the $23 \times \sqrt{3}$ reconstruction, but this is lifted as soon as the surface is brought to zero charge (Lindsay et al., 1992b). However, if the solution is not contaminated, the surface is otherwise quite flat. The surface maintains this morphology until an adsorbate forms near 0.2 V. However, this adsorbate is *not visible* in STM images, presumably being too tenuous to withstand the force of the STM tip. STM images in Figure 5.6 show a series of images taken over an Au(111) surface under 10 mM phosphate buffer as the potential is increased from 0 to 1.4 V (AgQRE). The area is dominated by two flat planes separated by a single atomic step of ~ 3 Å (Fig. 5.6A; the image has been levelled to enhance contrast). There are some small features (<1 Å high) on the surface that may be due to contamination.

As the potential is raised above 0.2 V (Fig. 5.6C), the images begin to blur. This loss of resolution continues up to 1.2 V (Fig. 5.6G). At 1.4 V, the image becomes unstable (Fig. 5.6H). When the potential is reduced to 0 V, the surface is found to be roughened (Fig. 5.6I), demonstrating that the reaction (at 1.4 V) is not reversible. The onset of the reaction under the tip is delayed somewhat (it occurs at 1.2 V in the bulk).

The onset of blurring at ~ 0.2 V is associated with a small peak in the voltammogram (see inset, Fig. 5.7). We studied the same system using an electrochemistry cell with AFM. The potential was controlled by a Pine Instruments AFRDE4 potentiostat using the AgQRE. Figure 5.7 shows an AFM topograph taken while the potential was cycled between $+0.7$ V and -0.7 V during the AFM scan (the contact force was 5 nN before potential control was applied).

A little over three full cycles were made during the acquisition of one AFM image. It is clear that a globular adsorbate structure forms on the surface when the potential exceeds 180 mV. It is removed at -50 mV. (We have levelled the image, because the raw data are perturbed by an interaction between the potentiostat and force sensor). This adsorbate is *not visible* in the STM images (Fig. 5.6), so the STM tip must be *ploughing through it to image the underlying gold*. Presumably this causes a disturbance of the tip that leads to the observed blurring when V_s exceeds 200 mV.

Although there is no evidence of an interaction below 180 mV between phosphate and Au(111) in the voltammogram, it is implicit in the lifting of the $23 \times \sqrt{3}$ reconstruction. Thus it appears that there is a potential window (0–0.18 V) over which the gold surface can be made positive in phosphate buffer

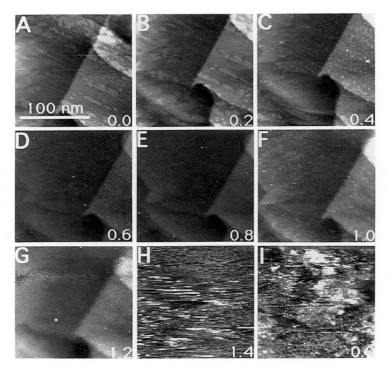

FIGURE 5.6 Scanning tunneling microscope images (raw data) of some atomically flat terraces on Au(111) under 10 mM phosphate buffer as the potential is raised in steps of 0.2 V from 0 V **A** to 1.4 V **H**, where the surface is too unstable to image. Returning to 0 V **I** shows the roughening caused by the reaction at 1.4 V. (From Lindsay et al., 1992b, Fig. 5. Reproduced from the *Biophysical Journal*, 1992, **61**, pp. 1570–1584, by copyright permission of the *Biophysical Society*.)

without forming dense adsorbates on the electrode. This is the region in which we have chosen to study adsorption of small DNA fragments.

5.5 LIGATED DNA OLIGOMERS ON AU(111)

5.5.1 Choice of DNA Samples

Our initial experiments (Tao et al., 1991) suggested that DNA might be denaturing on the electrode surface. Working with Rodney Harrington (University of Nevada, Reno) and Yuri Lyubchenko and Lyuda Shlyakhtenko (Institute of Molecular Genetics, Moscow) (Lindsay et al., 1992b), we have attempted to make a sample that consisted only of small DNA circles. Double helical, covalently closed, and with a circumference less than the persistence length of DNA, such circles should be immune to even the most brutal conditions at the electrode surface. Small circles can be made by ligating highly bent oligomers

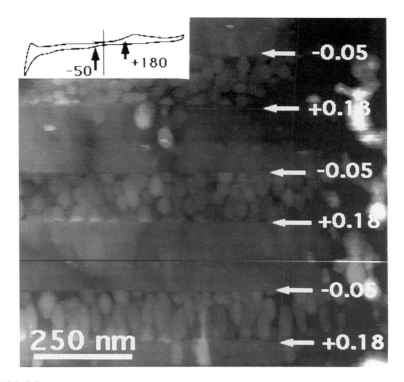

FIGURE 5.7 Atomic force microscope scan taken as the substrate potential is cycled between −0.7 V and +0.7 V under 10 mM phosphate buffer (the inset is a cyclic voltammogram for this system). Three electrochemical cycles occur as the image is collected. The image shows the formation of an adsorbate at +0.18 V that is removed at −0.05 V. (From Lindsay et al., 1992b, Fig. 6. Reproduced from the *Biophysical Journal*, 1992, 61, pp. 1570–1584, by copyright permission of the Biophysical Society.)

(this is the basis of one method for direct determination of DNA bending). The resulting ligation mixture contains both circles and linear DNA fragments. We failed to extract an adequate supply of circles from the dense polyacrylamide gel used for separating them from the linear fragments. Thus the experiments described here were carried out with the full ligation mixture that consists of both circles and linear fragments.

We chose the sequence

5′-   AAAAACCCCCAAAAAACCCCC,
3′-   GGGGTTTTTGGGGGTTTTTTG

which has a high intrinsic degree of bending according to the Trifonov wedge model and experimental determination of the wedge angles (Bolshoy et al., 1991). The smallest circle that can be made with this oligomer is a 7-mer (147 basepairs, yielding a diameter of ∼150 Å).

5.5.2 Synthesis and Characterization of DNA Samples

The oligomers were synthesized and purified (using HPLC) by E. Appella of the Molecular and Cell Biology Laboratory of the National Cancer Institute at the NIH. Details are given elsewhere (Roy et al., 1986; Joshua-Torr et al., 1988). Single-stranded samples were radioactively labeled with ^{32}P using T4 polynucleotide kinase. The complementary strands were annealed to give a four-base overhang for precise ligation. The duplexes were ligated using T4 DNA ligase. Full details of these procedures are given by Lyubchenko et al. (1991).

The result is oligomers of a length $21n + 4$ bases, some of which are circles. If we assume a B-DNA base stacking distance of 3.4 Å, this procedure yields a series of polymers of lengths $(71.4n + 13.6)$ Å. This calculation assumes that the four-base single-stranded ends are stacked as in B-DNA, but this stacking distance can double at low salt concentrations. The detection of ring closures was carried out with a two-dimensional gel analysis (Lyubchenko et al., 1991). Separation in the first dimension was performed in a 4% polyacrylamide gel in a TBE (90 mM Tris, 90 mM boric acid, 2 mM ethylenediaminetetraacetic acid (EDTA), pH 8.3) buffer. Separation in the second (perpendicular) dimension was in a 10% polyacrylamide gel in TBE. The result of this analysis is shown in Figure 5.8. (The smallest linear fragment on the lower right is a trimer; the gel was run this way in order to separate circles and long linear fragments clearly.)

STM studies were made using the original ligation mixture, which contained both linear and circular DNA molecules. These were purified after ligation by phenol extraction followed by gel chromatography on Sephadex G-50 columns. The 21 bp fragments were labeled with γ^{-32}P ATP followed by a large excess ($\sim 10^4$) of nonradioactive carrier material. Hence spot intensity on the gels is proportional to fragment concentration with essentially no length dependence. The first gel separation (not shown) gave spots of approximately equal area (within 50%) for trimers (T), dimers (D), and monomers (M), so the ratios of the concentrations of these fragments were approximately $1:1:1$ (T:D:M).

5.5.3 STM Imaging of DNA Adsorbates

Our first successful experiment used a solution of 30 μg/ml of the ligation mixture in 1.4 mM phosphate buffer (pH 6.5). The substrate was left at +100 mV (versus AgQRE) for 20 min under the solution. An image taken soon after lowering the tip is shown in Figure 5.9A. It is very streaky, in sharp contrast to images obtained on the bare substrate or on the substrate before the CE was connected. In this image $V_t = -100$ mV and $I_t = 1.0$ nA.

I_t was lowered to 0.26 nA, and an immediate improvement resulted (Fig. 5.9B). Many experiments since then have shown that there is a quite reproducible threshold for stable images of these adsorbates: I_t must be less than about 0.4 nA. We have imaged with currents as small as 0.05 nA, finding little effect as long as I_t is less than the threshold value. The small features in Figure 5.9B (~ 100 Å in length) are monomers ($n = 1$), as will be demonstrated below.

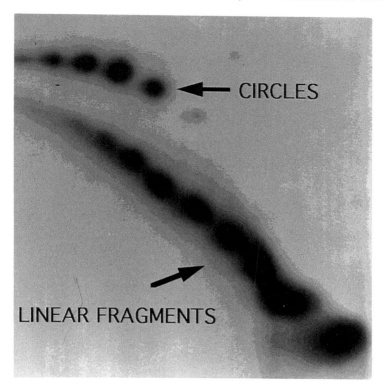

FIGURE 5.8 Two-dimensional gel analysis of the ligation products of the synthetic oligomer. The lower series starts at a trimer (right) and increases to an 11-mer. The smallest circle is a 7-mer. The first separation (not shown) gave approximately equal spots for the monomers, dimers, and trimers. (From Lindsay *et al.*, 1992b, Fig. 7. Reproduced from the *Biophysical Journal*, 1992, 61, pp. 1570–1584, by copyright permission of the *Biophysical Society*.)

The coverage shown in Figure 5.9 is rather typical. Many small linear fragments litter the surface, and spurious structure associated with decoration of the $23 \times \sqrt{3}$ reconstruction is not observed over much of the surface.

Our previous experiments (Tao *et al.*, 1991) had all been made at higher tunnel currents (which resulted in poor resolution at relatively low substrate potentials). This particular experiment at low potential and low tunnel current suggested that we had bound extended and isolated DNA molecules that were not denatured. The images were quite stable under repeated scanning.

5.5.4 What Holds the DNA in Place?

We have repeated runs like this a number of times with similar results. However, two negative results are important for understanding the deposition process.

Our first experiments with these samples used water with *no* added salt

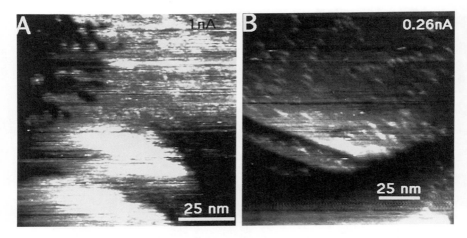

FIGURE 5.9 An Au(111) surface coated with the ligation products of the oligomers at a substrate potential of 0.1 V and a tunnel current of 1 nA **A** or 0.26 nA **B**. The same area is scanned in each case, but sample movement renders the high tunnel current image **A** unrecognizable (raw STM data). (From Lindsay et al., 1992b, Fig. 8. Reproduced from the *Biophysical Journal*, 1992, **61**, pp. 1570–1584, by copyright permission of the *Biophysical Society*.)

(since the circles remain stable without salt). If the attachment process is entirely electrostatic, we would expect to see the DNA circles attached to a bare gold substrate. We followed deposition procedures much like those described above (but with no salt) with little success. Noise in the images indicated that the interface was different from that observed without DNA present, but no recognizable images were obtained. *The presence of a salt coadsorbate appears to be essential to obtain stable STM images.* This indicates that electrostatic forces alone are inadequate for stabilizing the adsorbate (as argued elsewhere; Lindsay, 1992).

In a second negative experiment, 0.2 mM phosphate buffer was placed onto a substrate and the potential set to $V_s = 120$ mV. The surface was surveyed, and it appeared clean (with the $23 \times \sqrt{3}$ reconstruction lifted, as expected under phosphate buffer). The tip was withdrawn and the ligation mixture added to achieve a final concentration of 6 µg/ml. The tip was inserted from time to time to examine the substrate over the next 4 hr. Addition of the DNA caused a certain streakiness (as in the previous experiment) but no clear images were obtained. Thus, it appears that the DNA and salt must be adsorbed *together*. Salt alone forms a stable layer that excludes DNA. The time for the formation of this layer is less than that for DNA diffusion across the cell (Eq. 2). Thus we conclude that the population of fragments on the surface should be related to the population in solution by some function of the translational diffusion constants.

The reversibility of the adsorption is discussed in the next section.

5.5.5 Reversibility and the Role of the 23 × √3 Reconstruction

Regions can be found that contain unusual features, probably associated with decoration of the 23 × √3 reconstruction. We show a particularly interesting example in Figure 5.10. There, many chain-like features cover the surface. However, the chains do not display the periodicity expected for DNA: The repeat is closer to the ~70 Å period associated with the 23 × √3 reconstruction! However, this area also contains three of the microcircles and what appears to be an isolated linear chain in which the expected 36 Å repeat of the double helix is clearly resolved (arrow in Fig. 5.10B).

We use these images to demonstrate a number of points:

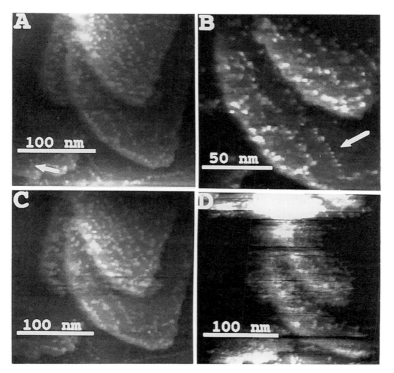

FIGURE 5.10 Scanning tunneling microscope scans (raw data) over an area of three atomically flat terraces on Au(111) at $V_s = 0.1$ V coated with some relatively high-molecular-weight ligated oligomers. The arrow in **A** points to a highly curved fragment. **B** is a zoom-in on the area scanned in **A** (the arrow points to a polymer displaying a 36 Å repeat). **C** is zooming out again; the image is almost identical to the first scan, demonstrating the stability of the adsorbate. In **D**, the same area is scanned again, but the substrate potential was switched to -0.1 V at the start of the scan, destabilizing the adsorbate. (From Lindsay *et al.*, 1992b, Fig. 9. Reproduced from the *Biophysical Journal*, 1992, 61, pp. 1570–1584, by copyright permission of the *Biophysical Society*.)

1. That the adsorbate is stable under repeated scanning
2. That the adsorbate destabilizes when the surface charge is made negative
3. That this destabilization occurs everywhere, including at the "questionable DNA" with the 70 Å repeat

Our control experiments demonstrate that, with appropriate cleanliness, features that resemble these "questionable DNA" *are almost never seen when DNA is not present in the electrolyte*. Taken together with this last observation, this implies that the "questionable DNA" *really are due to DNA adsorption* but with a contrast that is somehow modified by interaction with the $23 \times \sqrt{3}$ reconstruction.

Figure 5.10A shows three atomically flat terraces on the gold, each of which appears to be covered with DNA strands (the images show raw data). In Figure 5.10B we show the result of zooming in to one of the terraces. Most of the chain-like features do not show the expected periodicity for B-form helices (they might be the result of interaction with the $23 \times \sqrt{3}$ reconstruction as described earlier). However, at least one "molecule" on this scan (indicated by an arrow) has the correct width (of ~20 Å) and displays a 36 Å periodicity (close to that expected for B-DNA). Such images are quite rare in the runs we have conducted to date, suggesting that the DNA is usually modified on binding the substrate (by *modified* we mean retaining its double helical form, but with an unusual periodicity along the chain owing to electronic effects or physical interactions with salt, etc.).

Figure 5.10C shows the result of zooming out again over the same area. Careful comparison of Figures 5.10A,C reveals that the image has changed a little. Nonetheless, the overall stability is excellent. However, when the substrate potential was adjusted to -100 mV and the substrate rescanned (Fig. 5.10D), the adsorbate was immediately destabilized. Note that this applies to *all* the features on the substrate. Had the questionable DNA been gold islands associated with lifting of the $23 \times \sqrt{3}$ reconstruction, they would have remained stable for a considerable time (Tao and Lindsay, 1992).

The images in Figure 5.10 show raw data. Contrast across the steps is enhanced by high-pass and median filtering. The results of these operations are shown in Figure 5.11. The locations of the three microcircles are shown by arrows.

5.5.6 Statistical Analysis: Preferential Adsorption of Small Molecules

We have surveyed more than 100 images taken in five runs using this ligation mixture. Most of the images have three characteristics:

1. The overwhelming majority of fragments that are imaged are short (monomers or dimers).
2. The best images have most of the chains lying along the slow-scan (y) axis.

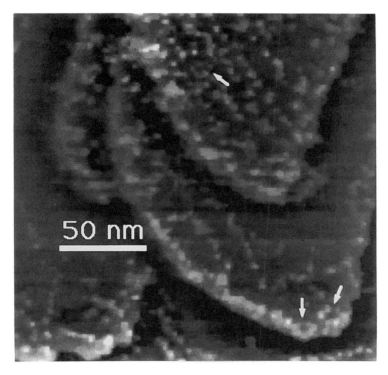

FIGURE 5.11 High-pass-filtered version of Figure 5.10A. Three DNA microcircles are pointed to by arrows.

Few of the fragments (in good images) make an angle of more than 45° with this axis.
3. The chain lengths are all somewhat longer than expected for B-DNA base stacking.

We have analyzed the fragment length distribution in detail (Lindsay *et al.*, 1992b) and present a summary of that analysis here.

We believe that the vertical alignment of the fragments is a consequence of the operation of the servo-system. We do see some images when fragments are aligned along the fast-scan direction (horizontal in these images), but the images are streaky. Image processing of noisy regions of the substrates often gives some hint of molecules aligned along the fast-scan direction. Thus the origin of the problem appears to be some instability as the tip is scanned rapidly along a molecule (as opposed to rapidly across it). Note that the same effect is not seen in adenine adsorbates, presumably because the tip interaction with the molecule is smaller.

Whatever its causes, this preferential imaging of molecules aligned with the y-axis exaggerates the effect of drift (which is much more of a problem in STM liquid cells than in UHV STM). This can be seen by considering the effective magnification factor (or demagnification factor for drift in the opposite direction) owing to drift. For a linear object aligned along the y-axis this is

$$M_d^y = (1 - V_d^y/V_s)^{-1}, \qquad (5.4)$$

where V_d^y is the drift velocity along the y-axis, and V_s is the scan velocity along this axis. Typical values for V_s are 5–25 Å/s, so rather small values for drift can result in infinite magnification for drift along the scan and a factor 0.5 decrease in magnification for drift against the scan. Thus the overall effect of drift is to displace the distribution to larger measured dimensions and to broaden it. It is not clear at this point whether this effect can explain the apparent shift of the measured length distribution to longer lengths. It seems unlikely, because we wait for drift to subside below about 20% from scan to scan before we record data. In these circumstances the slewing of the distribution should be small. Work is in hand to develop accurate methods to compensate for drift. Further experiments will shed light on this point.

A number of rather typical images are shown in Figure 5.12. In this particular run, DNA fragments were added to a final concentration of 32 μg/ml in 0.9 mM phosphate buffer and left to adsorb for nearly 1 hr (here $I_t = 0.26$ nA, $V_t = -300$ mV, and $V_s = 100$ mV versus AgQRE). Figure 5.12A shows a relatively large area of the surface that is littered with small, rodlike fragments. Figure 5.12B shows a scan at higher resolution over an area that contains a number of fragments that are quite bent (arrows point to the most extreme examples). Figure 5.12C shows a scan over an area where most of the fragments appear to be straight. In this run the great majority of fragments align along the y-axis (Fig. 5.9B shows fragments oriented at a much larger angle to the y-axis). We have composed a gallery of images taken from Figure 5.12C in the upper part of Figure 5.12D. Fragments 1 through 7 are representative of the shortest fragments seen. In this case they are 30% longer than the expected monomer dimensions. A shift rather like this is seen in most images (see later), and we designate these as *monomers*. On this basis, the fragment labeled 8 is a dimer and that labeled 9 is a trimer. We have also found a few circles (a total of eight in these five runs). This is too few to permit a statistical analysis, but we show three examples in the lower part of Figure 5.12D. Their dimensions are consistent with the minimum size ($n = 7$) expected for circle formation.

We have little data for circles, but if the discrepancy between circles and linear fragments is real (i.e., linear fragments appear somewhat longer than expected; circles appear similar or smaller than expected), then it may reflect a distortion of the molecule on binding the substrate. To make quantitative measurements, we waited until the changes in apparent dimensions with scan direction were less than about 20%. We then recorded a number of images that

5 Potentiostatic Deposition of Molecules for SXM   251

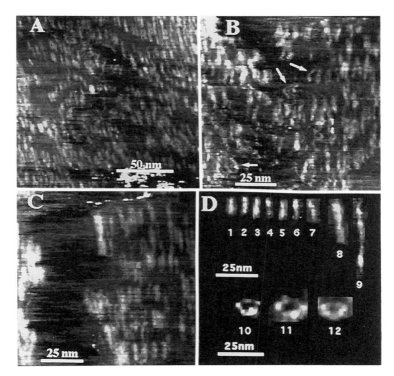

FIGURE 5.12 Scanning tunneling microscope scans (raw data) showing various linear fragments **A**, bent fragments **B** and multimers **C** made by ligating the 21-mer. The arrows in **B** point to some very bent polymers. All images are raw data. **D** (upper part and upper scale bar) shows a gallery of images taken from C. 1 through 7 are designated monomers (having measured lengths of 101, 96, 106, 106, 106 and 117 Å, respectively); 8 is a dimer (230 Å); and 9 is a trimer (370 Å). These dimensions are all about 30–50% longer than expected. The average width of the images presented here is ~32 Å. The lower part of the figure (refer to lower scale bar) shows three examples of circles. Their outside diameters are about 130 Å (10) and 160 Å (11 and 12); 11 and 12 have the dimensions expected for the smallest circles in the ligation mixture, while 10 is about 20% smaller than expected. (From Lindsay *et al.*, 1992b, Fig. 10. Reproduced from the *Biophysical Journal*, 1992, **61**, pp. 1570–1584, by copyright permission of the *Biophysical Society*.)

we processed by high-pass and median filtering to facilitate length measurement on individual fragments.

A histogram of the measured length distributions shows the expected quantization (Lindsay *et al.*, 1992b), but it is a little misleading because drift causes a run-to-run variation in lengths that smears out the length quantization that is so obvious in individual images (e.g., Fig. 5.12). Until we develop software to correct for drift, we are stuck with the somewhat subjective process of counting the number of monomers, dimers, and so forth, by identifying them by compari-

TABLE 5.1 Data for the Ratios of Trimers to Dimers to Monomers[a]

	Trimers	Dimers	Monomers
Number of fragments	22	33	119
Ratio to trimers	1	1.5	5.4
Ratio from gel data	1	1	1
$\dfrac{\rho_N \sqrt{D_N}}{\rho_3 \sqrt{D_3}}$	1	1.2	1.4

[a] Absolute numbers from STM images (first row), as normalized by the trimer population (second row), from gel data (third row), and as calcualted from equations 5 and 6 (fourth row).

son with their neighbors in the same run, as is done in Figure 5.12. Data are provided in Table 5.1.

The data confirm the impression given by the images, which is that small fragments are preferentially adsorbed. Analysis in the first gel dimension showed that trimers, dimers, and monomers were present in approximately equal concentrations. The population of monomers imaged on the substrate is enhanced fivefold!

Only a small part of this can be due to the difference in diffusion constants. We have shown above that the adsorbate layer stabilizes some time, T_0, after the substrate is made positive. Therefore, the number of molecules of species i (concentration ρ_i) that can diffuse (over a substrate area, A) into the double layer is

$$N_i = A\rho_i \sqrt{2D_t(i)T_0}, \qquad (5.5)$$

where we have used Eq. 5.2 and $D_t(i)$ is the appropriate translational diffusion constant. This short DNA can be treated as a rigid rod for which

$$D_t(p) = K \frac{\ln(p) + \gamma(p)}{1}, \qquad (5.6)$$

where K is a constant at a given viscosity and temperature; and $p = 1/2a$, where 1 is the fragment length and $2a$ is the DNA diameter (20 Å). $\gamma(p)$ can be calculated using the polynomials provided by de la Torre and Bloomfield (1981). We obtain ratios for $D_t(p)$ that are 1:1.28:1.82 (T:D:M). This simple model predicts that we should see the T:D:M in the ratios $\rho_3\sqrt{D_t(3)}:\rho_2\sqrt{D_t(2)}:\rho_1\sqrt{D_t(1)}$. Thus, for our approximately equal concentrations, the imaged ratios should be as listed in Table 6.1, and the observed excess cannot be accounted for by differences in diffusion constants.

The preferential adsorption process is not easily explained by obvious differences in large-scale configurational entropy (Muthukumar, 1987; Åkeson et al., 1989) because these fragments are all shorter than a persistence length for the polymer. Many other experiments in our laboratory have confirmed this preferential adsorption of smaller molecules. In particular, we have found that small amounts of single-stranded material clutter the substrate, complicating imaging of high molecular weight material. Purification of the sample appears to be an essential step. However, the molecular weight purification process itself must not introduce more contamination (i.e., polyacrylamide).

5.6 DISCUSSION AND CONCLUSIONS

We have shown how gold substrates can be prepared and imaged under potential control so as to display the $23 \times \sqrt{3}$ reconstruction quite routinely. We have shown how phosphate buffer lifts this reconstruction but does not cause unacceptable modification of the substrate in a potential window between 0 and 200 mV (versus AgQRE). We have also shown how small molecules (e.g., adenine) can be adsorbed and imaged with near-atomic resolution. These images illustrate the interesting electronic effects in the images associated with interactions between the adsorbate and the $23 \times \sqrt{3}$ reconstruction on Au(111).

Our experiments demonstrate that DNA molecules in phosphate buffer can be adsorbed onto an electrode and imaged by the STM with a high resolution. The resolution is demonstrated by the width of the images, which is typically 3 nm, showing that the instrumental broadening is limited to about 1 nm for double helical samples. This can be seen by examining Figure 5.12D. The images are typical of what can be done with a selected tip (i.e., if the images are much broader, they are usually improved on replacing the STM tip). Very high resolution can be obtained on occasion: An example is the molecule pointed to by an arrow in Figure 5.10B, which has an apparent width quite close to 2 nm.

Note, however, that the images of the stacked adenine bases show individual molecules quite clearly (Fig. 5.5). Therefore, there is *no intrinsic limitation preventing near-atomic resolution with this method.* The lower resolution on the double helical samples may be a consequence of their greater diameter, electronic properties or the way they pack into the salt coadsorbate layer.

Even given an ~1 nm resolution, one might expect that the helix pitch (~3 nm) would be more readily visible (i.e., the molecules would "look more like DNA"). In fact, we rarely see evidence of internal structure on short fragments. Longer fragments (i.e., Fig. 5.10) do appear to show internal structure, but it is not often near 3 nm. It would be possible to have quite narrow images with poor internal resolution if the tip is sharp, *but interacts strongly with the molecule once is makes contact.* Longer molecules may be more stable in this respect.

The details of the interaction with the substrate may be important. If the molecule is held more strongly at reactive sites associated with the $23 \times \sqrt{3}$ reconstruction present before the adsorption occurred, then this might be reflected in a variation in apparent contrast as the tip interacts differently with parts of the molecule that are bound differently. In any case, it may not be fruitful to pursue these points in the absence of a good theory for image contrast.

Small fragments are preferentially adsorbed. We have considered the effects of diffusion from the bulk, but it does not account for the enhanced adsorption of small fragments. Estimates of the interaction forces in an STM suggest that typical interatomic attractive forces are much too weak to stabilize an adsorbate (Lindsay, 1992). Our experiments are consistent with this in as much as simultaneous adsorption of phosphate and DNA are required to form a stable layer. The formation of such dense layers is beyond the scope of models based on the Poisson-Boltzmann equation and point ions.

If the loss of resolution (relative to the pure aqueous solution of small molecules) is due to complexes between the salt coadsorbate and the DNA, it is hard to see how this could be overcome, given the need for electrolyte to stabilize biologically relevant DNA structures. Indeed, it is conceivable that salt forms part of these structures. On the other hand, the method might yield images of single-stranded DNA (deposited with no salt) that permit the sequence to be read directly. However, a number of problems remain to be solved before long polymers could be sequenced rapidly.

In our view, the most remarkable aspect of the DNA-salt adsorbate is its uniformity. In the region from 0 to 180 mV (versus AgQRE) the gold appears to be atomically flat in both STM and AFM images taken under the phosphate buffer. Since there is no measurable current associated with the formation of the structure that lifts the reconstruction, the salt adsorbate must be thin (≤ 1 monolayer) and presumably quite uniform over the surface over the range of potentials and salt concentrations reported here. This suggests that the interface structure is controlled by the interface potential alone, being rather insensitive to the absolute amount of salt or the presence of a small amount of contamination (i.e., DNA molecules).

When DNA is present, it adsorbs with remarkable uniformity. Perhaps the best measure of this comes from the critical value of tunnel current for clear images, because this should be a very sensitive measure of the adsorbate thickness (because the tunnel current is such a strong function of distance). It is *always* near 0.4 nA. This suggests that the structure of the adsorbate is reproduced with high precision from run to run. Furthermore, the surface coverage of the added DNA particles is highly reproducible, if somewhat more length selective than expected.

At small positive charges ($0 < V_s < 180$ mV) the DNA appears to adsorb as an extended and undenatured double helix (sometimes in a B-conformation). This is a remarkable simplification of *ex situ* spreading techniques. The DNA is, of course, unstained and hydrated during imaging.

At high positive charges, the DNA appears to undergo conformational transitions. Whether this is due to the presence of a concentrated salt layer (as manifested in the adsorbate discovered in the AFM images), a consequence of the high positive charge on the electrode, or even selective adsorption of contamination is unclear at present. Nonetheless, the results indicate that this technique may be useful for studying denaturation and condensation processes.

We are engaged in a quantitative study of the interface charge in order to facilitate these developments.

At negative surface charges, the adsorbate destabilizes. It can be reformed to some extent by recycling the potential, but we have not investigated this thoroughly.

We have found that positive tip bias (with respect to the substrate) yields poor contrast. In general, contrast improves with increasing negative tip bias, but the effect is small compared with the role of tunnel current. We have explored tip bias effects up to $V_t = -600$ mV. Noise becomes a problem at higher negative bias (we have not removed air from our electrolytes).

All of the images we have obtained here show a rather simple set of morphologies for the DNA. The images are usually quite straight over small length scales. A number of oligomers are bent. Sometimes the bending is quite localized (Fig. 5.12B). Since the sequence we have used is believed to be highly bent in solution (Bolshoy et al., 1991), the images imply that *bonding to the substrate often straightens the oligomers.*

None of the images show the dramatic local kinking and bending we have seen many times in our earlier work that included *nonreversible* reactions (Lindsay et al., 1989a,b). We were clearly incorrect in attributing this local structure to DNA sequence (at least in any simple manner). We now believe that it was most likely a consequence of the inhomogeneous manner in which salt coreacted with the DNA and the substrate.

These data illustrate a further point, which is that a given sequence can adopt a variety of structures on the surface (straight, smooth bending, local kinking). We have only studied one other sequence to date, so it is not clear to what degree these arrangements are dictated by sequence.

The main experimental limitation on quantitative use of the method is microscope drift. Corrections can be made quite easily if drift is measured carefully by recording the appropriately timed sequence of images.

There are questions that remain unanswered, but it is clear that the procedure described here yields data that are at least as good as those obtained from electron microscopy (the best cryomicroscopy methods have detected the helix pitch; Fujiyoshi and Uyeda, 1981). It does so without staining and on molecules that are hydrated. The sample preparation and microscopy are relatively simple. The adsorption from solution is not yet understood on a quantitative basis, but the coverage is empirically predicable. The surface coverage is quite uniform, making statistical analysis relatively straightforward.

ACKNOWLEDGMENTS

The experiments with the ligation mixture were the result of an inspiring and productive collaboration with Rodney Harrington (University of Nevada, Reno) and Yuri Lyubchenko and Lyuda Shlyakhtenko (Institute of Molecular Genetics, Moscow). We are grateful to the following individuals, whose collaboration and assistance contributed in a vital way to the success of this work: Patrick Oden, who carried out the AFM experiments and the charge measurements on the gold substrates; Jim DeRose, who made and characterized substrates; Ettore Appella (National Institutes of Health), for synthesis and purification of oligonucleotides; Professor Seth Rose, for patient tutoring in aspects of DNA and electrolyte chemistry; Uwe Knipping and Jin Pan, for valuable help with computing; and Y. Li and L. Nagahara, for many helpful discussions and for assistance with some of the laboratory work. Support and advice has been received from the personnel of Angstrom Technology, Digital Instruments, Inc., and from John Graham (formerly of the Industrial Associates STM Program at ASU). Financial support is acknowledged through research grants from the NSF (Dir-89-20053); ONR (N00014-90-J-1455) to S.M.L.; and NIH (GM33455) and Hatch Projects 118 and 126 from the University of Nevada to R.E.H.

REFERENCES

Åkeson, T., Woodward, C., and Jönsson, B. (1989). Electric double-layer forces in the presence of polyelectrolytes. *J. Chem. Phys.* 91, 2461–2469.

Bard, A. J., and Faulkner, L. R. (1980). Electrochemical Methods: Fundamentals and Applications. Plenum, New York.

Barris, B., Knipping, U., Lindsay, S. M., Nagahara, L., and Thundat, T. (1988). Images of DNA fragments in an aqueous environment by scanning tunneling microscopy. *Biopolymers* 27, 1691–1696.

Bolshoy, A., McNamara, D., Harrington, R. E., and Trifonov, V. E. (1991). Curved DNA without A-A: experimental estimates of all 16 DNA wedge angles. *Proc. Natl. Acad. Sci. U.S.A.* 88, 2312–2316.

Clemmer, C. R., and Beebe, T. P. (1991). Graphite: A mimic for DNA and other biomolecules in scanning tunneling microscope studies. *Science* 251, 640–642.

de la Torre, J. G., and Bloomfield, V. A. (1981). Hydrodynamic properties of complex, rigid, biological macromolecules: Theory and applications. *Qt. Rev. Biophys.* 14, 81–139.

DeRose, J. A., Thundat, T., Nagahara, L. A., and Lindsay, S. M. (1991). Gold grown epitaxially on mica: conditions for large area flat faces. *Surf. Sci.* 256, 102–108.

Drake, B., Prater, C. B., Weisenhorn, A. L., Gould, S. A. C., Albrecht, A. R., Quate, C. F., Cannel, D. S., Hansma, H. G., and Hansma, P. K. (1989). Imaging crystals, polymers and processes in water with the Atomic Force Microscope. *Science* 243, 1586–1539.

Fujiyoshi, Y., and Uyeda N. (1981). Direct imaging of a double stranded DNA molecule. *Ultramicroscopy* 7, 189–192.

Garbarz, J., Lacaze, E., Faivre, G., Gauthier, S., and Schott, M. (1992). Disclocation networks in graphite: an STM study. *Philosophical Magazine* 65a, 853–862.

Heckl, W. M., Smith, D. P. E., Binnig, G., Klagges, H., and Maddocks, J. (1991). Two dimensional ordering of the DNA base guanine observed by scanning tunneling microscopy. *Proc. Natl. Acad. Sci. U.S.A.* 88, 8003–8005.

Joshua-Torr, L., Rabinovich, D., Hope, H., Frolow, F., Appella, E., and Sussman, J. L. (1988). The three-dimensional structure of a DNA duplex containing looped-out bases. *Nature (London)* 334, 82–84.

Lindsay, S. M., and Barris, B. (1988). Imaging deoxyribose nucleic acid molecules on a metal surface under water by scanning tunneling microscopy. *J. Vac. Sci. Technol. A* 6, 544–547.

Lindsay, S. M., Thundat, T., Nagahara, L., Knipping, U., and Rill, R. L. (1989a). Images of the DNA double helix in water. *Science* 244, 1063–1064.

Lindsay, S. M., Nagahara, L. A., Thundat, T., and Oden, P. (1989b). Sequence, packing and naometer-scale structure in STM images of nucleic acids under water. *J. Biomol. Struct. Dyn.* 7, 289–299.

Lindsay, S. M. (1992). Biological applications of the scanning probe microscope. In *Scanning*

Tunneling Microscopy: Theory, Techniques and Applications. (D. Bonnell, ed.) VCH Press, New York (in press).

Lindsay, S. M., and Sankey, O. F. (1992). Contrast and conduction in STM images of biomolecules. In *AIP Conference Proceedings 241: Scanned Probe Microscopy.* (H. K. Wickramasinghe, ed.), AIP, New York, pp. 125–135.

Lindsay, S. M., Lyubchenko, Y. L., Gall, A. A., Shlyakhtenko, L., and Harrington, R. E. (1992a). Imaging DNA molecules chemically bound to a mica surface. *SPIE Proc.* 1639, 84–90.

Lindsay, S. M., Tao, N. J., DeRose, J. A., Oden, P. I., Lyubchenko, Y. L., Harrington, R. E., and Shlyakhtenko, L. (1992b). Potentiostatic deposition of molecules for scanning probe microscopy. *Biophys. J.* 61, 1570–1584.

Lyubchenko, Y. L., Shlyakhtenko, L., Chernov, D., and Harrington, R. E. (1991). DNA bending induced by Cro protein as demonstrated by gel electrophoresis. *Proc. Natl. Acad. Sci. U.S.A.* 88, 5331–5334.

Lyubchenko, Y. L., Gall, A. A., Shlyakhtenko, L., Oden, P. I., Lindsay, S. M., and Harrington, R. E. (1993). Atomic force microscopy of long DNA: Imaging in air and under water. *Proc. Natl. Acad. Sci. U.S.A.* (in press).

Muthukumar, M. (1987). Adsorption of polyelectrolytes to a charged surface. *J. Chem. Phys.* 86, 7230–7235.

Nagahara, L. A., Thundat, T., and Lindsay, S. M. (1989). Preparation and characterization of STM tips for electrochemical studies. *Rev. Sci. Instrum.* 60, 3128–3130.

Nagahara, L. A., Thundat, T., Oden, P. I., Lindsay, S. M., and Rill, R. L. (1990). Electrochemical deposition of molecular adsorbates for in situ scanning probe microscopy. *Ultramicroscopy* 33, 107–116.

Roy, S., Weinstein, S., Borah, B., Nickol, J., Appella, E., Sussman, J. L., Miller, M., Shindo, H., and Cohen, J. S. (1986). Mechanism of oligonucleotide loop formation in solution. *Biochemistry* 25, 7417–7423.

Smith, T. (1980). The hydrophillic nature of a clean gold surface. *J. Colloid Interface Science* 75, 51–55.

Sonnenfeld, R., and Hansma, P. K. (1986). Atomic resolution microscopy in water. *Science* 232, 211–213.

Srinivasan, R., Murphy, J. C., Fainchtein, R., and Pattabiraman, N. (1991). Electrochemical STM of condensed guanine on graphite. *J. Electroanal. Chem.* 312, 293–300.

Tao, N. J., DeRose, J. A., Oden, P. I., and Lindsay, S. M. (1991). Imaging of molecules adsorbed onto electrodes under potential control by scanning probe microscopy. In *Proceedings of the 49th Annual Meeting of the Electron Microscopy Society of America.* (G. W. Bailey, ed.), San Francisco Press, San Francisco, pp. 376–377.

Tao, N. J., and Lindsay, S. M. (1991). Observations of the $23 \times \sqrt{3}$ reconstruction of Au(111) under aqueous solutions using scanning tunneling microscopy. *J. Appl. Phys.* 70, 5141–5143.

Tao, N. J., and Lindsay, S. M. (1992). Kinetics of a potential-induced $23 \times \sqrt{3}$ to $|x|$ transition of Au(111) studied by *in situ* scanning tunneling microscopy. *Surf. Sci. Lett.* 274, L546–L553.

Tao, N. J., Lindsay, S. M., and Lees, S. (1992). Measurements of the microelastic properties of biological material. *Biophys. J.* 63, 1165–1169.

CHAPTER 6
STM of DNA and RNA

Patricia G. Arscott and Victor A. Bloomfield
Department of Biochemistry
University of Minnesota, St. Paul, Minnesota

6.1 Introduction
    6.1.1 Need for High-Resolution Imaging
    6.1.2 Comparison of STM with Transmission Electron Microscopy
6.2 Experimental Methods
    6.2.1 Concentration and Composition of the Sample Solution
    6.2.2 Substrates
    6.2.3 Probes
    6.2.4 Sample Deposition
    6.2.5 Scanning Parameters
    6.2.6 Data Analysis
6.3 Results
    6.3.1 RNA
    6.3.2 DNA
    6.3.3 DNA Complexes
6.4 Discussion
    6.4.1 Theories of Image Formation
    6.4.2 Effects of the Probe and Substrate
    6.4.3 Effects of the Environment
    6.4.4 Future Possibilities
References

6.1 INTRODUCTION

6.1.1 Need for High-Resolution Imaging

In nucleic acids, small structural differences can have large consequences. A single base change can produce a lethal mutation. A slight bend in the backbone or the narrowing of a groove can determine whether a regulatory protein recognizes and binds to a specific site on a DNA molecule or whether an RNA molecule is active in transcribing and translating the genetic code. Cloning and sequencing technology have made it possible to compare the primary structures of nucleic acids and proteins on a large scale, but the detailed spatial information needed to relate primary to higher order structure and function has been lacking. The models that are available are based on X-ray and two-dimensional nuclear magnetic resonance (NMR) studies of a few select specimens and may represent only a fraction of the variety of structures that exist in a population, or in a single molecule, over time. Recent advances in scanning probe and scanning electron microscopies have shown that the fine structures of nucleic acids and proteins can be visualized at the angstrom level. The scanning tunneling microscope (STM), because of its sensitivity to surface texture, should prove especially useful in detecting binding sites and gauging the complementarity of surfaces in interacting systems.

6.1.2 Comparison of STM with Transmission Electron Miscroscopy

Details of surface structure can be reproduced more accurately by STM than by transmission electron microscopy (TEM), since there is little or no reason to coat the specimen with heavy metal. DNA and RNA are sufficiently thin and delicate that even a light coating of uranyl acetate, which has a grain size of only 4 Å, increases the diameter of the molecules by 20% or more and effectively limits resolution of surface features. The evidence thus far indicates that there is enough difference in effective conductivity or work function from one part of these molecules to another to produce the contrast for STM imaging. The voltage required to visualize both strands of the helix and finer features are on the order of 100 mV in STM as opposed to 100 kV in TEM. The risk of radiation damage and the need for a protective coating is therefore minimal.

In eliminating stains and shadows, however, there is some sacrifice of stability. Uncoated molecules tend to move under the force of the STM probe. Nucleic acids and proteins continually change conformation and orientation as they are being examined and are subject to dehydration and consequent changes in dimension unless humidity is carefully controlled. Although currently viewed as difficulties, these aspects of the scanning process hold promise as the biggest advantages of STM over TEM. Samples can be examined under water as well as in air and vacuum, which means that dynamic processes can be monitored under a wide range of experimental conditions.

There is still some question as to how thick a specimen can be imaged by STM. The upper limit is governed by the vertical range of the scan head. Whether an image is obtained at tip to surface separations greater than the tunneling distance (~10 Å) depends on the conductivity and compressibility of the specimen. The evidence thus far indicates that individual molecules and small aggregates of DNA and RNA can be imaged successfully, but it is not yet clear whether the same is true of thicker particles, such as the toroidal condensates and nucleosomes that represent the organized form of nucleic acids in some cells.

6.2 EXPERIMENTAL METHODS

6.2.1 Concentration and Composition of the Sample Solution

The objective in determining the appropriate concentration for STM samples is to reduce the search time for specimens on the substrate and to avoid problems of aggregation and clumping. In studying nucleic acids, the range of concentrations employed depends on the length and stiffness of the molecules. At high concentrations, tangling is a more significant problem, for example, for calf thymus DNA, which averages 20,000 basepairs and is close to 6.8 μm long, than it is for polynucleotides a few hundred basepairs long. Lee et al. (1989) obtained optimal results with full-length calf thymus DNA at 10–20 μg/ml, whereas others have used 1–20 mg/ml with sonicated fragments (Keller et al., 1989), plasmids (Cricenti et al., 1989a), and polynucleotides (Arscott et al., 1989).

In preparing nucleic acids for scanning in air and vacuum, care must be taken to minimize the effects of changes in salt concentration as the sample dries. The usual strategy is to use a low salt buffer, such as 10 mM NaCl, 1 mM Na cacodylate (Lee et al., 1989; Arscott et al., 1989), or one containing a volatile salt, such as ammonium acetate (Keller et al., 1989) or triethylammonium bicarbonate (G. Ruben, personal communication). A five times higher salt concentration can be tolerated when samples are to be scanned in solution (Barris et al., 1988; Lindsay and Barris, 1988), but even then it is important to realize that the distribution on the substrate cannot be completely controlled. Patches of salt are observed at the surface–water interface when samples are deposited electrochemically under certain conditions (Lindsay and Barris, 1988).

6.2.2 Substrates

The primary requirement for an STM substrate is that it be conductive and resistant to oxidation. Surface properties should permit an even spreading of sample solutions and aid in maintaining the stability of the specimen under the probe. Highly oriented pyrolytic graphite (HOPG) does not meet all of these criteria, but is the substrate most commonly used in working with biological molecules in air and vacuum. Single-stranded poly(dA) (Dunlap and Bustamante, 1989), and presumably other nucleic acids with exposed bases, are stabilized by hydrophobic interactions with the surface. The double-stranded forms and binding proteins that have a higher proportion of hydrophilic groups on the exterior adsorb weakly. Neither single- nor double-stranded nucleic acids adhere well enough to be scanned under water (Lindsay et al., 1988a; See also Chapter 5, this volume).

At least part of the appeal of HOPG is that it can be easily cleaned by stripping away the outer layers with sticky tape. Care must be taken, however, to remove each layer completely since curled strips and corrugated pieces left on the surface can be mistaken for biological helices and aggregates. In addition, the unsatisfied bonds at steps and cracks in the surface are reactive with some molecules and may contribute to the tendency of nucleic acids to accumulate and align at these sites. The low elastic modulus of HOPG leads to more serious concerns regarding measurements of adsorbate dimensions. In the course of a normal scan, the forces exerted by the probe can be large enough to deform the surface and lead to anomalous readings of both substrate and sample (Mamin et al., 1986; Yamada et al., 1988; Oden et al., 1990; Salmeron et al., 1991). Gold is currently the substrate of choice. Faceted spheres made from gold wire (Lindsay and Barris, 1988; Lindsay et al., 1988a,b), gold films epitaxially grown on mica (Lindsay et al., 1989b), and gold-plated aluminum (Cricenti et al., 1989a,b; Selci et al., 1990) are used with comparatively few problems.

Other substrates and ways of modifying surfaces continue to be explored. Among the more promising is a method to treat silicon surfaces with alkyl silanes that have reactive groups capable of cross-linking DNA to the substrate (Wilson et al., 1991). Chang and Bard (1990) propose deliberately making pits in HOPG with the probe to attract and hold molecules at specific sites. Size, shape, and distribution could be varied so as to easily distinguish them from nonconducting adsorbates. The idea is not without merit, but, as in other cases in which specimens are anchored, there is concern about distorting the biological structure.

6.2.3 Probes

Probes must also be conductive and resistant to oxidation, and they must be able to withstand the Joule heating accompanying a high tunneling current. Pt-Ir wire, in 9:1 or 8:2 proportions; Pt-Rh, 6:4; and pure Pt, Au, and W are typically used. For underwater work, it is necessary to insulate the probe (e.g., with glass, silicon, or wax) so that only the tip is exposed (Sonnenfeld and Hansma, 1986; Lindsay and Barris, 1988; Lindsay et al., 1988a; see also Chapter 5, this volume). High-resolution imaging of double-stranded DNA and RNA requires an elongated tip. The minor groove of B-form DNA, for example, is only 5.7 Å wide (Saenger, 1984). The more slender the probe, the less chance of generating multiple tunneling currents from the sides of the groove. Techniques have been developed for chemically etching probes to produce the desired size and shape (Fink, 1986; Bryant *et al.*, 1987), but surprisingly good results have been obtained with rough, irregular probes cut with shears and used as is.

6.2.4 Sample Deposition

The basic and most widely used method of preparing biological samples for scanning in air is simply to pipet 10–100 μl of sample solution onto a substrate and let it dry at room temperature. The drying process usually takes about 30 min, but may be expedited by placing the sample in a laminar flow hood or desiccator. Alternatively, the sample is lightly sprayed onto the substrate with an atomizer or the substrate is floated on a 100 μl drop of the sample deposited on Parafilm. Allen and colleagues (1991) suggest dipping the probe directly into the sample solution and spraying it onto the substrate by electrically vibrating the tip with a 10 μsec pulse as the scan begins. However, electrodeposition techniques are used to greater advantage in water than in air and are described in detail in Chapter 5, this volume.

Cricenti and coworkers' method of cross-linking DNA to a gold substrate (Cricenti *et al.*, 1989a,b; Selci *et al.*, 1990) has proved successful in air and may be adaptable for underwater scans as well. The DNA is treated with tris(1-aziridynl) phosphine oxide, which has three ethyleneimine groups capable of reacting with ribose moieties in the DNA backbone, as well as a phosphorus oxide group that is reactive with gold. Lyubchenko and coworkers (1991) used similar tactics in treating DNA with mercury, which binds to bases in the major groove and to S–H groups on graphite activated by oxidation and reaction with cystamine and 2-mercaptoethanol.

6.2.5 Scanning Parameters

Scanning parameters are essentially the same whether working in air, vacuum, or water. The microscope is operated in constant current mode, at setpoints ranging from 0.1 to 3.0 nA, with corresponding voltages ranging from 930 to 35 mV (Keller *et al.*, 1989; Arscott *et al.*, 1989). The scan rate depends on scan size and ranges from 100 to several thousand angstroms per second (Driscoll *et al.*, 1990; Dunlap and Bustamante, 1989). Jericho *et al.* (1990) recommend using a hopping mode to avoid the build up of lateral stress between the tip and sample. The probe is periodically raised and lowered so that most of the scan is carried out at a fixed distance above the sample. This is not an option with most commercial instruments and requires modification of the electronics.

6.2.6 Data Analysis

Both commercial and homemade microscopes are generally equipped with sophisticated software to facilitate measurement and manipulation of the digitized image. To identify specific features and to help elucidate the mechanism of image formation, the dimensions obtained by STM should be compared with molecular models based on crystallographic and NMR data whenever possible (Arscott et al., 1989; Driscoll et al., 1990). The pitch and diameter of the double-stranded DNA and RNA indicate the conformation and hydration state of the molecules and can be determined by tracking the vertical movement of the probe as it scans. The spacing of peaks along the helical axis corresponds to the pitch, or half pitch, and can be measured directly or by fast Fourier transform (Lee et al., 1989; Arscott et al., 1989). The latter method is faster and has the advantage of filtering nonrepetitive features, but does not give information about height or site-specific variations.

6.3 RESULTS

6.3.1 RNA

The first STM images of an RNA molecule (Lee et al., 1989) were obtained with poly(rA)–poly(rU), which almost always adopts the A form. The helical pitch of both natural and synthetic double-stranded RNA is 30.9 Å by fiber diffraction (Saenger, 1984). STM profiles of the probe's vertical versus translational movement along the helix indicate an average periodicity of 28.7 ± 1.1 Å. The small percentage difference in the two types of measurements is reproducible and attributable to a difference in hydration. Samples exposed to air for more than 2 hours show more shrinkage, with a maximum of about 19% after 10 hours. Center-to-center distances between helices in aggregates average 24.8 ± 0.8 Å, or slightly more than the molecule diameter of 21.3 Å (Saenger, 1984).

Lindsay and coworkers (1989a) made similar measurements of single-stranded RNA and poly(U) in solution. The periodicities obtained were variable but within the expected range. The average for the native RNA was 22 Å, and for the polynucleotide, 16 ± 2.5 Å. Both specimens appeared to be randomly coiled and kinked.

6.3.2 DNA

Double-stranded DNA is more structurally versatile than its RNA counterpart. The conformation observed depends on base sequence and chemical composition, as well as on the hydration and ionization state of the molecule. STM studies of double-stranded DNA in solution (Lindsay et al., 1989a) indicate that both natural and synthetic fibers exhibit the B-form periodicity expected with full hydration. Measurements of calf thymus DNA, poly(dAT)–poly(dAT), poly(dCG)–poly(dCG), and poly(dA)–poly(dT) yielded periodicities of 36 ± 5, 35 ± 4, 30 ± 3, and 33 ± 5 Å. The pitch of a typical DNA molecule at >92% humidity is 33.8 Å by fiber diffraction (Saenger, 1984). An STM image of DNA in the B form is shown in Figure 6.1.

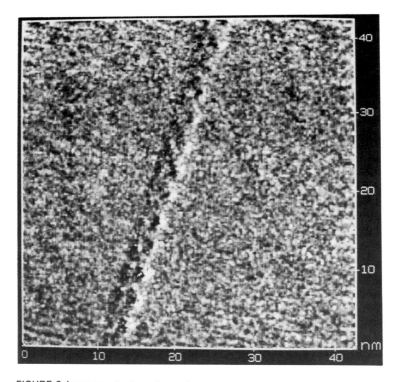

FIGURE 6.1 DNA in the B conformation.

Lee *et al.* (1989) noted that the B form is retained for some time after the sample is prepared for scanning in air. After 2 hr of drying under ambient conditions, the helical periodicity of calf thymus DNA is only 7% or so less than the pitch of a fully hydrated molecule, 31.5 ± 1.3 Å, compared with the standard 33.8 Å. The STM measurements were made on densely packed aggregates, which did not allow accurate determination of individual diameters, but center-to-center packing distances are about 30 Å, the same as in hydrated crystals. The B form has also been observed in closed circular plasmids (Travaglini *et al.*, 1987; Cricenti *et al.*, 1989a,b; Allison *et al.*, 1990) and an unusual parallel-stranded DNA (Li *et al.*, 1990).

B → A transitions occur with loss of hydration and are accompanied by a decrease in pitch and an increase in the diameter of the helix. Images made by Driscoll *et al.* (1990) of a 550 bp fragment of mouse B-cell DNA in ultrahigh vacuum show both grooves of the double helix as well as atomic-scale features of the phosphate-ribose backbone and bases. The helical pitch of 29 Å and other dimensions are within 1 Å of the values established by X-ray crystallography. In DNA containing regions of alternating purines and pyrimidines, loss of hydration results in B → Z transitions which are accompanied by an increase in pitch and a decrease in diameter. The pitch varies from 43.2 to 45.6 Å, depending on environmental conditions. STM images of poly(dG-me^{5}dC)–poly(dG-me^{5}dC) scanned in air show the left-handed twist and elongated contours of the Z helix (Arscott et al., 1989; Li et al., 1991). The periodicity of the helix determined by Arscott et al. (1989) was 42.1 Å, but there was some question as to whether the measurements represented the pitch or half pitch of the molecule (Arscott et al.,1990). Li and colleagues (1991) subsequently obtained a value of 45 Å, which is exactly the pitch determined by crystallography (Fujii et al., 1982).

Conformational transitions are associated with changes in the orientation of the bases, which cannot be visualized in an intact molecule. Only single-stranded "sticky ends" (Allen et al., 1991) and denatured regions (Biggar et al., 1990) are accessible to the probe. Dunlap and Bustamante (1989) obtained images of single-stranded poly(dA) which show the bases clearly outlined against the substrate. The imidazole and pyrimidine rings are distinguishable and the center-to-center distance between them is within 0.2 Å of the distance predicted by crystallography, 2.4 Å compared with 2.2 Å. Also, Cricenti et al. (1991) have tentatively identified the pyrimidine ring of cytosine in oligomers of adenine-cytosine.

6.3.3 DNA Complexes

Amrein et al. (1988, 1989) and Travaglini et al. (1988) were the first to visualize a DNA–protein complex by STM. Their work with the recA protein, which binds single-stranded DNA, is described in the Chapter 2, this volume.

Wang et al. (1990) reported studies of three proteins with different binding modes. Images of the DNA–VirE2 complex are consistent with previous data that suggest that the protein aligns alongside single-stranded DNA to form a protective coating (Sen et al., 1989). The second protein, C/EB, interacts with double-stranded DNA and is base sequence specific. The STM images show projections along the helix that are the appropriate length for the "leucine zipper" domain of the protein (Landschulz et al., 1988), but the rest of the molecule, lying in the grooves, is not distinguishable. Scans of the third protein, Z22 antibody, which preferentially binds to Z-DNA (Möller et al., 1982), revealed only large aggregates of uncomplexed protein and DNA.

Lu and coworkers (1991) obtained images of DNA polymerase I bound to a short fragment of DNA. The dimensions of the protein, which largely obscures the DNA, agree with crystallographic predictions.

6.4 DISCUSSION

6.4.1 Theories of Image Formation

Tunneling currents on the order of 10^{10} electrons/sec must pass between the probe and the surface to get a good image. Although it is generally believed that nucleic acids are insulators, high-resolution images of both DNA and RNA have been obtained with no additives other than the low levels of salt in the buffers. Keller *et al.* (1989) estimate that the magnitude of the tunneling current can be explained if electrons pass through the molecule in sequential steps, first through the backbone, then the core region occupied by bases, and then the backbone on the opposite side. Molecular orbital calculations indicate a possible pathway from sugars to bases and from one base to another (Otto *et al.*, 1983).

The σ-bonded sugar–phosphate chains present a greater intrinsic barrier to electron transfer than the conjugated π system of the bases. There is experimental evidence, however, to suggest that the backbone is more conductive than the bases. Purugganan *et al.* (1988) measured faster rates of electron transfer between metals bound in the outer regions of the helix than when bound internally (Purugganan *et al.*, 1988), and STM scans of A-DNA, made by Driscoll *et al.* (1990) indicate that barrier heights are correlated with topographical data over the backbone but not over bases exposed in the grooves. In Z-DNA, although the bases protrude into and occlude the major groove, they are apparently not seen by the probe. Arscott *et al.* (1989) obtained images of poly(dG-me^{5}dC)–poly(dG-me^{5}dC) in the Z form in which both grooves appear to be open. Although there was some ambiguity in interpretation of their data (Arscott *et al.*, 1990), later images of the same polynucleotide, made by Li and colleagues (1991), give no indication of bases in the major groove. These results suggest that the oriented waters and ions surrounding the phosphates may play a role in the transfer of electrons through the backbone chains as noted by Purugganan *et al.* (1988).

In the first STM image of DNA, made by Binnig and Rohrer (1984), the specimen looked as if it were lying below the surface of the substrate. Lindsay and colleagues (1988b, 1990) have since suggested that inverted images can be attributed to the pressure in the tunneling gap exerted by a blunt tip. They propose that the probability of electron transfer depends on the compressibility of the material in the tunneling gap. When there is less pressure, the specimen appears to be lying on top of the substrate, and the transfer coefficient may be related to the resonance energy of the tip–sample–substrate system. Their calculations indicate that energy shifts of up to 1 eV can be induced by the probe when the sample is near the middle of the tunneling gap. Estimates of the pressure required to bring the sample into resonance with the tip and substrate range from 0.1 to 10 gPa.

6.4.2 Effects of the Probe and Substrate

For electron microscopists accustomed to seeing individual (*i.e.*, unaggregated) molecules of DNA and RNA, it is surprising that few or none are encountered in the course of a normal scan. The most likely explanation is that they are swept aside by the probe, Biggar *et al.* (1990) clocked the movement of single strands of calf thymus DNA on HOPG at 10–20 Å/sec. The two-dimensional arrays and well-ordered bundles found in samples of both nucleic acids (Lee et al., 1989) may result from these types of large-scale movements bringing helices into close proximity. The DNA and RNA scanned in air and vacuum appear to be straighter and stiffer than could be expected from their calculated persistence length in solution. The rastering motion and pressure of the probe may help to straighten and align those that are weakly adsorbed and free to roll on the substrate. Considerably more curvature, kinking, and bending are observed in cases in which specimens are held in place by chemical fixation (Cricenti et al., 1989a,b; Selci et al., 1989) or electrostatic interactions with the substrate (Lindsay et al., 1989a).

Lindsay and colleagues (1990) calculate that the probe exerts pressures of 1–10 gPa, or 10^4–10^5 atmospheres, on the sample and suggest that vertical dimensions depend on the sample's compressibility. The diameter of a double-stranded DNA molecule is about 20 Å by fiber diffraction (Saenger, 1984). STM measurements of width are within the right range, but heights are about one-half the predicted value, only 9–12 Å (Lee et al., 1989; Keller et al., 1989; Driscoll et al., 1990). In contrast, the height of single-stranded poly(dA) is 8.4 Å by STM measurement instead of 5.7 Å. In the one case in which height and width are matched, the tip was retracted further than usual in order to pass over the protein capsid to which the DNA was attached (Keller et al., 1989). This suggests that with less pressure, dimensions are more accurate. However, dimensional changes of twofold are not expected even at these high pressures.

The compressibility of the substrate must also be taken into account in assessing the effects of pressure on the sample. The concavity of the surface directly beneath the tunneling tip has been documented in TEM images of HOPG (Lo et al., 1990). Irregularities in the surface lattice or, in the case of HOPG, in the orientation of layers beneath the surface (Oden et al., 1990) affect conductivity and hence the appearance and dimensions of the specimen. The roughness, hydrophobicity, and chemical reactivity of the surface determine how well the sample spreads and adheres. The single-stranded nucleic acids are more hydrophobic and adhere better to HOPG than do the double-stranded forms, for example. There are many variables, however, including sample concentration, the viscosity of the sample solution, and the presence of proteins and other solutes, that influence the sample–surface interaction. Both experimental and theoretical evidence indicate that the initial contact with the surface can induce conformational changes in both nucleic acids and proteins and may enhance the helicity and regularity of secondary and tertiary structures (Wattenbarger et al., 1990).

6.4.3 Effects of the Environment

Humidity, pH, and salt conditions must be preserved if results with nucleic acids and proteins are to be biologically trustworthy. The most obvious way to do this is to work under water, as recommended by Sonnenfeld and Hansma (1986), Lindsay and Barris (1988), and Lindsay et al. (1988a,b). Molecular motions pose a greater problem in water than in air and vacuum, however, and it is more difficult to attain the high resolution desired. The best alternative would be to scan in a controlled humidity chamber.

The most highly resolved images of double-stranded DNA made in air and vacuum show both strands of the helix, the major and minor grooves (Cricenti et al., 1989a; Selci et al., 1990; Arscott et al., 1989; Arscott and Bloomfield, 1990b; Driscoll et al., 1990, Li et al., 1991), and, in some cases, bases, sugars, and phosphates (Cricenti et al., 1989a; Selci et al., 1990; Driscoll et al., 1990). The shape and substructure of the bases can be discerned in at least two cases with single-stranded polynucleotides (Dunlap and Bustamante, 1989; Cricenti et al., 1990). Helical periodicities are well correlated with crystallographic data, and the observable differences are for the most part attributable to dehydration (Lee et al., 1989; Arscott et al., 1989).

TEM measurements indicate that the contour length of nucleic acids decreases linearly with a logarithmic increase in the ionic strength of the sample solution (Lang et al., 1987). One would assume that there is a similar relation in STM experiments, although there are other variables such as the surface properties of the substrate to be considered. Ionic effects may be directly related to the presence of the ions or to the displacement of water molecules and changes in the electrostatic potential of a given structure (Record et al., 1985). The effects may be specific or nonspecific, but cannot be very well controlled as the sample dries.

The binding of a protein to a DNA molecule will generally displace cations M^+ bound by counterion condensation near the DNA surface. If m cations are displaced on average, the reaction can be written

$$\text{Protein} + \text{DNA} \rightleftharpoons \text{Protein-DNA} + mM^+.$$

The thermodynamic equilibrium constant (neglecting activity coefficients) is

$$K_{eq} = [\text{Protein-DNA}][M^+]^m / [\text{Protein}][\text{DNA}].$$

The apparent observable equilibrium constant is

$$K_{app} = [\text{Protein-DNA}]/[\text{Protein}][\text{DNA}] = K_{eq}/[M^+]^m.$$

Typically, m is about 10, so a 10-fold increase in salt concentration $[M^+]$ can produce a 10^{10} decrease in K_{app} and thus in protein binding strength. An interaction that is strong and specific at low salt may be weak or nonexistent at high salt, which may account for the failure to find DNA–antibody complexes on the substrate even when their presence in solution is indicated by immunoassay. It may be necessary in some cases to cross-link nucleic acid and protein as was done with the DNA-VirE2 complex (Arscott et al., unpublished results).

6.4.4 Future Possibilities

The idea of sequencing nucleic acids by STM has sparked a great deal of interest. The images obtained by Dunlap and Bustamante (1989) and by Cricenti et al. (1990) demonstrate that it is possible not only to visualize individual bases but also to distinguish purines from pyrimidines. Both rings of the adenines in poly(dA) and the single ring of cytosine in oligo [$A_{15}C_{15}$ $(ACC)_5$] are visible. Whether there is sufficient difference in conductivity to distinguish an adenine from a guanine or a thymidine from a cytosine is still uncertain. With further advances in sample preparation and scanning technique, it may become feasible to visualize the constituents of the rings (the amino, carbonyl, methyl, and proton groups) that characterize each base. One could also tag the bases with specific ligands. Since there is no need to use heavy metals, as in TEM, a wide range of materials can be explored. It may be possible to find small enough ligands to fit into the grooves and allow sequencing of the native, double-stranded molecule. It seems unlikely that current methodology, which enables rapid sequencing of thousands of bases a day, will be supplanted by a technique that relies on visually identifying the bases one by one. The probable utility of sequencing by STM rather lies in the ability to determine ligand-binding sites with high precision and to engineer site-specific changes in the native structure.

REFERENCES

Allen, M. J., Tench, R. J., Mazrimas, J. A., Balooch, M., Siekhaus, W. J., and Balhorn, R. (1991). A pulse deposition method for scanning tunneling microscopy of deoxyribonucleic acid on graphite. *J. Vac. Sci. Technol.* B9, 1272–1275.

Allison, D. P., Thompson, J. R., Jacobson, K. B., Warmack, R. J., and Ferrell, T. L. (1990). Scanning tunneling microscopy and spectroscopy of plasmid DNA. *Scan. Microsc.* 4, 517–522.

Amrein, M., Stasiak, A., Gross, H., Stoll, E., and Travaglini, G. (1988). Scanning tunneling microscopy of recA–DNA complexes coated with a conducting film. *Science* 240, 514–516.

Amrein, M., Dürr, R., Stasiak, A., Gross, H., and Travaglini, G. (1989). Scanning tunneling microscopy of uncoated recA–DNA complexes. *Science* 243, 1708–1711.

Arscott, P. G., Lee, G., Bloomfield, V. A., and Evans, D. F. (1989). Scanning tunneling microscopy of Z-DNA. *Nature (London)* 339, 484–486.

Arscott, P. G., and Bloomfield, V. A. (1990a). Scanning tunneling microscopy in biotechnology. *Trends Biotechnol.* 8, 151–156.

Arscott, P. G., and Bloomfield, V. A. (1990b). Scanning tunneling microscopy of nucleic acids and polynucleotides. *Ultramicroscopy* 33, 127–131.

Arscott, P. G., Lee, G., Bloomfield, V. A., and Evans, D. F. (1990). Helical period of Z-DNA. *Nature (London)* 346, 706.

Barris, B., Knipping, U., Lindsay, S. M., and Nagahara, L. (1988). Images of DNA fragments in an aqueous environment by scanning tunneling microscopy. *Biopolymers* 27, 1691–1696.

Biggar, R. D., Major, S. A., McCormick, A. S., Spevak, E., and Vold, T. G. (1990). Scanning tunneling microscopy studies of single-stranded deoxyribonucleic acid. Preprint.

Binnig, G., and Rohrer, H. (1984). Scanning tunneling microscopy. In *Trends in Physics*. (J. Janta and J. Pantoflicek, eds.) European Physical Society, Petit-Lancy, Switzerland, pp. 38–46.

Bryant, P. J., Kim, H. S., Zheng, Y. C., and Yang, R. (1987). Techniques for shaping STM tips. *Rev. Sci. Instrum.* 58, 1115–1116.

Chang, H., and Bard, A. J. (1990). Formation of monolayer pits of controlled nanometer size on highly oriented pyrolytic graphite by gasification reactions as studied by scanning tunneling microscopy. *J. Am. Chem. Soc.* 112, 4598–4599.

Cricenti, A., Selci, S., Felici, A. C., Generosi, R., Gori, E., Djaczenko, W. D., and Chiarotti, G. (1989a). Molecular structure of DNA by scanning tunneling microscopy. *Science* 245, 1226–1227.

Cricenti, A., Selci, S., Felici, A. C., Generosi, R., Gori, E., Djaczenko, W. D., and Chiarotti, G. (1989b). Application of scanning tunneling microscopy to structural studies of DNA. *Helv. Phys. Acta* 62, 702–705.

Cricenti, A., Selci, S., Chiarotti, G., and Amaldi, F. (1991). Imaging of single-stranded DNA with the scanning tunneling microscope. *J. Vac. Sci. Technol.* B9, 1285–1287.

Driscoll, R. J., Youngquist, M. G., and Baldeschwieler, J. D. (1990). Atomic-scale imaging of DNA using scanning tunnelling microscopy. *Nature* 346, 294–296.

Dunlap, D. D., and Bustamante, C. (1989). Images of single-stranded nucleic acids by scanning tunnelling microscopy. *Nature (London)* 342, 204–206.

Fink, H.-W. (1986). Mono-atomic tips for STM. *IBM J. Res. Dev.* 30, 460–462.

Fujii, S., Wang, A. H.-J., van der Marel, G., van Boom, J. H., and Rich, A. (1982). Molecular structure of $(m^5dC-dG)_3$: The role of the methyl group on 5-methyl cytosine in stabilizing Z-DNA. *Nucleic Acids Res.* 10, 7879–7892.

Jericho, M. H., Blackford, B. L., Dahn, D. C., Frame, C., and Maclean, D. (1990). Scanning tunneling microscopy imaging of uncoated biological materials. *J. Vac. Sci. Technol.* A8, 661–666.

Keller, D., Bustamante, C., and Keller, R. W. (1989). Imaging of single uncoated DNA molecules by scanning tunneling microscopy. *Proc. Natl. Acad. Sci. USA* 86, 5356–5360.

Landschulz, W. H., Johnson, P. F., and McKnight, S. L. (1988). The leucine zipper: a hypothetical structure common to a new class of DNA binding proteins. *Science* 240, 1759–1764.

Lang, D., Steely, H. T., Jr., Kao, C.-Y., and Ktistakis, N. T. (1987). Length, mass and denaturation of double-stranded RNA molecules compared with DNA. *Biochim. Biophys. Acta* 910, 271–281.

Lee, G., Arscott, P. G., Bloomfield, V. A., and Evans, D. F. (1989). Scanning tunneling microscopy of nucleic acids. *Science* 244, 475–477.

Li, M.-Q., Zhu, J.-D., Zhu, J.-Q., Hu, J., Gu, M.-M., Xu, Y.-L., Zhang, L.-P., Huang, Z.-Q., Xu, L.-Z., and Yao, X.-W. (1990). Direct observation of parallel and antiparallel stranded DNA by scanning tunneling microscopy. In *Abstracts STM '90, Fifth Int. Conf. Scanning Tunneling Microscopy/Spectroscopy, and Nano I, First Int. Conf. Nanometer Scale Sci. & Technol* American Vacuum Society, Baltimore, p. 153.

Li, M.-Q., Zhu, J.-D., Zhu, J.-Q., Hu, J., Gu, M.-M., Xu, Y.-L., Zhang, L.-P., Huang, Z.-Q., Xu, L.-Z., and Yao, X.-W. (1991). Direct observation of B-form and Z-form DNA by scanning tunneling microscopy. *J. Vac. Sci. Technol* B9, 1298–1303.

Lindsay, S. M., and Barris, B. (1988). Imaging deoxyribose nucleic acid molecules on a metal surface under water by scanning tunneling microscopy. *J. Vac. Sci. Technol.* A6, 544–547.

Lindsay, S. M., Thundat, T., and Nagahara, L. (1988a). Imaging biopolymers under water by scanning tunneling microscopy. In *Biological and Artificial Intelligence Systems*. (E. Clementi and S. Chin, eds.) ESCOM B. V., Leiden, pp. 125–141.

Lindsay, S. M., Thundat, T., and Nagahara, L. (1988b). Adsorbate deformation as a contrast mechanism in STM images of biopolymers in an aqueous environment: images of the unstained hydrated DNA double helix. *J. Microsc.* 152, 213–220.

Lindsay, S. M., Nagahara, L. A., Thundat, T., and Oden, P. (1989a). Sequence, packing and nanometer scale structure in STM images of nucleic acids under water. *J. Biomol. Struct. Dyn.* 7, 289–299.

Lindsay, S. M., Thundat, T., Nagahara, L., Knipping, U., and Rill, R. L. (1989b). Images of the DNA double helix in water. *Science* 244, 1063–1064.

Lindsay, S. M., Sankey, O. F., Li, Y., and Herbst, C. (1990). Pressure and resonance effects in scanning tunneling microscopy of molecular adsorbates. *J. Phys. Chem.* 94, 4655–4660.

Lo, W., Spence, J. C. H., and Kuwabara, M. (1990). In situ electron microscope study of STM tip-surface interactions. In *Abstracts STM '90, Fifth Int. Conf. Scanning Tunneling Microscopy/Spectroscopy, and Nano I, First Int. Conf. Nanometer Scale Sci. and Technol.* American Vacuum Society, Baltimore, p. 192.

Lu, C.-D., Li, M.-Q., Qui, M.-Y., Yao, X.-W., Xu, Y.-L., Gu, M.-M., and Hu, J. (1991) Conformation of DNA–DNA polymerase I complex observed by scanning tunneling microscopy. *J. Biomol. Struct. Dyn.* 9, 233–238.

Lyubchenko, Y. L., Lindsay, S. M., DeRose, J. A., and Thundat, T. (1991). A technique for stable adhesion of DNA to a modified graphite surface for imaging by scanning tunneling microscopy. *J. Vac. Sci. Technol.* B9, 1288–1290.

Mamin, H. J., Ganz, E., Abraham, D. W., Thomson, R. E., and Clarke, J. (1986). Contamination-mediated deformation of graphite by the scanning tunneling microscope. *Phys. Rev.* B34, 9015–9018.

Möller, A., Gabriels, J. E., Lafer, E. M., Nordheim, A., Rich, A., and Stollar, B. D. (1982). Monoclonal antibodies recognize different parts of Z-DNA. *J. Biol. Chem.* 257, 12081–12085.

Oden, P. I., Thundat, T., Nagahara, L. A., Adams, G. B., Lindsay, S. M., and Sankey, O. F. (1990). Graphite superperiodicity observed with scanning tunneling microscopy. In *Abstracts STM '90, Fifth Int. Conf. Scanning Tunneling Microscopy/Spectroscopy, and Nano I, First Int. Conf. Nanometer Scale Sci. and Technol.* American Vacuum Society, Baltimore, p. 67.

Otto, P., Clementi, E., and Ladik, J. (1983). The electronic structure of DNA related periodic polymers. *J. Chem. Phys.* 78, 4547–4551.

Purugganan, M. D., Kumar, C. V., Turro, N. J., and Barton, J. K. (1988). Accelerated electron transfer between metal complexes mediated by DNA. *Science* 241, 1645–1649.

Record, M. T., Jr., Anderson, C. F., Mills, P., Mossing, M., and Roe, J.-H. (1985). Ions as regulators of protein-nucleic acid interactions *in vitro* and *in vivo*. *Adv. Biophys.* 20, 109–135.

Saenger, W. (1984). *Principles of Nucleic Acid Structure*. Springer-Verlag, New York.

Salmeron, M., Ogletree, D. F., Ocal, C., Wang, H.-C., Neubauer, G., Kolbe, W., and Meyers, G. (1991). Tip-surface forces during imaging by scanning tunneling microscopy. *J. Vac. Sci. Technol.* B9, 1347–1352.

Selci, S., Cricenti, A., Felici, A. C., Generosi, R., Gori, E., Djaczenko, W., and Chiarotti, G. (1990). Molecular structure of organic compounds observed by high resolution scanning tunneling microscopy. *J. Vac. Sci. Technol.* A8, 642–644.

Sen, P., Pazour, G. J., Anderson, D., and Das, A. (1989). Cooperative binding of *Agrobacterium tumefaciens* VirE2 protein to single-stranded DNA. *J. Bacteriol.* 171, 1573–2580.

Sonnenfeld, R., and Hansma, P. K. (1986). Atomic resolution microscopy in water. *Science* 232, 211–213.

Travaglini, G., Rohrer, H., Amrein, M., and Gross, H. (1987). Scanning tunneling microscopy on biological matter. *Surf. Sci.* 181, 380–390.

Travaglini, G., Rohrer, H., Stoll, Z., Amrein, M., Stasiak, A., Sogo, J., and Gross, H. (1988). Scanning tunneling microscopy of recA–DNA complexes. *Physica Scripta* 38, 309–314.

Wang, L., Arscott, P. G., and Bloomfield, V. A. (1990). In *Abstracts STM '90, Fifth Int. Conf. Scanning Tunneling Microscopy/Spectroscopy, and Nano 1, First Int. Conf. Nanometer Scale Sci. and Technol.* American Vacuum Society, Baltimore, p. 156.

Wattenbarger, M. R., Chan, H. S., Evans, D. F., Bloomfield, V. A., and Dill, K. A. (1990). Surface-induced enhancement of internal structure in polymers and proteins. *J. Chem. Phys.* 93, 8343–8351.

Wilson, T. E., Murray, M. N., Ogletree, D. F., Bednarski, M. D., Cantor, C. R., and Salmeron, M. B. (1991). Scanning tunneling microscopy at high gap resistances and on chemically modified silicon surfaces. *J. Vac. Sci. Technol.* B9, 1171–1176.

Yamada, H., Fujii, T., and Nakayama, K. (1988). Experimental study of forces between a tunnel tip and the graphite surface. *J. Vac. Sci. Technol.* A6, 293–295.

SFM in Biology

CHAPTER 7

Investigation of the Na,K-ATPase by SFM

Hans-Jürgen Apell
Department of Biology, University of Konstanz
Konstanz, Germany

Jaime Colchero
Department of Physics, University of Konstanz
Konstanz, Germany

Achim Linder
Department of Biology, University of Konstanz
Konstanz, Germany

Othmar Marti
Department of Physics, University of Konstanz
Konstanz, Germany

7.1 The Structure and Role of Biomembranes
7.2 Investigation of Biological Membranes by SFM
7.3 Na,K-ATPase
    7.3.1 Structure
    7.3.2 Isolation
    7.3.3 Crystallization
7.4 Sample Preparation for SFM
    7.4.1 Binding of Membrane Fragments onto Mica Surface
    7.4.2 Removal of Surface Contaminations
    7.4.3 Manipulation of Surface Properties
    7.4.4 Water Cell
7.5 Imaging Techniques
    7.5.1 Tip-Sample Interactions
    7.5.2 Scanning Friction Force Microscopy
7.6 Results
    7.6.1 Topography
    7.6.2 Forces
    7.6.3 Friction
    7.6.4 Comparison of Preparations

7.7 Outlook
Appendix: Model of the Tip-Sample Interaction
References

7.1 THE STRUCTURE AND ROLE OF BIOMEMBRANES

Living organisms are all organized by the same general principle: They exist in the form of cells, from single cells to aggregates of several and up to a multitude of cooperating cells with specialized functions. Nevertheless, each of the cells consists of a small aqueous volume with diameters from 1 μm to a few millimeters.

This volume is surrounded by and pervaded with membranes that form the boundaries of the cell and internal cellular compartments. Membranes consist of a bimolecular layer of lipid molecules and a broad variety of proteins that can span the membrane or be attached to the surface (Fig. 7.1). The thickness of the biomembranes ranges from approximately 4 nm in domains of pure lipid to 15 nm in regions where bulky proteins protrude into the aqueous environment. In addition, many proteins are decorated with sugars ("glycosylated"). To fortify structural features of cells, organisms have developed additional outer membranes or cell walls that are constructed of proteins, sugars, or mineral components to conserve specific shapes or to protect the cells against osmotic stress.

The plasma membrane of a cell separates the cytoplasm from the environment. As a consequence, the ionic concentration and the composition of other substrates in the intracellular aqueous phase are maintained across the membrane such that metabolic processes can take place under optimized conditions.

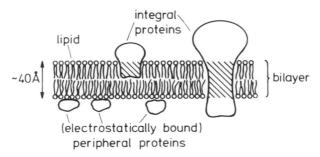

FIGURE 7.1 Schematic structure of a biological membrane. The membrane consists of amphiphilic lipid molecules that form a bilayer structure with their polar head groups facing the aqueous compartments. Into and onto the membrane, proteins are bound. Some proteins are anchored in the membrane by hydrophobic parts of the surface (hatched regions). Other proteins are attached to lipid head groups by electrostatic interactions.

Therefore, a continual entry and exit of matter has to happen. On the one hand, uncontrolled diffusion of molecules has to be minimized. Indeed, the lipid bilayer is a high diffusion barrier for most of the substances solved in the aqueous phases. On the other hand, specific transport processes have to be performed to guarantee the uptake of nutrients and necessary cofactors into the cell and the export of metabolic decomposition products, as well as toxic materials that appear in the cytosol of the cell. Proteins take care of the transport processes. The energy to perform these functions is stored in the transmembrane electrochemical potential of mainly the concentration gradients of Na^+ and K^+ ions and the electrical membrane potential.

Since the 1920s there has been much investigation of membranes, components, and functions of membranes. Beginning with Gorter and Grendel (1925), the model of bimolecular layers of lipids was brought up and is now well accepted since Singer and Nicolson (1972) published the fluid mosaic model of cell membranes. Although lipid molecules are more or less ordered in a membrane, the structure of a membrane is not rigid (at least in the physiological range of temperatures). Biological membranes are thought to be in a liquid-crystalline state. This could be proved by experimentally determined diffusion coefficients for lipid molecules in the range of $D \sim 10^{-7} - 10^{-8}$ cm^{2}s^{-1} and for proteins in membranes of $D \sim 10^{-9}$ cm^{2}s^{-1}. This high fluidity indicates a significant softness of biomembranes that has to be taken into account when membranes are scanned by methods contacting the surface. The physical chemistry of membranes has become textbook knowledge and can be retrieved easily (e.g., Silver, 1985).

Membrane proteins with a broad variety of functions, from transport of different classes of substances to voltage-and concentration-gradient generating ion pumps and to information-transducing devices, have been identified. The sizes of proteins range from a few ten thousand to a million daltons. During the processes of their actions, proteins may perform peristaltic motions ("conformational changes") or may interact with other proteins. Hence their spatial coordination is of crucial importance.

For the understanding of molecular mechanisms of proteins, therefore, their structures and arrangements in membranes are of crucial importance. Besides experimental techniques such as electron microscopy, X-ray diffraction studies, and nuclear magnetic resonance (NMR) methods, as well as theoretical approaches, have been applied. From known amino acid sequences the protein topologies are constructed by determination of hydropathy plots and compared with the results of specific labeling and specific proteolytic treatments (Eisenberg, 1984; Jennings, 1989).

Since structure-function relationships are of great importance to understand biological processes, it is of interest to adapt new techniques to investigate the structure of proteins with nanometer resolution under physiological conditions in which they are capable of performing their functions. Therefore it is promising to apply scanning force microscopy to investigate biological membranes containing native proteins.

7.2 INVESTIGATION OF BIOLOGICAL MEMBRANES BY SFM

Scanning force microscopy (SFM) is not restricted to conducting surfaces. This advantage was used almost immediately to scan organic surfaces such as crystals of amino acids (Gould *et al.*, 1988) or organic monolayers (Marti *et al.*, 1988). It was then obvious to extend the method to biological systems. Initial materials were blood cells and bacteria (Gould *et al.*, 1990). The purple membrane, which consists of crystalline areas of a single proton pumping protein, bacteriorhodopsin, was investigated extensively. Its three-dimensional structure in the membrane has been well characterized by various techniques. Therefore it is ideally suited to test the capabilities of SFM. Bacteriorhodopsin has been scanned in dried form on lysine-coated glass, on mica (Worcester *et al.*, 1990; Gould *et al.*, 1990), and in buffer solutions (Butt *et al.*, 1990). The findings were in agreement with results of electron diffraction experiments and documented that the SFM can be a useful tool with which to image structures of biological membranes. A second integral membrane protein investigated by SFM was Na,K-ATPase (Apell *et al.*, 1992). This protein and its function and structure are described in detail in this chapter.

Besides membrane surfaces of living organisms, model systems of biological membranes can be investigated. They are obtained by transferring monomolecular films of phospholipids onto solid surfaces by the Langmuir-Blodgett technique. The surface structure of pure lipid membranes or asymmetric proteolipid (Fab lipid)–containing membranes has been revealed by SFM (Egger *et al.*, 1990; Weisenhorn *et al.*, 1990).

Larger structures such as unicellular algae, so-called diatoms, have also been successfully imaged with SFM. The walls of their shells are coated by layers of polysaccharides, proteins, and lipids and are sculptured with patterns in the range of approximately 100 nm. Scanning these shells at room temperature and in aqueous environments resulted in images known otherwise only from scanning electron microscopy (Linder *et al.*, 1992).

The following section is concerned with the preparation and investigation of Na,K-ATPase–containing membrane fragments by SFM. A series of methodological and technical aspects are discussed, and their implications for SFM imaging are presented.

7.3 Na,K-ATPase

Na,K-ATPase is found in the cytoplasmic membranes of virtually all animal cells, and it is responsible for active transport of Na^+ ions out of the cell and K^+ ions into the cell. Low sodium and high potassium concentrations in the cytoplasm are essential for basic cellular functions such as excitability, secondary active transport, and volume regulation (Läuger, 1991).

7.3.1 Structure

The enzyme consists of two subunits, α and β, formed by different polypeptide chains. The α-subunit contains approximately 1000 amino acids, with a molecular mass of about 112 kDa (Shull et al., 1985) and performs all catalytic functions (Jørgensen and Andersen, 1988). The β-subunit is a glycoprotein consisting of approximately 300 amino acids and 50 carbohydrate residues (Shull et al., 1986), with a molecular mass of about 35 kDa (excluding the carbohydrate). The role of the β-subunit is not yet clear. However, the proteins are enzymatically active and transport competent only when α- and β-subunits are inserted into the membrane at the same time and are able to interact during the processes of folding and arranging the amino acids. They form heterodimers of one α- and one β-subunit (Ackermann and Geering, 1990). The minimal functional unit of the protein is still an open question. Besides the $\alpha\beta$-structure, mainly the $(\alpha\beta)_2$ is claimed to form the transport competent form of the ion pump (Nørby and Jensen, 1991). From a Fourier analysis of tilted specimen of negative-stained $(\alpha\beta)_2$-crystals, a three-dimensional reconstruction of the Na,K-ATPase has been derived (Maunsbach et al., 1988). The proteins seem to protrude at least 4 nm on the cytoplasmic side of the lipid bilayer and about 2 nm on the extracellular. Approximately 40% of the mass of the protein is disposed within the hydrophobic core of the membrane, about 40% is located on the cytoplasmic side, and about 20% is on the extracellular side. The enzymatic machinery is situated in the cytoplasmic part of the α-subunit.

Na,K-ATPase can exist in two principal conformations, E_1 and E_2, each of which can be stabilized by different ionic compositions of the buffer. The E_1 conformation is maintained preferentially in the presence of (cytoplasmic) Na^+ ions and ATP in absence of Mg^{2+}. Under these conditions up to three Na^+ ions are bound to selective ion-binding sites facing the cytoplasm. Addition of Mg^{2+} as a necessary cofactor leads to a phosphorylation of the enzyme and to a conformational transition to an E_2 state. The E_2 conformation can be stabilized by buffers containing a few millimolar of K^+ ions and inorganic phosphate. Then the ion-binding sites are facing the extracellular medium and have a high affinity for K^+ ions. The digestion of the protein by proteases (trypsin or α-chymotrypsin) in both conformations results in a different pattern of digestion products, indicating significant rearrangements of the quaternary structure of the α-subunit during conformational changes (Jørgensen and Andersen, 1988).

7.3.2 Isolation

Although all animal cells are equipped with Na,K-ATPase, tissues that are specialized for Na transport are especially rich in this protein. Therefore, outer medulla of mammalian kidney, salt glands of sea birds, rectal glands of sharks, or excitable tissues like nerve cells are used to isolate this protein.

The isolation and purification of the Na,K-ATPase from rabbit renal tissue was performed by a slightly modified procedure described by Jørgensen (1974) and Apell et al. (1985). The outer medulla is homogenized, and a differential centrifugation yields a fraction of plasma membrane vesicles with Na,K-ATPase. Treatment of these vesicles with the detergent sodium dodecylsulfate removes selectively and quantitatively membrane proteins other than the Na,K-ATPase and part of the membrane lipids. After a second centrifugation step a preparation of flat, purified membrane fragments is obtained. They have diameters of 0.4–1 μm and contain up to $\sim 10^4$ $\alpha\beta$-protomers per micrometer squared as determined from electron microscopic studies (Deguchi et al., 1977). The membrane preparations have typical enzymatic activities of 2000 μmol released inorganic phosphate per milligram of protein per hour. These preparations can be stored at $-70\,°C$ for long times without changing their properties.

7.3.3 Crystallization

A standard technique with which to gain information on protein structure is the electron microscopic investigation of proteins that have been treated to form a two-dimensional crystal. Image processing of electron microscopic pictures of stained crystalline areas have resolutions on the order of a few angstroms, depending on the size of the crystalline areas. This technique has been applied to Na,K-ATPase for more than 10 years (for review, see Maunsbach et al., 1991). To grow crystals, Na,K-ATPase in purified membrane fractions may be incubated with various combinations of ligands and ion concentrations under different conditions of temperature, pH, and time. Crystallization of the protein is possible in both the E_1 and E_2 conformations and as monomers or dimers in the unit cell, depending on the composition of the crystallization buffer (Maunsbach et al., 1991). Besides vanadate or phosphate, phospholipase A_2 has been used to induce the crystallization (Mohraz et al., 1985). It has been shown that crystallization occurs as the result of hydrolysis and solubilization of the phospholipids of the membrane fragments.

To prepare Na,K-ATPase crystals for SFM investigations, a suspension of 1 mg/ml of protein in freshly prepared purified membranes was treated with buffer containing 10 mM Tris-HCl, 1 mM CaCl$_2$, 1 mM MgCl$_2$, 5 mM H$_3$PO$_4$, and 60 μg/ml phospholipase A_2 at pH 7.3. The reaction mixture was transferred into a dialysis tube. The pore size excluded molecular masses above 14 kDa. The external medium had the 1000-fold volume of the internal volume; thus the decomposition products of the phosholipase were diluted. To avoid bacterial growth in the solution, azide was added to the buffer. After 2 days the preparation was transferred into a dialysis tube with a pore size of 300 kDa, and the buffer was replaced by a calcium-free one for another 1–2 days. This step reduced the amount and activity of the phospholipase. The dialyzed membrane fragments finally were concentrated by centrifugation (20,000 g, 15 min). All procedures were performed at 4°C. The residual enzymatic activity of the crystallized protein was determined to be approximately 20 μmol inorganic phosphate per milligram of protein per hour. The almost complete loss of activity could be explained by the digestion of the essential lipid environment of the ion pump or by blocking of the protein in a particular conformation when crystal structures are formed.

A transmission electron microscopic picture of a crystallized membrane preparation is shown in Figure 7.2. The image was taken from a freshly crystallized membrane sample. The treatment with phospholipase A_2 lasted 48 hr, and then the preparation was washed, concentrated by centrifugation, and deposited on a carbonized copper grid. Negative staining was performed with a 1% solu-

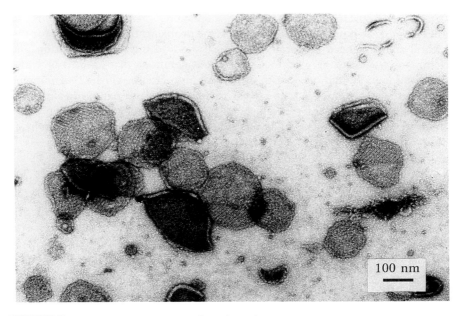

FIGURE 7.2 Electron microscopic image of membrane fragments containing Na, K-ATPase (negative staining with 1% uranyl acetate). The membrane fragments have been incubated with phospholipase A_2 for 48 hours and have sizes of 100 to 500 nm. On the surfaces two different types of domains can be distinguished: Uncrystallized areas with randomly distributed particles, which are identified as proteins, sometimes in groups of 3 or 4; and a crystallized pattern with rows of dimers. The unit cell has a typical size of 15 × 6 nm² and an inner angle of 65°. Conspicuous is the buckling edge of the fragments, which is extremely rich in particles.

tion of uranyl acetate. On the images of the membrane fragments different structures could be discriminated: uncrystallized areas with randomly distributed particles, crystallized domains with rows of dimers or sometimes trimers, and buckled edges of the fragments that were extremely rich in particles. The dimensions determined for the unit cell were in good agreement with published data: $a = 16$ nm, $b = 6$ nm, and $\gamma = 65°$ (Mohraz et al., 1985). The sizes of the crystalline areas depended on the duration of the phospholipase A_2 incubation. They appeared after approximately 24 hr and increased relative to the total area. After 6 days, only crystalline structures remained. The areas with randomly distributed particles can be interpreted as lipid bilayers with uncrystallized proteins. Rather frequently oligomers of two to four particles were found. Characteristic was the finding of invaginations and holes in the membrane, while the edges of fragments were formed by bulges of protein particles. These findings have been also reproduced with SFM (Apell et al., 1992).

7.4 SAMPLE PREPARATION FOR SFM

Scanning the surface of Na,K-ATPase–containing membrane fragments (or other membrane-bound proteins) causes some problems that are specific for this method and that need preparative investigations and technical precautions: (1) Membrane fragments are obtained as aqueous solutions that have to be transferred and immobilized on smooth surfaces appropriate for SFM; (2) contaminations that cover surface features have to be avoided or removed; and most important, (3) changes of the sample due to the scanning forces have to be minimized in order to achieve good contrast and to reduce mechanical damage of the soft biological material.

7.4.1 Binding of Membrane Fragments onto Mica Surface

Solutions of membrane fragments were diluted with distilled water or buffer to obtain protein concentrations of 0.1–0.3 mg/ml. A drop of this solution was placed onto freshly cleaved muscovite mica (potassium aluminosilicate, thickness 0.15 mm) for 10–60 s. Then the mica was rinsed carefully with buffer and transferred to the SFM.

This procedure led to a coverage of the mica surface of 20–60% by membrane fragments. The adhesion was an all-or-none process. When fragments had "bound" to the surface, they could not be moved on the surface without serious damage. Since Mg^{2+} and Ca^{2+} ions improved the speed and yield of the adhesion to the mica, we rinsed the freshly cleaved mica with buffer containing a moderate concentration of Mg^{2+} ions (5–10 mM) and removed the residual liquid before adding the fragments. This treatment presumably replaced the K^+ ions in the surface layer of the green mica by Mg^{2+} ions and therefore enabled a more effective adhesion process and improved crystal stability.

7.4.2 Removal of Surface Contaminations

Scanning of adhered crystallized or uncrystallized fragments without further procedures detected much debris in addition to the membrane fragments. The contaminations could be classified in two groups: on the one hand, coarse particles with diameters of about 50 nm; and on the other hand, a thin, contrast-reducing film that covered the fragments and the mica surface almost homogeneously. The first type of deposited material presumably consisted of small protein–lipid aggregates. Freezing and thawing of fragments increased the fraction of these minifragments from about 40% to more than 80% coverage on the mica surface. This type of contamination was increased also in preparations that had been stored at 4°C for more than 1 week and in preparations with freshly crystallized fragments that had been kept at room temperature for more than 3 hr. The second type of debris had the appearance of a film of soft material that covered all surfaces. It prevented the observation of the mica structure between membrane fragments. This material could have its origin in the degradation products of the phospholipase treatment such as fatty acids. A similar problem was the removal of additional enzymes from the preparation procedure (e.g., phospholipase or neuraminidase).

The amount of debris can be reduced in different ways. First, the mica-adsorbed fragments should be rinsed with a large excess of buffer. However, this treatment is not sufficient to achieve good contrast and to make the mica structure visible between the fragments. To get rid of small-sized or dissolved material, the following procedure has been developed as a standard technique: Solutions of (crystallized) membrane fragments were dialyzed for several days in a tubing with a molecular mass cut-off of 300 kDa in order to remove the phospholipase. This preserves the fragments in a given state of the crystallization process. A subsequent two- to threefold repetition of centrifugation as described above, discarding of the supernatant, and homogenization of the soft pellet at 4°C yields membrane preparations with good contrast.

It has been found that fragments adhere also to debris structures on mica, thus producing distortions of the scanned fragment surfaces. This process is controlled by the Mg^{2+} and Ca^{2+} concentrations in buffer containing the fragments. Since Mg^{2+} is essential for the stability of the crystallized fragments, their concentrations can be decreased only for a short period of time. Fragments that have been suspended overnight in buffer containing 1 mM phosphate and no Mg^{2+} exhibit less contaminated images, but adhere only slowly to the mica surface.

On the other hand, we have found that the large crystallized fragments will sediment within a day out of a suspension to the bottom of a vial that contained 2 ml of buffer. This property was used as a purification procedure. Samples containing 10 μg protein were placed on 1 ml of standard buffer to sediment overnight. The fragments collected from the pellet were so clean that the atomic structure of mica could be resolved between the fragments.

7.4.3 Manipulation of Surface Properties

Compared with solid-state surfaces, biological materials are soft, and membranes especially can be described as two-dimensional liquids. Thus, modified techniques are required to prevent mechanical damage and maintain maximal contrast.

Mechanical damage can be reduced by different means. Fragments can be stabilized by adding covalent cross-linking agents or certain cations that are able to link the constituents of the sample to each other or to the substrate. Near a phase transition point between liquid and solid state of the biological membranes, cooling can also stiffen the sample. Nevertheless, all these methods have significant influence on the properties of biomolecules. Therefore, the ideal approach is minimizing the imaging forces by using the sharper tips that are now available. Because of the smaller curvature angle, the attractive van der Waals forces and Coulomb forces are much lower, and a smaller contact area results in a reduced mechanical load and higher contrast.

In the case of ATPase crystals, it was not possible to achieve good images with forces below 1 nN. In the following section different approaches to stabilize the samples for SFM are discussed.

7.4.3.1 COOLING

ATPase crystals tend to dissociate in suspensions at room temperature. However, crystals adsorbed to mica are stable for hours even without cooling. Reducing the temperature of the samples to 3°C with an SFM placed in an absorber refrigerator increased their stability, but the resolution could not be increased to reveal more features than the experiments performed at 20°C. Actually, the contrast seemed to be slightly lower at 3°C. Presumably, the standard SFM tips were too blunt, so they could not intrude into the spaces between the protein rows due to their enhanced rigidity. Since the transition temperature of the lipid mixture in the fragments was lower than $-20°C$, cooling to temperatures in the range of 0°C did not lead to a significantly increased stability of the membrane fragments. The slight effect hardly compensated for the additional trouble with condensing water and thermal drifts. At temperatures below 0°C, antifreezing agents had to be added to the buffer. In the case of Na,K-ATPase, high ion concentrations could not be used to reduce the freezing temperature, because high ionic strength alters the conformation of the protein. With sugars instead, temperatures could be achieved only in the range of $-4°C$ due to the limited solubility of the sugars. Beyond the reduction of the freezing temperature it is known that dissolved sugar molecules change the physical properties of membranes (Crowe *et al.*, 1988). Cryoprotectants like methanol or glycols are known to alter the membrane structures; therefore, they are not suitable for our purposes.

Glycerol was less interfering. We have used buffers with 46% glycerol, which allowed us to image ATPase crystals at $-20\,^\circ$C in the deep-freeze compartment of our refrigerator. However, glycerol entered the membrane phase and lowered its phase transition temperature. This compensated at least partly for the cooling effect. In summary, the images taken at $-20\,^\circ$C were comparable with images at room temperature in the absence of antifreezing agents.

7.4.3.2 FIXATIVES

The standard fixation agent used in electron microscopy is glutaraldehyde. This compound links amino groups of the proteins to each other. Due to its high partition coefficient for the aqueous phase, it produces cross reactions between groups at the surface of the proteins and therefore modifies the features of interest. As an alternative, one can use aromatic azides that have a high partition coefficient for the lipid phase and link intramembrane parts of proteins when irradiated with ultraviolet light (McBeath and Fujiwara, 1984). A disadvantage of this method is the low absorption of the agent in the lipid phase. The necessary high light intensities at an excitation wavelength of 280 nm are expected to induce photo damage of the tryptophane residues of the protein. Agent-induced conformational changes of the protein cannot be excluded. Covalent linking of the proteins to mica is virtually impossible, since the O—Si—O bonds of the mica layers are extremely inert. Different laminar substrates that are sufficiently reactive to allow chemical modification have not been tested so far. Surface modifications have to be sufficiently homogenous to maintain atomically flat surfaces. The cross-linking reactions have to be specific enough to avoid binding of debris from the environment.

7.4.3.3 IONIC CONDITIONS

Certain cations such as Ca^{2+}, Mg^{2+}, Fe^{3+}, and UO_2^{2+} are known to alter the viscosity of biomembranes or to induce phase separation of some lipids. High concentrations of Mg^{2+} and Ca^{2+} are known to inhibit the function of ATPase and lead to precipitation reactions with the phosphate. We have imaged uncrystallized membrane fragments in buffer containing 10 mM $CaCl_2$. In contrast to Ca^{2+}-free buffers, it was possible to resolve rows of proteins on the fragments even with blunt tips. However, the rows were less regular and the distance between them was larger than in crystallized fragments (~ 13 nm). Adsorption of the membrane fragments was much quicker than in buffer without Ca^{2+}. Hence, at least part of the higher stability is due to higher forces between the mica and the fragments. On the other hand, Ca^{2+} is known to enhance the forces between the tip and the sample (Israelachvili, 1992). In crystallized samples, Ca^{2+} has not shown any additional effect. Since Mg^{2+} and Ca^{2+} ions improve adhesion to the mica, we rinse the freshly cleaved mica with standard buffer and remove the residual liquid before adding the fragments (see Section 7.4.1). K^+ concentrations above 100 mM have been found to destroy the E_2-crystals within a few minutes.

Exchanging the phosphate buffer for water enhanced the contrast in SFM images. This finding indicates that the ionic composition at the membrane–buffer interface was crucial for the contrast mechanism. The influence of imaging forces is discussed in Section 7.6.2.

7.4.4 Water Cell

To scan Na,K-ATPase–containing membrane fragments, an SFM from Park Scientific Instruments (Mountain View, California) was used, modified by a self-designed water cell. Several problems have to be overcome before using SFM in aqueous solutions. At least three specifications have to be fulfilled: (1) The laser beam has to be guided through air–cell and cell–buffer interfaces in a way that it is reflected from the cantilever to the position-sensitive photo diode, (2) the approach of the cantilever to the mica surface should be observable to make the coarse positioning simple, and (3) the evaporation of water has to be minimal in order to maintain defined ionic conditions. All of these requirements could be satisfied with the design of a water cell that can be used in combination with a standard SFM. A schematic drawing is shown in Figure 7.3 and described in the following.

The laser beam enters and leaves the buffer through small glass cubes. The glass surfaces are perpendicular to the laser beam, which avoids disturbing refraction effects. Glass has been chosen to obtain an undisturbed image of the laser beam on the photodiode and because it is a material that can be cleaned appropriately. The distance between the glass cubes and the cantilevers is small to minimize the effects of particles in the buffer passing through the light path. The cantilever can be watched through a piece of plexiglass that is passed into an opening of the water cell (and that can be removed if required). The view onto the mica has an angle of less than 10° relative to the surface. Under these conditions, it is possible to observe the image of the cantilevers as a reflection on the mica in spite of the low difference between the refractive indices of mica and water. This facilitates the manual approach of the tip to the mica within a distance of approximately 2 μm. The loss of water by evaporation is low due to the small surface between aqueous phase and air and can be compensated for by refilling a few microliters of distilled water once an hour. The total aqueous volume of the cell is approximately 100 μl. To clean the water cell after use, it is rinsed with water and ethanol. With this procedure salt and organic materials are removed and the accumulation of debris on the optical surfaces is prevented.

7.5 IMAGING TECHNIQUES

It is evident that for imaging soft surfaces like biological samples the interaction between tip and sample has to be kept as small as possible to avoid distortions or

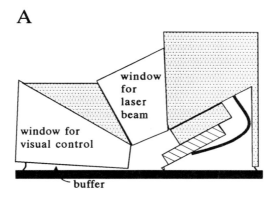

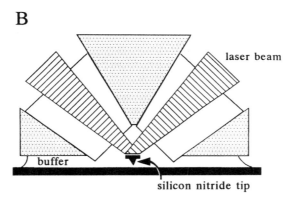

FIGURE 7.3 Schematic set-up of the water cell used in combination with an SFM from Park Scientific Instruments to scan images of membrane fragments containing Na,K-ATPase. **A:** Longitudinal view of the cell: The main body of the cell has been manufactured from acrylic glass (dotted areas); the chip with the silicon nitride tip is pressed to a guiding holder by a stainless-steel spring. Two cylindrical windows are mounted from above to allow the laser beam to be reflected on the upper surface of the cantilever. A third glass body is mounted on the front side of the cell to allow visual control of the approach of the cantilever to the sample surface. The total length of the cell is 20 mm. **B:** Transverse cross section illustrating the optical path of the laser beam through the water cell. The size of the scanning tip has been oversized drastically for clarity. The thickness of the glass windows is 5 mm, and the width of the cell 10 mm.

even damage. Typical imaging forces in air lie in the range of (several) hundred nanoNewtons (1 nN = 10^{-9} N). Using the water cell to image in aqueous buffers reduced the interaction by two to three orders of magnitude and led to promising structural resolutions of protein-containing membrane fragments. Besides the reduction of the interaction, imaging in liquids allows the investigation of biological samples in their natural environment. Moreover, experiments may be made with functional units to investigate, for example, the dynamics of proteins. Some principal considerations of tip-sample interactions are presented in the next section.

7.5.1 Tip-Sample Interactions

In air, samples are covered by a liquid film of water and hydrocarbons; therefore the leading interaction is given by the meniscus formed between tip and sample. The attractive force produced by the Laplace pressure of this liquid bridge may be as high as several hundred nanonewtons. If the sample and the whole cantilever are immersed in a liquid, this meniscus is absent and the leading force is controlled by the van der Waals interaction:

$$F = \frac{A \cdot R}{6z^2}, \tag{7.1}$$

where A is the Hamaker constant, R is the tip radius, and z is the tip-sample separation.

The Hamaker constant A depends on the optical properties of the interacting media, including the medium between tip and surface, and is usually smaller in liquids than in air (e.g., $A = 13.2 \times 10^{-20}$ J for mica–air/vacuum–mica, and $A = 2.2 \times 10^{-20}$ J for mica–water–mica) (Israelachvili, 1992). Therefore the interaction between tip and sample in liquids is further reduced. Typical adhesion forces in water are 0.2 nN for a Si_3N_4 tip on a mica substrate (Weisenhorn et al., 1992). The interaction force between tip and sample is typically two or three orders of magnitude less in liquid than in air. Since forces larger than about 1 nN can distort or even damage biological samples, imaging in liquids is fundamental for biological applications.

If physiological buffer is used, ions are present in the medium between the tip and the sample. Since biological samples are usually charged, this gives rise to an additional electrostatic interaction, the electrical double-layer interaction (Israelachvili, 1992). Essentially this interaction can be understood by the following mechanism: The surface charges repel equal charges in the electrolyte and attract opposite charges. Therefore close to the surface a net charge distribution ρ is induced that generates an electrical potential ϕ according to

$$\Delta\phi(z) = -\frac{\rho(z)}{\varepsilon\varepsilon_0}.$$

It can be shown that for small potentials (<25 meV $\sim kT$),

$$\phi(z) = \phi_0 \cdot e^{-\frac{z}{d}}$$

where

$$\phi_0 = \frac{\sigma \cdot d}{\varepsilon\varepsilon_0}$$

is surface potential,

$$d = \frac{1}{e}\sqrt{\frac{\varepsilon\varepsilon_0 kT}{2I}}$$

is Debye length,

$$I = \frac{1}{2}\sum_i n_i z_i^2$$

is Ionic strength, n_i is concentration, z_i is valence of the i-th ion, and e is elementary charge. The Debye length d is independent of the surface charge, while the potential and the force increase linearly with this charge.

In this chapter we assume that the interaction between the tip and the biological sample is described by the van der Waals interaction and some strong repulsive force at small distances, together with the just-described so-called electrical double-layer forces. Steric and solvation interactions, which are due to the discrete nature of the liquid, are therefore neglected (Israelachvili, 1992).

When an SFM tip approaches the sample, different potentials and distances have to be taken into account. It is important to note that it is not possible to adjust the tip-sample distance z directly by piezoelectric movement; only the distance Δ between sample and the base of the cantilever can be set experimentally. It follows that the forces versus distance curves as recorded experimentally do not directly correspond to the tip-sample interaction. However, this interaction can be calculated from the force versus distance curves as explained in the Appendix. Furthermore, the Appendix presents a simple model for the tip-sample interaction and a discussion of stability conditions.

7.5.2 Scanning Friction Force Microscopy

The newly developed scanning friction force microscopy (SFFM) allows one to measure simultaneously forces normal to the surface as well as in lateral direction. Since a detailed description of the instrument and the method has been given in Chapter 1, this volume, we only describe some features of lateral forces that are necessary to understand the presented experiments. In SFM, imaging occurs usually without wear. We can qualitatively understand friction between tip and a solid surface as follows.

While the tip is dragged over the rigid surface, physical bonds produced by the adhesion between the tip and the solid surface are continuously broken. In this process energy is dissipated and a mean lateral force is exerted onto the cantilever. At high resolution the atomic modulation of this lateral force has been observed on mica and on graphite. If the lateral force is plotted versus the lateral tip displacement during one scan (forward and backward scanning has to be included) a hysteresis is observed. The area enclosed by this curve is proportional to the energy dissipated in this process. We note that friction is strongly related to the adhesion of tip and sample and increases with higher normal loading. If the substrate is not a solid but a two-dimensional liquid, as in the case of biological membranes, the friction process is different. In this case the molecular constituents of the substrate that may adhere to the tip are free to move within the two-dimensional liquid, and thus the lateral force is produced by the viscosity of the two-dimensional liquid. Friction is then expected to increase with increasing scan speed.

7.6 RESULTS

7.6.1 Topography

At low resolution (image size 7×7 μm^2), we found that the mica substrate was covered by membrane fragments with sizes ranging from 0.25 to 1 μm (Fig. 7.4A). At high scanning forces (>20 nN), the membrane fragments could not be moved, but the surface structure was significantly changed when the same area was scanned again. Therefore, we concluded that the adhesion of the entire fragments onto the mica was rather high.

At medium resolution (image size 1×1 μm^2), fragments showed two different domains (Fig. 7.4B). One type of domain was completely flat, had a thickness of approximately 4 nm, and was assigned to a pure lipid membrane without protein. The second domain was 12 nm thick, in agreement with the electron microscopic images of Na,K-ATPase crystals. It had a granular surface structure.

At high resolution (image size 300×300 nm^2) in standard buffer and with "sharp" tips, the fragments showed a crystalline structure (Fig. 7.5). This image has been Wiener filtered to enhance the visibility of the crystalline structure. Rows could be identified that extended from the upper right to the lower left corners. The distance between them was 10 nm. Along the axis in the other direction, at an angle of approximately 75° to the first one, the distance between the rows was about 8 nm. The size of the corrugations between the rows depended strongly on the quality of the tips. With the sharpest tips we have used thus far, corrugations of about 1 nm have been observed. This was one-half the value that has been described for the extracellular protrusions of the protein from electron microscopic investigations. This discrepancy can be explained by the blunt shape of the tips. The dimensions of the features corresponded well with the predicted values.

With state of the art tips currently available (spring 1992), on clean samples images are obtained as shown in Figure 7.6. These fragments are partially covered by a thin film of about 0.5 nm thickness, which sometimes can be removed by the scanning motion of the tip. Part of this film is shown in the middle and on the border in Figure 7.6A. The protein was hardly resolved through this film. Figure 7.6B shows a close-up into the structure of the fragment to present more details. The long-range order of proteins in this image was rather low. The size of the single bumps that we considered to be (dimers of) the proteins was 10×12 nm; the z-corrugation was 0.4 nm. On some proteins in the high-resolution image a small dimple of about 0.2 nm was detected, which could be due to the

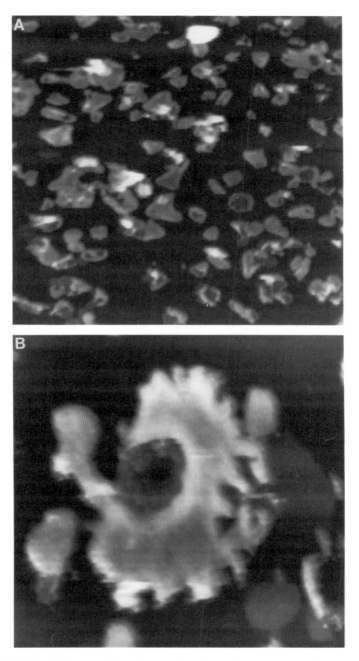

FIGURE 7.4 **A**: Typical low-resolution image of membrane fragments containing Na,K-ATPase. Image size is 7×7 μm^2; the gray-scale corresponds to a corrugation of about 25 nm. Fragment size varies between 0.25 and 1 μm. **B**: Medium-resolution image of a membrane fragment containing Na,K-ATPase. Two different domains are clearly resolved: a lower, flat one of about 4 nm height and a higher, structured domain of about 12 nm height. The first domain is composed only of lipids; the second domain contains the proteins. Image size is 1×1 μm^2; the gray-scale corresponds to a corrugation of about 20 nm.

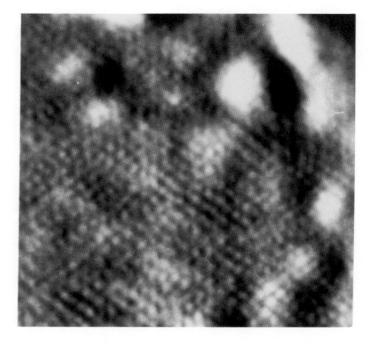

FIGURE 7.5 High-resolution image of the protein domain of a membrane fragment. To enhance the crystal structure this image has been Wiener-filtered. The unit cell of the crystal is 8 × 10 nm, with an angle of 75°. Image size is 250 × 250 nm².

dimer structure of the protein. We hope to confirm this result with a new generation of cantilevers with ultrasharp tips of 10 nm radius (e.g., Park Scientific Instruments, United States; Nanotechnology, Germany; Olympus, Japan).

7.6.2 Forces

To investigate the tip-sample interaction and its local variation, we measured force-distance curves on the fragments as well as on the mica substrate. In the first series of experiments, these curves were measured with high indentation loading (about 100 nN), and the pull-off force (adhesion) was determined. The adhesion of the tip on the mica substrate increased from about 3 nN at the first measurement to about 35 nN after the fourth. Subsequent measurements showed a variation between about 15 and 35 nN. On the membrane fragments the adhesion was found to be about 3 nN. On the membrane fragments the tip did not snap off the sample suddenly as observed with rigid surfaces, but released gradually. Even after 100 nm retraction the tip was still weakly bound by the membrane fragment with a force of ~0.5 nN. This behavior suggested that the membrane fragment or some of its constituents were lifted from the substrate

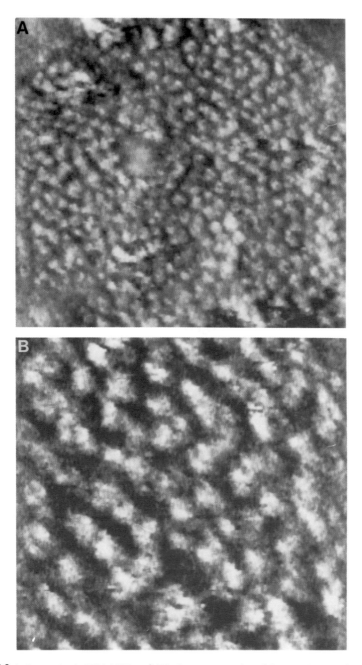

FIGURE 7.6 **A:** Image size is 300 × 300 nm². The long-range order of the proteins is rather low, but single (dimers of) proteins are clearly resolved. Parts of a thin film (thickness about 0.5 nm) that covered the fragments is shown in the center as well as on the borders of this image. **B:** Image size is 125 × 125 nm²; the gray-scale corresponds to a corrugation of about 1 nm. The size of single (dimers of) proteins is 10 × 12 nm; their corrugation about 0.4 nm. On some proteins, a small dimple of about 0.2 nm is resolved.

while the tip was retracted. On the other hand, the increase of pull-off force after successive measurements on the mica substrate indicates that some modification of the tip had taken place. Possible modifications may have been due to either geometric change (e.g., breaking of the tip) or to adhesion of molecules from the sample. Since the curvature radius of the used tips was rather large, the tips should have been mechanically stable. We therefore suggest that adsorbates from the substrate, presumably lipids, adhered to the tip.

A scan taken after force-distance measurements with high indentation loads showed severe damage to the fragments. Therefore standard measurements were taken with indentation loads lower than 5 nN. In this case the adhesion forces were strongly reduced and ranged between 3 and 5 nN on the mica substrate and less than 0.5 nN on the membrane fragments.

When the tip approached the surface of the membrane fragments, we observed a repulsive force in the distance range of 2–10 nm. From the experimental curves $F(\Delta)$, we calculated curves of $F(z)$ as described in the Appendix and found an exponential decay length with a Debye length $d = 2.4 \pm 0.4$ nm and a force $F_0 = 0.8 \pm 0.2$ nN. This agreed reasonably with the decay length of 1.7 nm that was calculated from the ionic strength of the buffer used in this experiment (1 mM Mg^{2+}, 5 mM Tris, 3 mM H_3PO_4, pH 7.2).

Over the mica substrate we found a small repulsive force until a height of about 2.0 nm was reached. Then the tip was unstable and snapped onto the surface. The corresponding force discontinuity was $\Delta F = 0.4$ nN. On the membrane fragments such a discontinuity was smaller than 0.2 nN.

Information about tip-sample interaction may also be obtained from careful interpretation of topographic images. Figure 7.8 shows topographic and friction images where the force was decreased during the scan from 1 nN at the beginning (bottom of the image) until the tip snapped off the surface (upper part of the image). As discussed later, the brighter areas in the friction image corresponded to points where the friction was negligible, (i.e., the tip did not adhere onto the substrate). At low imaging force (<0.25 nN) the tip scanned about 2.5 nm above the substrate, 4 nm over the lipid phase of the fragments, but still in contact with the protein phase. Imaging between 0 and 4 nm over the lipid and between 0 and 2.5 nm over the mica substrate was not stable. In fact, to the right of the membrane fragment in Figure 7.8 the tip jumps on and off the surface. At this low force, the *apparent* height of the membrane fragment was 5.5 nm for the lipid phase and about 10 nm for the protein phase, measured from a level 2.5 nm above the mica substrate where the tip found a stable z-position. The resolution at this low force was significantly smaller than at higher forces ($F > 0.5$ nN). We interpret these results as follows.

As explained in the Appendix, a physically stable tip-sample position requires a positive effective compliance:

$$k_{\text{eff}} = \frac{\partial^2 V_{\text{surf}}}{\partial z^2} + c = \frac{\partial^2 V_{\text{edl}}}{\partial z^2} + \frac{\partial^2 V_{\text{vdW}}}{\partial z^2} + c = k_{\text{edl}} + k_{\text{vdW}} + c.$$

7 Investigation of the Na,K-ATPase by SFM 295

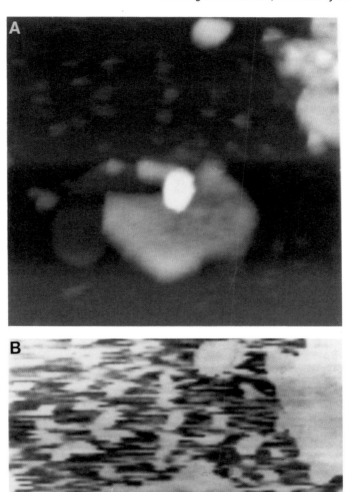

FIGURE 7.7 Topographic (A) and lateral force (B) images of a membrane fragment containing Na,K-ATPase. Image size is $1.5 \times 1.5 \; \mu m^2$; the gray-scale corresponds to a corrugation of about 20 nm (left) and a lateral force of about 40 nN (right). The normal force was 2.5 nN.

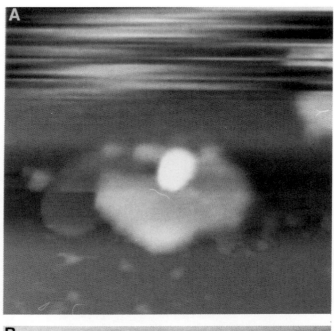

FIGURE 7.8 Images taken at the same location as those in Figure 7.7 a short time later. The normal force was decreased from about 1 nN (bottom) until the tip retracted from the surface (top). Note that in the central part of the image (low normal force) the tip does not scan in contact with the lipid domain and the substrate, but about 2.5 nm over the substrate and 4 nm over the lipid domain. Image size is $1.5 \times 1.5\ \mu m^2$; the gray-scale corresponds to a corrugation of about 20 nm (**A**) and a lateral force of about 30 nN (**B**).

The complicance term k_{edl} from the electrical double-layer interaction is positive. From the measured values F_0 and d we obtain $k_{edl}(z) = 0.3 \, e^{-z/d}$ nN. The compliance term k_{vdW} from the van der Waals interaction is negative (see the Appendix for sign convention) and may produce an instability if the van der Waals interaction is strong enough. The Hamaker constants across water between the tip (Si_3N_4) and the different materials in our experiment are $A_{mica} = 4.1 \cdot 10^{-20}$ J, $A_{protein} = 3.6 \cdot 10^{-20}$ J, and $A_{lipid} = 2.1 \cdot 10^{-20}$ J. Assuming a typical curvature radius of 100 nm for the tip, we obtain from Eq. 7.1 a compliance $k_{vdW} = -0.2$ nN/nm at a distance $z = 2.0$ nm over the mica substrate. At the same distance $k_{edl} = +0.1$ nN/nm, and, since the compliance of the cantilever was 0.2 nN/nm, it seems reasonable that the unstable point was about 2 nm over the surface. Since the Hamaker constant $A_{protein}$ is only slightly lower than A_{mica}, the unstable point over the protein phase should also be at a height of about 2 nm. However, experimentally we obtain a value of about 1 nm. We attribute this to an additional interaction, probably due to steric repulsion from the sugar trees on the proteins.

The interpretation of the instabilities in the topographic images is somewhat different. These images were taken at constant force with the feedback set to repulsive imaging mode; therefore instability will occur whenever $k_{surface} < 0$ (see Appendix). We therefore conclude that $k_{surface}$ is negative below 2.5 nm over the mica substrate and below 4 nm over the lipid phase.

We think that we qualitatively understand the main features of the tip-sample interaction. However, further measurements with higher precision are needed for quantitative results.

7.6.3 Friction

Figures 7.7 and 7.8 show a set of topography and lateral force images. For each pair, topography and lateral force were measured simultaneously; Figure 7.7 was taken with a force of 2.5 nN, and shortly thereafter Figure 7.8 was taken while the imaging force was slowly reduced from 1 nN until the tip released from the surface. The most evident feature in the lateral force images was the inhomogeneity of the substrate. The lateral force was low on the fragment as well as on the debris that partly covered the substrate. This can be seen directly in Figure 7.8: After the tip retracted from the substrate, no lateral force acted on the cantilever; therefore, this grey level of the image corresponded to zero lateral force. Analyzing the data in Figure 7.7, we found lateral forces ranging from 40 to 25 nN on the dark regions of the substrate and about 3 nN on the protein phase of the membrane fragments. Within our detection limit (about 1 nN for lateral forces) we could not detect any lateral forces on the lipid phase of the membrane fragment. Vanishing lateral forces have been observed also on some areas of the protein phase. This is probably due to soft adsorbed material. When the force normal to the surface was below 1 nN, we found lateral forces between 20 and 30 nN on the mica substrate and lateral forces below 1 nN on the membrane fragment.

Some of the bright structures on the substrate also appeared in the topographic images and showed a corrugation between 2 and 4 nm; therefore, we think that these structures were composed of lipid debris from preparation. Comparing Figures 7.7 and 7.8, we found that these structures were moved by the scanning motion of the tip.

On the one hand, lateral forces on the mica substrate were surprisingly high compared with the low imaging force. Under standard conditions the force normal to the surface did not exceed 5 nN, even if the additional adhesion force was taken into consideration. Therefore lateral forces on the substrate are higher at least by a factor of four. On the other hand, the lateral forces on the lipid phase are very small. This might be due to the fact that the lipid bilayer is a two-dimensional liquid. While the lipids exposed to the mica substrate are presumably bound, the lipid molecules facing the buffer may move freely across the lipid phase of the membrane fragments. The lateral force on the tip is then produced by the viscosity of the two-dimensional lipid phase and is considerably smaller than conventional friction. Since SFFM is a very new field, no additional data of SFFM in electrolytes are available for comparison.

7.6.4 Comparison of Preparations

A goal of the investigation of Na,K-ATPase–containing membrane fragments is the determination of the protein shape as a function of external parameters. Therefore, it is necessary to optimize the conditions for the membrane preparations to obtain the highest possible resolution. In the hitherto performed investigations we focused our interest on crystalline preparations, since they are easier to identify under nonoptimum circumstances. In the following section, the effects of different preparations on the image performance are presented.

Figure 7.9 shows part of a freshly crystallized fragment imaged in standard buffer containing 1 mM Mg^{2+} and 5 mM phosphate, pH 7.2. The size of the scanned area was 250×250 nm^2 and the dimensions of the unit cell of the two-dimensional crystals were 8×10 mn^2, with an angle of about 75° between the rows. This angle compares reasonably well with the findings of Mohraz et al. (1985). The buckling of the fragment in the upper right corner of the image was caused by underlying debris adsorbed to the mica.

In Figure 7.10A a similar fragment that had been imaged in distilled water (scanned area 400×400 nm^2) is shown. The lack of phosphate and Mg^{2+} made the crystals instable, which gave rise to a larger distance of about 15 nm between the protein rows. After replacing distilled water with standard buffer it was not possible to resolve regular structures on the same fragments. This indicated that the divalent cations were of crucial importance for the interaction between tip and sample. If divalent ions are not necessary to maintain proper conditions for the protein, they should be removed for the sake of higher resolution.

Fragments that have not been treated with phospholipase exhibited no crystalline features on their surface in standard buffer. Na,K-ATPase could be

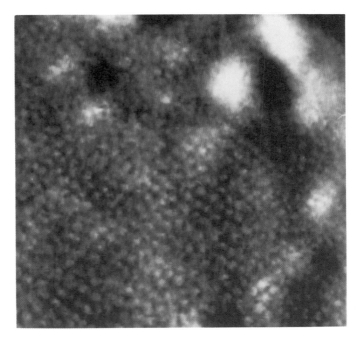

FIGURE 7.9 Scanning force microscopy image of membrane fragments containing crystallized Na,K-ATPase in buffer containing 1 mM Mg^{2+} and 5 mM phosphate, pH 7.2. The size of the scanned area is 300×300 nm^2, and the dimensions of the unit cell of the two dimensional crystals were 8×10 nm^2, with an angle of 75° between the rows. These structures are interpreted as dimers of the Na,K-ATPase in accordance with Mohraz *et al.* (1985). The buckling of the fragment in the upper right corner of the image has been caused by underlying debris adsorbed to the mica.

detected as single bumps in an almost smooth surface. These fragments were very susceptible to imaging forces. However, when kept in buffers containing 10 mM Ca^{2+}, surface structures could be detected that were stable against the scanning process. As presented in Figure 7.10B, the fragment had domains showing some order; the distance between the rows was about 15 nm. This finding indicated that the phase separation and/or transition of lipids induced by divalent cations seemed to be more significant for structural features than the imaging forces. Another intriguing effect can be seen from Figure 7.10B. The fragments in this experiment were taken from a sample that had been frozen for a couple of weeks at −74°C and then thawed before use. Almost all fragments exhibited irregular shapes such as invaginations and holes. We attribute these structural artifacts to the growth of ice crystals during the freezing process. Therefore, we normally used only freshly isolated membrane preparations.

Storage of crystallized membrane fragments in buffer at 4°C for 1 week reduced the image quality. Figure 7.11A shows part of a fragment taken from such a preparation. To enhance the image quality by reducing the debris, the

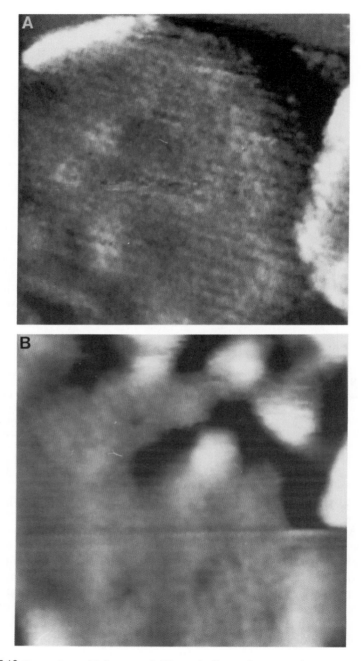

FIGURE 7.10 Comparison of influences of different buffer conditions on the SFM images taken from membrane fragments containing Na,K-ATPase. **A:** Crystallized protein structures scanned in distilled water. The lack of phosphate and Mg^{2+} ions induces the disappearance of the crystalline structure with time. Early effects of this process is an increase of the size of the unit cell (10×15 nm^2). The size of the scanned area is 400×400 nm^2. **B:** Membrane fragments stored and scanned in buffer containing 10 mM Ca^{2+} ions, which induce a kind of order in the protein domains of the membrane fragment. The torn appearance of the fragment at the upper right part was found rather frequently in this frozen and thawed preparation. The size of the scanned area is 300×300 nm^2.

membrane-containing solution was centrifuged and resuspended before scanning. In spite of this treatment, the crystalline structure was obscured by an adhering film of soft material. No crystalline structure was resolved at normal interaction forces (1 nN); at higher forces the crystal structure could be revealed, but was altered. Rescanning of the same area resulted in different structures. The rows had a horizontal distance of 12 nm and a vertical distance of 11 nm (size of the scanned area was 300 × 300 nm^2). Compared with an image from freshly prepared crystals, as shown in Figure 7.9, it is obvious that storage should be avoided to maintain high resolution.

Removing sugar residues from the extracellular face of the protein by neuraminidase treatment yielded crystals with enhanced contrast. Figure 7.11B shows a (partially) deglycosylated fragment imaged with an average tip in Tris buffer containing 4 mM phosphate and 1 mM Mg^{2+}. The rows have a horizontal distance of 11 nm and a vertical distance of 12 nm (size of the scanned area was 370 × 340 nm^2). Even in a preparation that had been stored for more than 1 week and that contained much debris, nearly each fragment showed regular patterns at a higher contrast than in a comparable preparation as presented in Figure 7.11A. Apparently the debris adhered less readily to crystals without the sugar tree containing neuraminic acid.

7.7 OUTLOOK

In SFM, image quality is determined by the mechanical properties of the sample, by the tip-sample interaction, and by the geometry of the tip. The ionic strength is a fundamental parameter for the total surface potential, since it determines the strength and the range of the electrical double-layer interaction. By appropriate choice of the ionic strength and the surface charge, the tip-sample interaction can be minimized. Experimentally, the charge density on tip and sample can be controlled in an electrochemical cell. Further experiments are needed to achieve a better understanding and control of the processes governing tip-sample interactions. These experiments should finally lead to improved imaging conditions.

Buffer composition, especially ionic strength, may affect tip-sample interaction, as well as the physical properties of biomembranes, and thereby influence image quality. In principle, SFM-related techniques such as force versus distance curves, SFFM, and local stiffness measurements can determine whether the effect on image quality is produced by changes in tip-sample interactions or by modifications of the properties of membrane fragments. Moreover, we think that these techniques will open new possibilities for locally characterizing the biomembrane–buffer interface.

The interaction between tip and sample increases linearly with the radius of the tip (Eq. 7.1). Therefore sharper tips will not only increase the accessibility of narrow corrugations but also lower imaging forces. Moreover, a substantial reduction of the imaging forces should reduce sample damage. To resolve crystals of Na,K-ATPase with a standard tip, imaging forces are necessary that

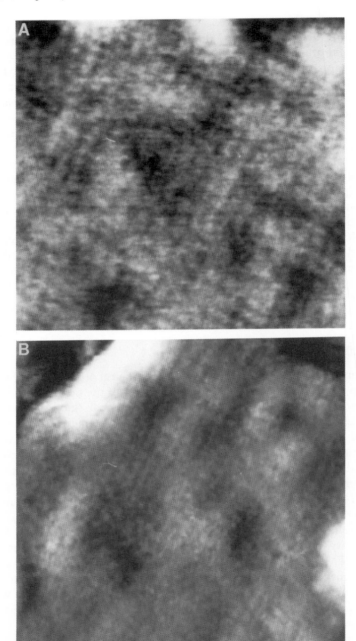

are in the range of forces that are able to destroy the crystal structure. Tips with a curvature radius of 10 nm have become available and open the opportunity to improve resolution at lower forces.

Further enhancement of resolution should be achieved by cooling the membrane fragments or cross-linking the proteins (on cost of their functionality). Finally, image processing by averaging a large number of scans of the same area will increase the signal-to-noise ratio, which is rather high in our images. If we can prove that the sample is not changed appreciably by the imaging forces, SFM and SFFM will yield a realistic image of the surface of Na,K-ATPase–containing membrane fragments. Our goal is to enhance the resolution to be able to distinguish between the E_1 and E_2 conformations, which may lead to a better understanding of the structure–function relation of the pump.

We hope that these investigations on Na,K-ATPase may provide basic knowledge in the application of SFM and SFFM to other membrane-bound proteins.

APPENDIX: MODEL OF THE TIP-SAMPLE INTERACTION

The interaction between tip and surface can be modeled as follows (see Fig. 7.12). The surface potential as well as the harmonic potential from the cantilever act on the tip; therefore the effective potential is

$$V_{\text{eff}}(z, \Delta) = V(z) + \frac{c}{2}(\Delta - z)^2, \qquad (7.2)$$

where z is the tip-sample distance, Δ the distance between the sample and the base of the cantilever, $V(z)$ is the surface potential, and c is the compliance of the cantilever. For a fixed distance Δ, the net force on the tip has to vanish:

$$-F = \frac{\partial V_{\text{eff}}}{\partial z}(z, \Delta) = \frac{\partial V}{\partial z}(z) - c(\Delta - z) = 0 \Rightarrow z_{\text{eq}}(\Delta). \qquad (7.3)$$

This is an implicit equation for the equilibrium distance z_{eq} and is a function of Δ If this is to be a stable point, the compliance (i.e., curvature) of this effective

---

FIGURE 7.11 Influence of different treatments of membrane fragments containing Na,K-ATPase on the SFM images. A: Influence of the storage duration of crystallized membrane fragments in buffer. The image was taken after 1 week in standard buffer at 4°C. At image forces as applied in Figure 7.9 no surface structure could be resolved. When increased to forces in the range of 10 nN unit cells of the size of 11 × 12 nm² have been observed. The size of the scanned area is 300 × 300 nm². B: Influence of sugar removal from the extracellular face of the protein by neuraminidase treatment. Although this preparation has been stored also for 1 week at 4°C, the crystalline structure could be obtained clearly (the size of unit cells: 11 × 12 nm²). The size of the scanned area is 370 × 340 nm².

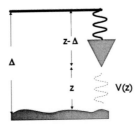

FIGURE 7.12 Schematic model of an SFM setup: The tip is influenced by the harmonic potential of the cantilever, $c/2\, d^2$ and by the surface potential $V(z)$. The distance Δ between the sample and the base of the cantilever is set experimentally. For each distance Δ, the tip-sample distance z and the cantilever deflection $d = \Delta - z$ adjusts itself such that the total force on the tip is zero.

potential has to be positive:

$$k_{\text{eff}} = \frac{\partial^2 V_{\text{eff}}}{\partial z^2}(z, \Delta) = \frac{\partial^2 V}{\partial z^2}(z) + c > 0. \tag{7.4}$$

For a logarithmic potential, $V(z) = A \cdot \ln(z)$, these equations can be solved analytically as a model for an attractive potential:

$$V_{\text{eff}}(z, \Delta) = A \cdot \ln(z) + \frac{c}{2}(\Delta - z)^2$$

$$-F_{\text{tot}} = \frac{A}{z} - c(\Delta - z) \Rightarrow z^{\pm}(\Delta) = \frac{\Delta}{2}\left(1 \pm \sqrt{1 - \frac{4A}{c\Delta^2}}\right) \tag{7.5}$$

$$k_{\text{eff}} = -\frac{A}{z^2} + c > 0 \Rightarrow z_{\text{eq}} > \sqrt{\frac{A}{c}} = z_{\text{inst}}.$$

It can be shown that only the positive root z^+ leads to a stable point. At a distance less than z_{inst} the tip jumps toward the surface without finding a stable point. For a physical potential with a long-range attractive and a short-range repulsive interaction, the tip reaches a stable point on the surface. In this case the minimum compliance $k_{\text{min}} = \partial^2 V / \partial z^2$ of the potential $V(z)$ does not exceed some finite value, and no unstable point occurs for cantilevers with $|k_{\text{min}}| < c$. For softer cantilevers the lever jumps and a hysteresis is observed in the $F(\Delta)$ curve. In this case, Eq. 7.3 yields three physical meaningful equilibrium distances: z_1, z_2, and z_3. The first two correspond to $z^{\pm}$ as described above, while z_3 corresponds to the stable position on the surface.

We should note that, apart from the physical instability described above, another type of instability might occur if images are obtained as usual with a feedback loop. The sign of the feedback has to be set either for the attractive mode ($k_{\text{surf}} < 0$) or for the repulsive mode ($k_{\text{surf}} > 0$). In the first case, the feedback has to *approach* the tip toward the sample if the cantilever is bent away

7 Investigation of the Na,K-ATPase by SFM

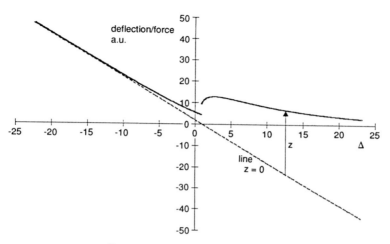

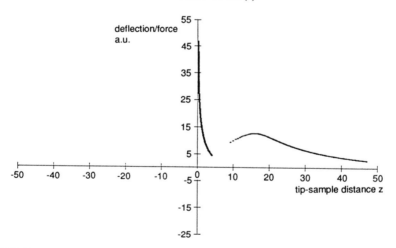

FIGURE 7.13 A: Force versus (sample-base of the cantilever) distance curve $F(\Delta)$ as typically recorded experimentally. As the tip is approached toward the surface (smaller Δ), it feels the surface potential and is deflected. When $d^2V/dz^2 \leq -c$ the cantilever jumps onto the surface. The tip now feels a strong repulsive surface potential (Pauli-repulsion, mechanical contact). Therefore, as a distance Δ is further reduced, the tip-sample distance z is almost zero and barely changes with increasing load. Piezoelongation Δ and cantilever deflection d are equal. The $F(\Delta)$ curve asymptotically approaches a line that defines $z = 0$. For each point $F(\Delta)$ the tip-sample distance z is graphically given by the horizontal spacing between this line$_{z=0}$ and the curve $F(\Delta)$. **B:** Force versus (tip-surface) distance curve $F(z)$ calculated from $F(\Delta)$ (Fig. 7.5A) as explained in the text. Essentially, each point $F(\Delta)$ is plotted against its horizontal distance to the line$_{z=0}$.

from the sample; in the repulsive mode, however, the tip has to be *retracted* if the cantilever is bent away from the sample to reestablish the preset force. For a positive slope of $F_{\text{tip-sample}} = -\partial V/\partial z$, the feedback has to be set to attractive mode and for a negative slope, to repulsive mode. To summarize, the tip will be physically unstable if $k_{\text{surf}} < -c$, whereas the feedback will be unstable if $k_{\text{surf}} < 0$ when set to the repulsive mode.

From a measured $F(\Delta)$ curve, an $F(z)$ curve is calculated as follows (Fig. 7.13A,B): After contact on a hard surface, the tip-sample distance is approximately zero and barely changes with increasing load. Therefore piezoelongation and the cantilever deflection are equal. The $F(\Delta)$ curve asymptotically approaches a line; this line defines the position $z = 0$ (see Fig. 7.13). If force is measured by the deflection d of the cantilever (which means that the vertical axis of the force vs. distance curve has a length scale), then the tip-sample position z corresponds to the horizontal distance between this line$_{z=0}$ and the $F(\Delta)$ curve:

$$z(\Delta) = d(\Delta) - \text{line}_{z=0}(\Delta), \tag{7.6}$$

with $d(\Delta) = \dfrac{F(\Delta)}{c}$. Force versus distance curves $F(z)$ are now obtained from the measured $F(\Delta)$ data by rescaling the horizontal axis with the calculated z-position according to

$$F(z) = c \cdot d[z(\Delta)]. \tag{7.7}$$

Mathematically this is a (nonlinear) transformation of the vertical axis, which shifts each point $\{\Delta, F(\Delta)\}$ to $\{z(\Delta), F[z(\Delta)]\}$. An instability of the system leads to a discontinuity in the $F(\Delta)$ curve. This discontinuity corresponds to some jumping distance, and no $F(z)$ data are obtained in this region (Fig. 7.13B).

REFERENCES

Ackermann, U. and Geering, K. (1990). Mutual dependence of Na,K-ATPase α- and β-subunits for correct posttranslational processing and intracellular transport. *FEBS Letters* 269, 105–108.

Apell, H. -J. Marcus, M. M., Anner, B. M., Oetliker, H., and Läuger, P. (1985). Optical study of active ion transport in lipid vesicles containing reconstituted Na,K-ATPase. *J. Membr. Biol.* 85, 49–63.

Apell, H. -J., Colchero, J., Linder, A., Marti, O., and Mlynek, J. (1992). Na,K-ATPase in crystalline form investigated by scanning force microscopy. *Ultramicroscopy* 42–44, 1133–1140.

Butt, H. -J., Downing, K. H., and Hansma, P. K. (1990). Imaging the membrane protein bacteriorhodopsin with atomic force microscopy. *Biophys. J.* 58, 1473–1480.

Crowe, J. H., Crowe, L. M., Carpenter, J. F., Rudolph, A. S., Aurell Winstrom, C., Spargo, B. S., and Anchordoguy, T. J. (1988). Interaction of sugars with membranes. *Biochim Biophys. Acta* 947, 367–384.

Deguchi, N., Jørgensen, P. L., and Maunsbach, A. B. (1977). Ultrastructure of the sodium pump: comparison of thin sectioning, negative staining, and freeze fracture of purified, membrane-bound (Na$^+$,K$^+$)-ATPase. *J. Cell Biol.* 75, 619–634.

Düzgünes, N., and Papahadjopoulos, D. (1983). Ionotropic effects on phospholipid membranes: calcium magnesium specifity in binding, fluidity and fusion. In *Membrane Fluidity in Biology*. (R. C. Aloia, ed.) pp. 187–216. Academic Press, Inc., New York.

Egger, M., Ohnesorge, F., Weisenhorn, A. L., Heyn, S. P., Drake, B., Prater, C. B., Gould, S. A. C., Hansma, P. K., and Gaub, H. E. (1990). Wet lipid-protein membranes imaged at submolecular resolution by atomic force microscopy. *J. Struct. Biol.* 103, 89–94.

Eisenberg, D. (1984). Three-dimensional structure of membrane and surface proteins. *Annu. Rev. Biochem.* 53, 595–623.

Gorter, E., and Grendel, F. (1925). On bimolecular layers of lipoids on the chromocytes of blood. *J. Exp. Med.* 41, 439–443.

Gould, S. A. C., Drake, B., Prater, C. B., Weisenhorn, A. L., Maune, S., Hansma, H. G., Hansma, P. K., Massie, J., Longmire, M., Elings, V., Dixon Northern, B., Mukergee, B., Peterson, C. M., Stoeckenius, W., Albrecht, T. R., and Quate, C. F. (1990). From atoms to integrated circuit chips, blood cells, and bacteria with the atomic force microscope. *J. Vac. Sci. Technol.* A8, 369–373.

Gould, S. A. C., Marti, O., Drake, B., Hellemans, L., Bracher, C. E., Hansma, P. K., Keder, N. L., Eddy, M. M., and Stucky, G. D. (1988). Molecular resolution images of amino acid crystals with the atomic force microscope. *Nature (London)* 332, 332–334.

Israelachvili, J. (1992). *Intermolecular & Surface Forces*. (2nd Ed.) Academic Press, Orlando, FL.

Jennings, M. L. (1989). Topography of membrane proteins. *Annu. Rev. Biochem.* 58, 999–1027.

Jørgensen, P. L. (1974). Isolation of ($Na^+ + K^+$)-ATPase. *Methods Enzymol.* 32, 277–290.

Jørgensen, P. L., and Andersen, J. P. (1988). Structural basis of E_1-E_2 conformational transitions in Na,K-pump and Ca-pump proteins. *J. Membr. Biol.* 103, 95–120.

Läuger, P. (1991). *Electrogenic Ion Pumps*. Sinauer Associates, Sunderland, MA, pp. 168–225.

Linder, A., Colchero, J., Apell, H.-J., Marti, O., and Mlynek, J. (1992). Scanning force microscopy of diatom shells. *Ultramicroscopy* 42–44, 329–332.

Marti, O., Ribi, H. O., Drake, B., Albrecht, T. R., Quate, C. F., and Hansma, P. K. (1988). Atomic force microscopy of an organic monolayer. *Science* 239, 50–52.

Maunsbach, A. B., Skriver, E., and Hebert, H. (1991). Two-dimensional crystals and three-dimensional structure of Na,K-ATPase analyzed by electron microscopy. In *The Sodium Pump: Structure, Mechanism, and Regulation*. (J. H. Kaplan and P. De Weer, eds.) The Rockefeller University Press, New York, pp. 159–172.

Maunsbach, A. B., Skriver, E., Söderholm, M., and Herbert, H. (1988). Three-dimensional structure and topography of membrane-bound Na,K-ATPase. *Progr. Clin. Biol. Res.* 268A: 39–56.

McBeath, E., and Fujiwara, K. (1984). Improved fixation for immunofluorescence microscopy using light-activated 1,3,5-triazido-2,4,6-trinitobenzene (TTB). *J. Cell Biol.* 99, 2061–2073.

Mohraz, M., Yee, H., and Smith, P. R. (1985). Novel crystalline sheets of Na,K-ATPase induced by phospholipase A_2. *J. Ultrastruct. Res.* 93, 17–26.

Nørby, J. G., and Jensen, J. (1991). Functional significance of the oligomeric structure of the Na,K-pump from radiation inactivation and ligand binding. In *The Sodium Pump: Structure, Mechanism, and Regulation*. (J. H. Kaplan and P. De Weer, eds.) The Rockefeller University Press, New York, pp. 173–188.

Shull, G. E., Schwartz, A., and Lingrel, J. B. (1985). Amino-acid sequence of the catalytic subunit of the ($Na^+ + K^+$)-ATPase deduced from complementary DNA. *Nature (London)* 316, 691–695.

Shull, G. E., Lane, L. K., and Lingrel, J. B. (1986). Amino-acid sequence of the β-subunit of the ($Na^+ + K^+$)-ATPase deduced from a DNA *Nature (London)* 321, 429–431.

Silver, B. L. (1985). *The Physical Chemistry of Membranes: an Introduction to the Structure and Dynamics of Biological Membranes*. Allen and Leuvin, Boston.

Singer, S. J., and Nicolson, G. L. (1972). The fluid mosaic model of the structure of cell membranes. *Science* 175, 720–731.

Weisenhorn, A. L., Drake, B., Prater, C. B., Gould, S. A. C., Hansma, P. K., Ohnesorge, F., Egger, M.,

Heyn, S. -P, and Gaub, H. E. (1990). Immobilized proteins in buffer imaged at molecular resolution by atomic force microscopy. *Biophys. J.* 58, 1251–1258.

Weisenhorn, A. L., Maivald, P., Butt, H. -J., and Hansma, P. K. (1992). Measuring adhesion, attraction, and repulsion between surfaces in liquids with an atomic force microscope *Phys. Rev.* B45, 11226.

Worcester, D. L., Kim, H. S., Miller, R. G., and Bryant, P. J. (1990). Imaging bacteriorhodopsin lattices in purple membranes with atomic force microscopy. *J. Vac. Sci. Technol.* A8, 403–405.

CHAPTER 8

SFM and Living Cells

J. K. H. Hörber
EMBL, Heidelberg, Germany

W. Häberle, F. Ohnesorge, G. Binnig
IBM Research Division
IBM Physics Group Munich
Munich, Germany

H.G. Liebich, C. P. Czerny
Tierärztliche Fakultät der
Ludwig-Maximilians-Universität München
München, Germany

8.1 Introduction
8.1 Instrumentation
8.3 Preparation of Cells
    8.3.1 *Bacillus coagulans*
    8.3.2 Erythrocytes
    8.3.3 Cell Cultures and Virus Propagation
8.4 Results and Discussion
    8.4.1 Structures of a Bacterial Cell Membrane
    8.4.2 Structures of the Erythrocyte Membrane
    8.4.3 Membrane Structures of Cultured Monkey Kidney Cells
    8.4.4 Surface Processes on Virus-Infected Cells
8.5 Conclusions
References

8.1 INTRODUCTION

Scanning probe techniques like scanning tunneling microscopy (STM) or scanning force microscopy (SFM), which are basically surface science instruments (Binnig et al., 1982, 1986), have shown that they can be developed into powerful tools with which to study biological structures (Binnig and Rohrer, 1984), in particular the dynamics thereof (Drake et al., 1989). Observations with the STM of artificial lipid films (Fuchs et al., 1987; Smith et al., 1987) and with integrated proteins (Hörber et al., 1988) demonstrated very early that even membranes might be accessible to these instruments. This was confirmed by SFM observations of lipid films (Egger et al., 1990; Meyer et al., 1991). Therefore the first attempts to study natural membranes (Worcester et al., 1988; Guckenberger et al., 1989; Jericho et al., 1990; Hörber et al., 1991) with SFM and STM were soon made, but showed that the reliable preparation of such membranes on substrates is quite difficult, because numerous surface effects at the substrate influence preparation and imaging by introducing considerable identification problems.

One way to avoid at least some of the problems at the surface of the sample is to coat it with homogeneous conducting material. Some groups tried such and other methods borrowed from electron microscopy to improve the results (Travaglini et al., 1987; Stemmer et al., 1987; Zasadzinski et al., 1988). Nevertheless the preparation of artificial and natural membranes on well-defined solid supports is rather difficult, and it is not obvious to what extent they can actually show the normal cell membrane structures. Therefore, it is tempting to deal with whole cells and thus to make use of nature's optimized technique to stabilize membranes. In this case, of course, many problems arise from the complexity of the structure, leading again to identification problems. First attempts to investigate whole cells have been made with both STM (Ruppersberg et al., 1989; Dai et al., 1991; Ito et al., 1991) and SFM (Gould et al., 1990; Butt et al., 1990; Kasas and Celio, 1992), with various techniques to affix the cells to a substrate.

A rather different approach to investigate cell membranes with a specially developed SFM emerged in 1989 when the first reproducible images were made of the outer membrane of a living cell, fixed only by a pipet in its normal growth medium (Häberle et al., 1989). In this way the cell can be kept alive for days while imaging. This makes studies of live activities and kinematics, in addition to the application of other cell physiological measuring techniques, possible. With this step in the development of scanning probe instruments the capability of optical microscopy to investigate the dynamics of biological processes of cell membranes under physiological conditions could be extended into the nanometer region. But it is not this imaging capability that might be the most attractive feature of this technique in the future; it is the ability to measure interaction forces that determine the dynamic processes between macromolecules that may lead to new and essential insights.

At the moment, structures as small as about 10 nm can be resolved. This gives access to processes such as the binding of labeled antibodies and endo- and exocytosis. As with the integrated tip of the cantilever forces are applied to the investigated cell membrane, the mechanical properties of cell surface structures are also involved in the imaging process. On the one hand, this fact mixes topographic and elastic properties of the sample in the images, and on the other hand, it provides additional information about cell membranes and their dynamics in various situations during the life of the cell, if these two aspects in the SFM images can be separated. For this, independent data are needed, such as topographic data from electron microscopy or from modulation techniques by SFM.

The following section of this chapter is concerned with the instrumental set-up that is necessary to image living cells by SFM. Furthermore, the preparation of various cells for SFM imaging is described.

We present results on the regular surface layer (S-layer) of *Bacillus coagulans*. Next, the structure of erythrocyte membrane, the effect of changing osmotic pressure to this membrane, and the binding of labeled antibody are shown. Finally, we demonstrate the ability of SFM to ellucidate relevant dynamic processes on living cells on the example of cultured monkey kidney cells upon their infection with pox viruses.

8.2 INSTRUMENTATION

Figure 8.1a shows the schematic arrangement of the SFM with a tunneling detection above the objective of an inverted optical microscope (Nikon). The sample area can be observed from below through a planar surface defined by a glass plate with a magnification of ×600–1200 and from above by a stereomicroscope with a magnification of ×40–200. The illumination is from the top

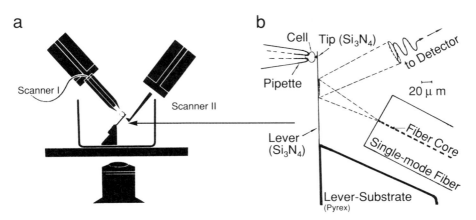

FIGURE 8.1 Schematic drawing of the SFM set-up on top of an inverse optical microscope. **a:** With tunneling detection; **b:** with optical detection (in detail).

through the less well-defined surface of the aqueous solution. In order not to block the illumination, the tunneling sensor and the scanned micropipet point toward the focal plane at an angle of 45°. The lever is mounted in a fixed position within the liquid slightly above the glass plate. Its long axis is parallel to the glass plate, but its two other axes are rotated by 45°. In this way the lever is deflected in a direction parallel to the tunneling tip.

A short pipet of 0.8 mm diameter is pulled to about 2–4 μm on one end, in the way used for the production of patch clamp pipets (Sakmann and Neher, 1983). It is mounted on the piezotube scanner and coupled with a fine and flexible teflon tube, through which the pressure in the pipet can be adjusted by a piston or water pump. The pipet is fixed at an angle that allows imaging of the cell but without the danger of touching the pipet with the lever. All these components are located in a container of 50 μl volume. The glass plate above the objective of the optical microscope forms the bottom of this container.

Before the cells are added the tunneling tip is brought into tunneling contact with the lever already in the solution. The electropotentials of all contacting parts have to be adjusted such that the ion current to the tip and hopefully also the electrochemistry are minimized. After adding several microliters of the cell suspension, a single cell can be sucked onto the pipet and fixed there by maintaining low pressure in the pipet. The other cells are removed through the pumping system. The fixed cell is placed close to the SFM lever by a rough approach with screws and finally positioned by the piezoscanner. When the cell is in close contact with the tip of the lever, scanning of the capillary with the cell attached leads to position-dependent deflections of the lever. The maximum possible scan width is 3.6 μm. The levers used are microfabricated silicon and silicon nitrite triangles with 100 μm length and with a spring constant of 0.12 N/m (and were provided courtesy of Shinya Akamine, Stanford University, and Mike Kirk, Park Scientific Instruments).

The piezotube holding the tunneling tip is part of a feedback loop that stabilizes the tunneling current to a value in the range of 0.5–1 nA. The changes in the feedback voltage on this piezo are a measure for the deflection of the lever. This signal is used by an Arlunya to create a video signal, with respect to the x,y movement of the pipet, which is recorded on videotape. Single pictures are low pass Fourier filtered to reduce the signal noise and displayed by a Silicon Graphics Iris workstation.

To increase the long-term stability of the instrument in order to observe cells for many hours, we can switch to optical detection (Häberle et al., 1992). For certain results, especially for experiments with the erythrocytes, the tunnel probe was used to sense the lever motion. The other results were obtained by optical detection of the lever deflection, which was sensed by measuring the deflection of a reflected laser beam with a bi-cell photodiode, as has become common practice in the SFM field. In our set-up, however, no lenses are used. Instead, an optical fiber is brought very close to the lever and positioned in three dimensions by the manipulator built for the tunneling tip. This configuration is shown in Figure 8.1b.

8.3 PREPARATION OF CELLS

8.3.1 Bacillus Coagulans

Bacteria cells were placed with 1 drop of solution on a quartz glass slide, which had previously been coated with polylysine to improve the fixation of the cells to the substrate. While imaging, the sample was kept immersed in buffer solution. The SFM used for the experiments with bacteria cells on glass slides was a conventional type SFM described in Ohnesorge et al. (1992). The cells of *Bacillus coagulans* E38-66 were provided courtesy of D. Pum (Center for Ultrastructural Research, Vienna).

8.3.2 Erythrocytes

Fresh human erythrocytes were used for the experiments after briefly cleaning them by washing three times with a physiological NaCl solution and concentrating them again in a centrifuge at 1200 rpm for 10 min. The concentrate was diluted again to 1:1000 before adding about 10 μl to the physiological solution in the container in which the cell is observed by the SFM. Higher salt concentrations were achieved by adding grains of salt to obtain a concentration of up to 250 mM.

For the experiment of binding molecular structures to the erythrocyte surface, concanavalin A (Sigma) was used, which is not blood group specific but has an affinity for terminal mannosyl and glycosyl residues. It was labeled with 20 nm colloidal gold particles and used in 10^{-3} mM concentrations.

8.3.3 Cell Cultures and Virus Propagation

The permanent African green monkey kidney cell line MA-104 was grown in monolayer cultures in Earl's minimal essential medium (EMEM) containing 5% fetal calf serum (FCS) and antibiotics. Stocks of the neurovirulent vaccinia virus strain Munich 1 were prepared by infecting the MA-104 cells at a multiplicity of 1 $TCID_{50}$/cell and adsorption for 60 min at 25°C (Czerny et al., 1989, 1990). The virus medium was EMEM with 1% FCS. One day later a cytopathogenic effect (CPE) had developed up to 90–100% cytolysis. The infected cultures were freeze-thawed ($-70/25$°C), intensively sonicated in a Branson sonifier, and centrifuged at 1000 g at 4°C for 10 min to remove cell debris. The supernatants containing the virus were collected and used as the test virus. Infectivity titers were determined on Falcon 96-well microplates (Becton Dickinson, Heidelberg)

314   J. K. H. Hörber et al.

and calculated by the 50% endpoint method of Spearman and Kaerber ($TCID_{50}$) 7 days after incubation at 37°C (Mayr et al., 1974).

8.4 RESULTS AND DISCUSSION

8.4.1 Structures of a Bacterial Cell Membrane

The outer cell envelope of bacteria of the strain *Bacillus coagulans* E38-66 is formed by a regularly structured protein surface layer (the S-layer). Electron micrographs of freeze-etched preparations of whole cells reveal an oblique lattice with the parameters a = 9.4 nm, b = 7.4 nm, g = 80° Sleytr et al., 1989. Membrane fragments can be recrystallized and show the same two-dimensional lattice. This has been shown by electron microscopy on stained preparations (Pum et al., 1989) and by SFM on uncoated membrane sheets that were kept in buffer solution while imaging (Ohnesorge et al., 1992).

In Figure 8.2a the regular structure of an oblique lattice is visible. To form a closed envelope around the blimp-shaped bacteria cell, there have to be lattice defects especially near the poles of the cell. A so-called wedge dislocation as previously observed by electron microscopy (Pum et al., 1989) can also be seen in the SFM images (Fig. 8.2b).

The lattice structure appears, however, to be almost a factor of two too large. This effect can be observed occasionally in STM and SFM images because

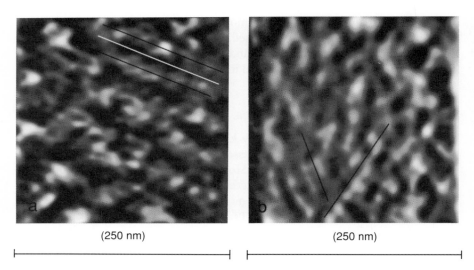

(250 nm)   (250 nm)

FIGURE 8.2 Structure of a bacterial S layer observed with the SFM at a scan size of 250 nm with a: the regular lattice structure and b: wedge dislocations.

of friction. Insufficient adhesion to the substrate or the elasticity of the cell itself combined with friction between tip and sample results in a lateral displacement that reduces the effective scan size. This, of course, would also lead to distortion, for instance of the angle between the lattice vectors, which are not observed in the images.

On the other hand, the lattice can be modulated such that a superstructure containing two unit cells appears, as observed on membrane fragments of this bacterial membrane (Ohnesorge et al., 1992). Also, electron microscopic images show a tendency to form superstructures. Therefore it seems possible that the SFM images of the ordering of the surface protein structure of whole bacterial cells show twice the expected periodicity.

But at the moment there is still a great deal to be learned, especially about artifacts produced by the scanning tip, before such measurements can be discussed in detail. Nevertheless, these measurements demonstrate that the protein lattice of the bacterial cell can be imaged by SFM. The resolution is high enough for future investigations of *in situ* antigen and antibody reactions on living cells.

8.4.2 Structures of the Erythrocyte Membrane

Figure 8.3a shows the erythrocyte membrane detected by the SFM at a scan size of 1 μm. The large structures seen in this image are comparable with those observed on dried blood cells (Gould et al., 1990). In all images, long linear as well as many circular ringlike structures are visible at all scales. In experiments with concanavalin A attached to 20 nm gold colloids, large blobs become visible. Figure 8.3b shows the cell membrane before and a few minutes after adding the gold-labeled concanavalin. These images demonstrate that the distribution of glycoproteins is not uniform on the cell surface.

Some large-scale dynamics can be seen on erythrocyte membranes by raising the salt concentration. Figure 8.3c shows the cell membrane under normal conditions in a physiological salt solution and the membrane after raising the salt concentration to about 250 mM NaCl. The surface became rough on the horizontal scale of 100 nm, which indicates a certain folding of the membrane. The structures that are not pushed inward by the osmotic pressure are of the size and form expected for the shell structure of erythrocytes composed of 200 nm long spectrin tetramers joined at their ends by junctional complexes of 35 nm long actin microfilaments, which consist of an average of five spectrin fibers radiating from each complex (Darnell et al., 1986).

Similar to STM in which the images are always a result of combined local spectroscopic and topographic features, the SFM images reflect topographic configurations as well as local compressibility. The linear structures in these pictures might therefore also be due to such local variations in the stiffness of the cell surface. Hence, the observed structures can reflect the cytoskeleton under-

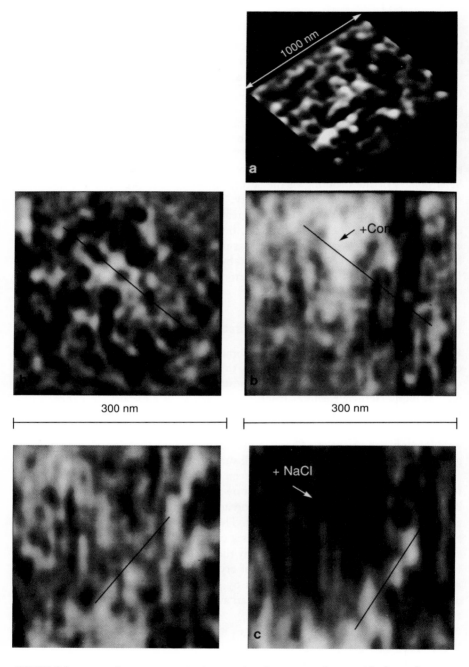

FIGURE 8.3 Scanning force microscopic pictures of erythrocyte membranes with observed corrugation amplitudes of 2–5 nm. a: Scan size 1 μm; b: 300 nm × 300 nm membrane area before adding concanavaline A, labeled with 20 nm gold colloids, and the same area after adding the concanavaline; c: Before and after rising the salt concentration up to 250 mM, scan size 300 nm.

neath the membrane which is a rather stiff molecular network that gives the cell its characteristic form.

8.4.3 Membrane Structures of Cultured Monkey Kidney Cells

The surfaces of monkey kidney cells, a cultured type used for various investigations, looks, as expected, different from those of erythrocytes. The main structures, aside from the marked folding, are reminiscent of mountain ranges.

Some of the structures seen are very stable as the applied force is raised, while others tend to disappear. Some details in Figure 8.4a show high contrast only at high loading forces. For high loading forces the samples become more deformed depending on the local compressibility and the images become more dominated by the compressibility aspect of the imaging mechanism, whereas high contrast at low loading forces can be more closely related to topography.

The apparent shadowing of pronounced structural details in SFM images occurring particularly on strongly corrugated biological sample surfaces (i.e., 10–100 nm) is a scan direction-dependent artifact due to phase shifts in the imaging system caused electronically and mechanically. At this point it should be noted that the convolution of a nonsymmetric tip geometry also influences the appearance of structural details, which has to be controlled by selecting well microfabricated imaging tips.

Force microscopy detects normal forces acting between the sample and the tip and thus obtains the topography information. Depending on the length of the tip, lateral forces might also deform the lever and/or the sample and therefore might contribute to some extent to the images (Fig. 8.4b). These forces can distort the image or can cause stable surface features to move or can even be so destructive that they tear off structural details (Ohnesorge *et al.*, 1992).

In the SEM images the monkey kidney epithelial cell surface is characterized by numerous fingerlike longitudinal extensions, called *microvilli* and *microplicae* (Fig. 8.4c). These cytoplasmic structures make the cells' absorptive surface area 5–10 times greater than it would be otherwise. These surface cell structures are also involved in cell phagocytosis. The plasma membrane that covers these extrusions is highly specialized and contains a surface coat of polysaccharide and enzymes. Under the plasma membrane these cells possess a cortical layer of actin filaments and actin-binding proteins that are cross-linked into a three-dimensional network closely connected to the surface membrane. These filaments have a structural role: they can pull on the membrane and create the changes seen in the SFM by means of the dynamics of the surface protrusions (Fig. 8.4d).

The relatively stable arrangements of the actin filaments are responsible for their relatively persistent structure. But these surface actin filaments are not permanent. During phagocytosis or cell movement, rapid changes of shape occur at the cell surface. These changes depend on the transient and regulated poly-

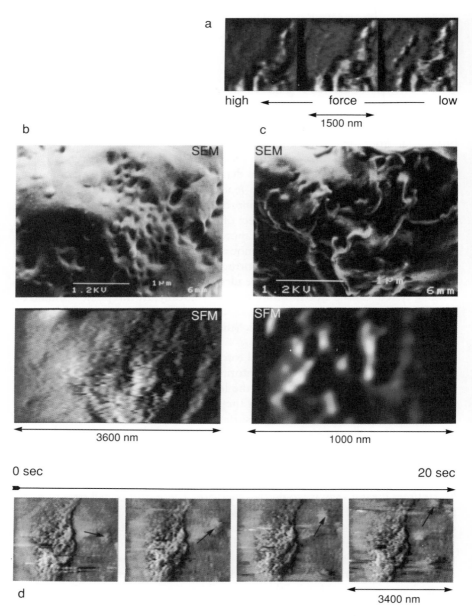

FIGURE 8.4 Surface structures of a monkey kidney cell in normal growth solution. **a:** Area of 1.5 μm with different loading forces between 10^{-10} and 10^{-7} N. **b:** Area of 3.6 μm with friction effects on structures also seen in SEM images. **c:** Microvilli seen with the SEM and SFM. **d:** Area of 3.5 × 3 μm showing stable structures in the left parts of the images and dynamic processes in the right parts.

merization of cytoplasmic-free actin or the depolarization during the breakdown of the actin filament complex.

The time scale of cell surface changes on a larger scale observed with the SFM was about 1–2 hr at room temperature. Except for small structures some 10 nm in size, everything is quite stable for 1–2 hr. More and larger structures are rearranged only 10–15 min after this period has elapsed.

8.4.4 Surface Processes on Virus-Infected Cells

With optical SFM detection it is possible to image a cell for many hours. The longest session we had with a monkey kidney cell was 44 hr. In this experiment we infected the cell after 2 hr of observation with vaccinia viruses and observed it until the cell finally died.

The idea to investigate virus-infected cells developed because large viruses such as pox viridae might be of the right size to be observed easily when attacking a cell or when the progeny viruses leave the cell. We chose the vaccinia type as they are the most unlikely to infect humans.

This virus is 200–300 nm in size and has a quite complex protein cover. Certain proteins are known to be responsible for its adsorption to the cell membrane without needing special binding sides. They also are involved in the penetration of the cell membrane and the disassembly of the virus to release its DNA. The reproduction of this type of virus is located in the cytoplasm.

In our experiments dealing with cells infected by viruses we always started by observing one cell for 1 or 2 hr to determine its normal behavior and structure. The cell is kept in the normal growth solution in order to change the environment as little as possible. As we apply only light suction with the pipet, the cell is affixed quite similar to its normal growing situation.

The highly concentrated viruses are kept in the same solution so that by adding them to the container in which the cell is observed by the SFM the environment of the cell is not changed except for the virus. We add the viruses such that it takes them 2–3 min to reach the cell by diffusion. In our first series of eight experiments we observed that the behavior of the cell membrane changed drastically at that point. The lever no longer detected a well-defined cell surface as the sharp onset of a repulsive force when the sample approaches the lever with a certain force. Hardly any contrast can be seen in the images. This situation lasted for several minutes, in some cases only less than a minute, before the cell membrane became stiff again, even stiffer than before.

Later experiments did not confirm this effect, although we endeavored to work under identical conditions. Therefore the dependences of this effect are not yet clear, but we consider it a worthwhile topic of a future investigation.

During the 2–3 hr following infection no drastic changes were observed with our SFM, but then blobs began to form along chainlike structures and grew for several minutes before suddenly disappearing (Fig. 8.5a). The blobs, which are 50–200 nm, are smaller than the virus itself and may either appear and

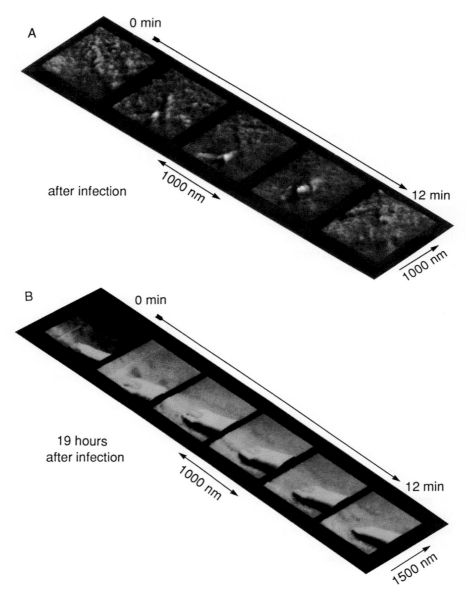

FIGURE 8.5 a: Exocytotic process seen 3 hr after infection. b: Pox virus leaving the cell at the end of a microvillus 19 hr after infection.

disappear again six or eight times at the same place on the cell surface, or the process is seen only once.

It is known that it takes 2–3 hr for all the steps necessary before the final virus assembly can start (Moss, 1986), so we think these exocytotic processes may be connected to the reaction of the cell to the production of virus proteins or related processes. The connection to the chainlike structures supports the assumption that they belong to the stabilizing actin structures, as these are also known to be involved in transport processes at this size (Stokes, 1976).

After 6–8 hr when the first progeny virus are ready, the cell starts to build more microvilli, small tubes of 0.1–0.5 μm diameter and 0.5–2 μm length. These are known to be stabilized also by actin structures belonging to an internal transport system (Stokes, 1976). As Stokes showed in 1976 with electron microscopic pictures, the virus can attach itself to this transport system and use it to leave the cell. In this case a Golgi apparatus along the way makes a double lipid layer around the virus, enabling it to leave the cell by exocytosis in some cases through these microvilli. Stokes made images of microvilli 20 hr after infection, showing the viruses just leaving the cell.

The structures in Figure 8.5b that we observed with the SFM are extremely similar to the electron microscopy images of Stokes. But the electron microscopic images are merely snapshots, whereas with the SFM we observed the event like a movie. The exocytosis of such enveloped viruses is rather seldom; only a certain percentage leave the cell in this way. With electron microscopic snapshots one has to be lucky to fix the sample at the right moment, while with the SFM we had to be lucky to observe the right microvillus at the right time.

8.5 CONCLUSIONS

In these studies it was demonstrated that with a technique using basically force microscopy, the door to a new way of investigating cell surfaces locally is opened. It was shown that it is possible to image the surface of living cells with a resolution of a few nanometers. The method of fixing a cell to a pipet is flexible enough to allow integration of and combination with well-established cell physiological techniques of manipulations used in investigations of single cells. The structures observed might be partially related to known features of membranes, but detailed structural analyses have to be left to further investigations. Most important than the observations of structures with this new technique is the possibility of conducting dynamic studies of living organisms on this scale. This might give access to the observation of immune reactions, the forming of cell pores, as well as endo- and exocytosis in various cell types, the details of which are still speculative. In general this technique makes studies of the evolution of cell membrane structures possible and provides information that brings us closer to understanding not only the "being" of these structures but also their "becoming."

REFERENCES

Binnig, G., Rohrer, H., Gerber, C., and Weibel, E. (1982). Surface study by scanning tunneling microscopy. *Phys. Rev. Lett.* 49, 57-60.
Binnig, G., and Rohrer, H. (1984). Scanning tunneling microscopy. In *Trends in Physics 1984*. (J. Janta and J. Pantoflicek, eds.) Europen Physical Society, pp 38-46.
Binnig, G., Quate, C. F. and Gerber, C. (1986). Atomic force microscope. *Phys. Rev. Lett.* 56, 930-933.
Butt, H. J., Wolff, E. K., Gould, S. A. C., Dixon Northern, B., Peterson, C. M. and Hansma, P. K. (1990). Imaging cells with the atomic force microscope. *J. Struct. Biol.* 105, 54-61.
Czerny, C.-P., Mahnel, H., and Hornstein, O. J. (1989). Prüfung der Immunität gegen Orthopockenenviren an der weissen Maus mit Vaccinia Virus. *Vet. Met.* B36, 100-112.
Czerny, C.-P., and Mahnel, H. (1990). Structural and functional analysis of orthopoxvirus epitopes with neutralizing monoclonal antibodies. *J. Gen. Virol*, 71, 2341-2352.
Dai, J. W., Jliao, Y. K., Dong, Q., Su, Y. X., and Lin, K. C. (1991). The surface structure of natural membrane of macrophages in water as studied by scanning tunneling microscopy. *J. Vac. Sci. Technol.* B9(2), 1184-1188.
Darnell, J., Lodish, H., and Baltimore, D. (1986). *Molecular Cell Biology*. Scientific American Books, New York.
Drake, B., Prater, C. B., Weisenhorn, A. L., Gould, S. A. C., Albrecht, T. A., Quate, C. F., Cannell, D. S., Hansma, H. G., and Hansma, P. K. (1989). Imaging crystals, polymers, and processes in water with the atomic force microscope. *Science* 243, 1586.
Egger, M., Ohnesorge, F., Weisenhorn, A. L., Heyn, S. P., Drake, B., Prater, C. B., Gould, S. A. C., Hansma, P. K., and Gaub, H. E. (1990). Wet lipid-protein membranes imaged at submolecular resolution by atomic force microscopy. *J. Struct. Biol.* 103, 89-94.
Fuchs, H., Schrepp, W., and Rohrer, H. (1987). STM investigations of Langmuir-Blodgett films. *Surf. Sci.* 181, 391-393.
Gould, S. A. C., Drake, B., Prater, C. B., Weisenhorn, A. L., Manne, S., Hansma, H. G., and Hansma, P. K. (1990). From atoms to integrated circuit chips, blood cells, and bacteria with the atomic force microscopy. *J. Vac. Sci. Technol.* A8(1), 369-373.
Guckenberger, R., Wiegräbe, W., Hillebrand, A., Hartmann, T., Wang, Z., and Baumeister, W. (1989). Scanning tunneling microscopy of a hydrated bacterial surface protein. *Ultramicroscopy* 31, 327-332.
Häberle, W., Hörber, J. K. H., and Binnig, G. (1989). Force microscopy on living cells. *J. Vac. Sci Technol.* B9(2), 1210.
Häberle, W., Hörber, J. K. H., Ohnesorge, F., Smith, D. P., and Binnig, (1992). In situ investigations of single living cells infected by viruses. *Ultramicroscopy* 42-44, 1161-1167.
Hörber, J. K. H., Lang, C. A., Hänsch, T. W., Heckl, W. M., and Möhwald, H. (1988). Scanning tunneling microscopy of lipid films and embedded biomolecules. *Chem. Phys. Lett.* 145(2), 151.
Hörber, J. K. H., Schuler, F. M., Witzemann, V., Schröter, K. H., Müller, H. G., and Ruppersberg, J. P. (1991). Imaging of cell membrane proteins with a scanning tunneling microscopy. *J. Vac. Sci. Technol.* B9(2), 1214.
Ito, E., Takahashi, T., Hama, K., Yoshioka, T., Mitzutani, W., Shimizu, H., and Ono, M. (1991). An approach to imaging of living cell surface topography by scanning tunneling microscopy. *Biochem. Biophys. Res. Commun.* 177(2), 636-643.
Jericho, M. H., Blackford, B. L., Dahn, D. C., Frame, C., and MacLean, D. (1990). Scanning tunneling microscopy imaging of uncoated biological material. *J. Vac. Sci. Technol.* 8, 661-666.
Kasas, S., and Celio, M. R. (1992) Application of scanning-probe microscopes in biology. *Biophys. J.* (in press).
Marti, O., Ribi, H. O., Drake, B., Albrecht, T. R., Quate, C. F., and Hansma, P. K. (1988). Atomic force microscopy of an organic monolayer. *Science* 239, 50-52
Mayr, A., Bachmann, P. A., Bibrack, B., and Wittmann, G. (1974). *Virologische Arbeitsmethoden Band I: Zellkulturen-Bebrütete Hühnereier-Versuchstiere*. Gustav Fischer Verlag, Stuttgart.

Meyer, E., Howald, L., Overney, R. M., Heinzelmann, H., Frommer, J., Güntherodt, H.-J., Wagner, T., Schier, H., and Roth, S. (1991). Molecular-resolution images of Langmuir-Blodgett films using atomic force microscopy. *Nature* (London) 349, 398–399.

Moss. B. (1986). Replication of poxviruses. In: *Fundamental Virology*. (B. N. Fields and D. M. Knape, eds.) Raven Press, New York. pp. 637–655.

Ohnesorge, F., Heckl, W. M., Häberle, W., Pum, D., Sara, M., Schindler, H., Schilcher, K., Kiener, A., Smith, D. P. E., Sleytr, U. B., and Binnig, G. (1992). SFM Studies of antibody binding to S-layers, and of protein crystals. Ultramicroscopy 42–44, 1236–1242.

Pum, D., Sara, M., and Sleytr, U. B. (1989). Structure, surface charge, and self-assembly of the S-layer lattice from *Bacillus coagulans* E38–66. *J. Bacteriol.* 177, 5296–6001.

Ruppersberg, J. P., Hörber, J. K. H., Gerber, C., and Binnig, G. (1989). Imaging of cell membraneous and cytoskeleton structures with a scanning tunneling microscope. *FEBS Lett.* 257(2), 460–464.

Sakmann, B., and Neher, E. (1983). *Single-Channel Recording*. Plenum Press, New York.

Sleytr, U. B., and Messner, P. (1989). Self-assembly of crystalline bacterial cell surface layers (S-layer). In *Electron Microscopy of Subcellular Dynamics*. (H. Plattner, ed.) CRC Press, Inc., Boca Raton, FL, pp. 13–31.

Smith, D. P. E., Bryant, A., Quate, C. F., Rabe, J. P., Gerber, C., and Swalen, J. D. (1987). Images of a lipid bilayer at molecular resolution by scanning tunneling microscopy. *Proc. Natl. Acad. Sci. U.S.A.* 84, 969–972.

Stemmer, A., Reichelt, R., Engel, A., Rosenbusch, J. P., Ringger, M., Hidber, H. R., and Güntherodt, H.-J. (1987). Scanning tunneling and scanning transmission electron microscopy of biological membranes. *Surf. Sci.* 181, 394–402.

Stokes, G. V. (1976). High-voltage electron microscope study of the release of vaccinia virus from whole cells. *J. Virol.* 18(2), 636–643.

Travaglini, G., Rohrer, E., Amrein, M., and Gross, H. (1987). Scanning tunneling microscopy on biological matter. *Surf. Sci.* 181, 380–390.

Worcester, D. L., Miller, R. G., and Bryant, P. J. (1988). Atomic force microscopy of purple membranes. *J. Microsc.* 152, 817–821.

Zasadzinski, J. A. N., Schneir, J., Gurley, J., Elings, V., and Hansma, P. K. (1988). Scanning tunneling microscopy of freeze-fracture replicas of biomembranes. *Science* 239, 1013–1015.

INDEX

A

Acetone, 219
α-Chain, 184
AChR, *see* Nicotinic acetycholin receptor channel
Actin filament, 131, 134, 317, 320
Adatom, 10
Adenine, 219, 239–241, 270
Adhesion, 77, 288–289, 292
A-DNA, 267
Adsorbate mobility, 206–207
Adsorption, 134, 137–138
α-Elastin, 187
AgQRE, *see* Silver quasireference electrode
α-Helical domain, 179
Air-drying, 139, 143, 197
4-n-Alcyl-4'-cyanobiphenyl, 210
n-Alkane, 215–216
Alkyl, 214, 216–218
Alkylcyanobiphenyl, 210–214
Alkyl derivatives, 216–219
Alkyl silane, 262
Au(111), 197, 233, 236, 238–239
Azobenzene, 220

B

Bacillus coagulans, 313
Background plane removal, 99–101, 145, 160, 162
Bacterial cell membrane, 314–315
Bacteriophage, 155–157
Ballistic electron emission microscopy, 58–59
Barrier height, 5, 7, 13, 133, 147, 187, 196, 217
Base stacking, 239–240, 244
B-DNA, *see also* DNA, B-form, 244
Behenic acid, 217
Benzene, 207–208
Besocke-type microscope, 32
Biomembrane, 276–278
Biotin-substituted thiol, 202
Bivalent cations, 139
Bloch wave vector, 9
β-Reverse turn, 179
Brillouin zone, 50
β-Spiral conformation. 179
Buffer, 197, 237, 288, 241, 261, 313
  amphoteric, 197–198
  phosphate, 241
Building vibrations, 21
Bungee cord, 22

C

Cadmium arachidate, 220
Calorimetric measurement, 216
Cantilever spring, 66–67
  compliance, 67
  resonance frequency, 66–67
Cantilever spring deflection detection, 69–81
  by interferometry, 69–74, 79, 80
  by optical lever method, 74–78, 79, 81, 312
  by tunneling, 68–69, 79, 80, 311–312
Capacitance measurement, 83–84
Capsomere, 155, 157
Carbon film, 137
Carbon monoxide, 207–208
Carboxyl, 217
8CB, *see* 4-n-Octyl-4'-cyanobiphenyl
CdA, *see* Cadmium arachidate

325

CE, see Counter electrode
Cell, 312–314
Cell culture, 313–314
Chemisorbed atom, 10
Chemisorbed organic molecules, 207
Chemisorption, 137, 207–209
CITS, see Current imaging tunneling spectroscopy
Coarse sample positioning, 32–33
Collagen, 183–187
Collagen fiber, 184
Concanavalin A, 313
  gold-labeled, 315–316
Conducting solid, 221
Constant current mode, 27
Constant force mode, 60
Constant height mode, 27
Constant z-mode, 60
Contact mode, 60
Contamination resist, 33
Contrast, 149–152, 169–170, 217, 239–241, 260–261
Control electronics, 25–31
Cooling, 284–285
Copper (100) surface, 208
Correlation averaging, 163–164
Coulomb force, see also Electrostatic force, 284
Counter electrode, 93, 232
Cross-correlation function, 163
Cross-linking, 263
Cryogenic STM, 35
Crystalline film, 206
Current imaging tunneling spectroscopy, 29–30
Current-to-voltage converter, 25, 26, 222–223
Cyclic voltammograms, 93
Cytosine, 219, 270

Display, 160–162
Distortions, 37–38, 160–162
  linear, 37–38, 160–162
  nonlinear, 37–38
Distortions correction, 101–103, 160–162
dI/ds, 27–28, 133
dI/dV, 27–28
DMF, see Dimethylformamide
DMPC, see Dimyristoylphosphatidylcholine
DNA, see also A-DNA, B-DNA, Z-DNA, 135, 153, 155, 219, 233–235, 242–248, 260–270
  B→Z transition, 265
  backbone, 237, 263, 267
  bending, 243
  B-form, 264
  circles, 242–243
  circular, 244
  conformation, 264
  double-helical, 230, 242
  double-stranded, 264
  ligation, 244
  linear, 244
  major groove, 263, 269
  minor groove, 262, 269
  oligomer, 242–248
  pitch, 264–266
  protein-binding, 269
  repeat, 247
  synthesis, 244
DNA complex, 266, see also RecA–DNA complex
DNA polymerase I, 266
DNA–VirE2 complex, 266, 269
Double-layer region of potential, 232
Drift, 250

D

Data analysis, 39–44, 160–168, 263–264
  real space, 39–41
  spatial frequency domain, 42–44
Debye length, 231, 289
Decay length of the potential, 63–64
Decoration, 141, 156–157
Dehydration, 139–140, 260, 269
Density of states, 8, 9, 10, 16, 219
Diatoms, 278
Difference frequency, 94–96
Digital image averaging, 162–165
Dimethylformamide, 189, 191
Dimethylsulfoxide, 219
Dimyristoylphosphatidylcholine, 160–161

E

Eddy current damping, 21
Effective mass, 18
Effective potential, 303
Eigenfrequency, see Resonance frequency
Einstein–Smoluchowski relation, 234
Elastin, 187–189
Electrical double-layer interaction, 288, 297
Electrochemical immobilization, 139
Electrochemistry, 92–93
Electrochemistry cell, 231–232, 235
Electrodeposition, 196–197, 263
Electrolyte, 233, 238, 248, 288
Electromagnetic drive, 23
Electron beam–heated gun, 137, 141

Electronic states, 8, 230, 239
Electron tunneling, 3–11
Electrostatic force, 83–84, 246, *see also* Coulomb force
Epitaxial adsorption, 180
Erythrocyte membrane, 315–317
Erythrocytes, 313
Evanescent waves, 89–90
Exocytosis, 321
Exocytotic process, 319, 321

F

Fast Fourier Transform, 42
Fatty acid, 217
Feedback, 25–31
  analog, 25
  digital, 25
Feedback loop, 25, 26
Fermi distribution, 6, 9
Fermi energy, 5, 7, 13
Fermi statistics, 6
FFT, *see* Fast Fourier Transform
Fiber diffraction, 264, 268
Fiberoptic Interferometer, 73
Field-emission, 147
Filter, 39–44
  convolution averaging, 39
  convolution low pass, 39
  gradient, 40–41
  high pass, 160
  Laplacian, 39–41
  median, 40, 41
Filtering, 33–44
  fourier space, 42–44
  real space, 39–41
Fine approach, 33–34
Fixative, 285
Flame annealing, 197
Fluorescent labeling, 130
Force, 61–66, 287, 292–297
Force-distance curve, 292, 305
Fourier filtration, 163
Fourier space, 42, 161
Fourier transform, 42, 161, 180, 191
Four-segment photodiode, 77
Freeze-drying, 139–140
Freeze-etch unit, 140
Freeze-fracture replica, 142
Friction, 77, 82, 289–290, 294, 297–298, 315, 318
Friction mode, 60

G

Gap resistance, *see also* Tunneling resistance 207, 212
Gel analysis, 244–245
Gliadin, 183–184
Glow discharge, 138
Glycerol, 139, 285
Glycol, 284
Glycoprotein, 279, 315
Gold, *see also* Au(111) 137, 197, 209, 262, 263
Graphite, *see also* Highly oriented pyrolytic graphite, 48–49, 81, 219
Graphite basal plane, 209, 210, 216
Guanine, 219, 270

H

Hamaker constant, 288, 297
Hamiltonian, 4, 8, 9
Harmonic oscillator, 21
Harmonic potential, 303
Height measurement, 165–166
Helical structure, 163, 189, 191
Hexagonally packed intermediate layer, 145, 147–150, 157–160, 166–168
High-frequency noise, 39
Highly oriented pyrolytic graphite, 133–136, 180, 183, 189, 209, 261–262, 268
High molecular weight subunit protein, 179–183
High resolution electron microscopy, 216
High vacuum apparatus, 137
HMW subunit protein, *see* High molecular weight subunit protein
HOPG, *see* Highly oriented pyrolytic graphite
Hopping mode, 263
HPI layer, *see* Hexagonally packed intermediate layer
Hydration shell, 140
Hydrophobic support, 139
Hysteresis, 37

I

IHP, *see* Inner Helmholz plane
Image potential, 8
Image processing, 37–48, 160–168
Imaging force, 287
Immunoglobulin G, 201
Indium-tin-oxide, 221
Inner Helmholz plane, 231

Index

Interaction potentional, 61–63
Interferometry, 69–74
  heterodyne, 69–73
  homodyne, 69–73
Ionic conditions, 285–286
Ionic conductivity, 169
I(V) curve, 27
I-V spectroscopy, 193, 196
I/V converter, see Current-to-voltage converter

J

Jellium, 10

L

Langmuir–Blodgett film, 220–221
Langmuir–Blodgett technique, 217, 220, 278
Lateral force, 76–78, 82, 289, 295–298, 317
LB, see Langmuir–Blodgett
Least-squares approximation, 99
Lennard–Jones potential, 62
Light mixing, 94–97
Lipid bilayer, 277
Lipid membrane, 290
Liquid-crystal, 179, 191, 210–215
Local experiment, 94–97
Local-potential, 53–56
Low-temperature experiment, 52

M

Magnesium acetate, 139
Magnesium chloride, 145
Magnetic dipol, 86
Magnetic force, 85–88
Magnetostrictive material, 23
mCB, see 4-n-Alcyl-4′-cyanobiphenyl
Mean free path, 58–59
Membrane, 276–278
  fluid mosaic model, 277
Metal-coating, 141–142, 153, 170
Metal replica, 142, 160–161
Mica, 81–82, 137, 282, 297
Microfibrils, 184–188
Micropipet, 312
Microplicae, 317
Microvilli, 317, 321
Molecular dynamics, 217–218
Molecular orbital calculation, 267
Molecular structure research, 171–172
Molecular-ordered monolayer adsorption, 179

Molybdenum disulfide, 210, 212–213, 219
Monkey kidney cell, 313, 320
  membrane structure, 317–318
Monolayer, 179, 180, 189, 191, 206, 234
Mo S2, see Molybdenum disulfide
Multiple tip, 81, 131–132

N

Na,K-ATPase, 278–303
  α-subunit, 279
  β-subunit, 279
  crystallization, 280
  E1 conformation, 279, 280
  E2 conformation, 279, 280
  electron microscopic image, 281
  enzymatic activity, 280, 281
  isolation, 279–280
  structure, 279
Naphtalene, 208
Near-field optical scanning microscope, see Scanning near-field optical microscope
Nematics, 210
Nicotinic acetylcholin receptor channel, 199–200
Noise, 162–163
$1/f$ Noise, 57
Noise potentiometry, 57–58
Nomarski Interferometer, 74
Noncontact mode, 60
Nucleic acid, 262, 270
  double-stranded, 262
  sequencing, 270
  single-stranded, 262

O

4-n-Octyl-4′-cyanobiphenyl, 210–211
OHP, see Outer Helmholtz plane
Operating medium, 35–36
Optical grating, 83–84
Optical table, 22
Optical tunneling, 89–90
Outer Helmholtz plane, 233

P

Paraffin, 215–216
PBLG, see Poly(g-benzyl-L-glutamate)
Periodic data, 44
Phospholipid liquid crystal, 160–161
Phthalocyanine, 208

Physisorbed organic molecules, 207
Piezoactuator, 23
Piezoceramic material, 25
  conversion ratio, 25
  hysteresis, 25
  sensitivity, 25
Piezoelectric material, 23
Piezoscanner, 23–25
Piezotranslator, see Piezoactuator
Piezotripod, 24
Platinum carbon film, 133, 137–138, 141, 142
Platinum iridium carbon film, 141–142
Platinum (111) surface, 208
Pocket-sized STM, 22
Poly(dA), 262, 266
Poly(g-benzyl-L-glutamate), 191–193
Polyhead, 155–157, 164–165
Polyimide, 220
Polylysine, 313
Polynucleotide, 261
Polypeptide, 188–193
Polytyrosine, 189–191
Potential of zero charge, 232
Pox virus, 319
Preamplifier, see also Current-to-voltage converter, 130
Probe, 262–267
Proteins, 128–129, 139–140, 169, 178, 193, 199, 201
  elongated, 179–193
  globular, 193–202
  membrane, 199–201, 277
Pt-C film, see Platinum carbon film
Pt-Ir-C film, see Platinum iridium carbon film
Pulses to the tunneling voltage, 133–134, 222
Purine, 219, 270
Purple membrane, 143–144, 146, 148–152, 166, 278
Pyrimidine, 219, 270

Q

Quality factor, 19

R

Raleigh's method, 18
RE, see Reference electrode
RecA–DNA complex, 143, 145, 147–148, 153, 155–156
Reciprocal lattice, 163
Reciprocal-lattice vector, 9, 11

$3 \times \sqrt{23}$ reconstruction, 236, 238–239, 247–248
Reference electrode, 93, 232
Regularly structured protein surface layer, 314–315
  electron microscopy, 314–315
Relative humidity, 148
Relief reconstruction, 166–168, 171
Repulsive force, 170, 294
Repulsive mode, see Contact mode
Resonance frequency, 17–22
Rest gas, 33
Rhodium crystal, 207–208
RNA, 260, 264
  double-stranded, 264
  single-stranded, 264
RNA polymerase holoenzyme, 196–199

S

Salt coadsorbate, 246
Sample deposition, 137–139, 263
Sample preparation, 133–142, 282–286
Scanning electron microscope, 33, 130, 318
Scanning force microscope, 78–81, 311–313
  special design considerations, 78–80
Scanning force microscopy, 60–88, 278
  selected experiments, 81–88
  theory, 61–65
Scanning friction force microscopy, see also Friction mode, 289–290
Scanning near-field optical microscope, 89–90
Scanning noise microscopy, 56–58
Scanning optical microscopy, 88–92
Scanning optical tunneling microscope, see Scanning near-field optical microscope
Scanning probe microscope, 2
Scanning transmission electron microscope, 137
Scanning tunneling microscope, 2, 17–36, 129–130
  mechanical design, 17–20
  miniaturization, 97
Scanning tunneling microscopy, 2–59
  high voltage, 145–153
  resolution, 11–16
  theory, 2–17
Scanning tunneling optical microscope, see Scanning near-field optical microscope
Scanning tunneling potentiometer, 53–56
Schottky barrier, 59
Schottky emission, 147
Schrödinger equation, 3
Self-assembly, 207, 209, 220

SEM, *see* Scanning electron microscope
SFFM, *see* Scanning friction force mircroscopy
SFM, *see* Scanning force microscopy
Signal-to-noise ratio, 163–164, 168, 171–172
Silicon surface, 262
Silicon wafer, 221
Silver quasireference electrode, 232
S-layer, *see* Regularly structured protein surface layer
Small-angle X-ray scattering, 193
Smectics, 210
SNOM, *see* Scanning near-field optical microscope
SOM, *see* Scanning optical microscopy
Sound isolation, 22
Spectral noise density, 56–57
Spraying technique, 139
Spring system, 21
STEM, *see* Scanning transmission electron microscope
STM, *see* Scanning tunneling microscopy
Stray capacitance, 25
Streptavidin, 202
Substrate, 133–137, 261–262
Superconductor, 52
Superhelix, 184
Surface contamination, 283
Surface potential, 303
Surface profile, 83–84
SXM, *see* Scanning Probe Microscope

T

TCNQ, *see* Tetracyanoquinodimethane
TEM, *see* Transmission electron microscope
Tetracyanoquinodimethane, 221
Tetrathiofulvalene, 221
TFE, *see* Trifluoroethanol
Thermal drift, *see also* Drift
Thermal expansion coefficient, 23
ω-thiol, 209
Three-dimensional reconstruction, 171, 279
Thymine, 219, 270
Tip, 292, 294
  artifact, 131–132
  conditioning, 133–134
  preparation, 30–32, 131–133
  radius, 11, 12, 292, 294, 297
Tip-sample distance, 150, 153
Tip-sample interaction, 130–132, 288–289, 294, 303–306
Tip-sample separation, 11, 12

Tomographic technique, 171
Top view display, 37, 47, 162
Tracking force, 60
Transfer Hamiltonian method, 8–10
Translational diffusion constant, 252
Transmission coefficient, 5
Transmission electron microscope, 130, 166–168, 171, 260, 268, 281
Transverse piezoelectric effect, 23
Trifluoroethanol, 179, 183
Tropoelastin, 187
TTF, *see* Tetrathiofulvalene
TTF–TCNQ, 221
Tube scanner, 24
Tungsten tip, 131–133
Tunnel current, *see* Tunneling current
Tunnel gap, 3
Tunnel junction, 4, 6
Tunneling barrier, 3, 5
  reflectivity, 5
  resonance, 8
  transmissivity, 3, 5
Tunneling current, 3–8, 130, 143, 147, 169, 193, 207, 222–223, 254, 266–267
Tunneling current density, 6, 7
Tunneling current filament, 11, 13
Tunneling resistance, 8
Tunneling spectroscopy, 16, 27–31
Tunneling tip, *see also* Probe, 130–133, 223, 237
Two-dimensional crystal, *see also* Two-dimensional lattice, 179, 180
Two-dimensional lattice, 163
Two-dimensional projection, 171
Two-segment photodiode, 76

U

UHV, *see* Ultrahigh vacuum
Ultrahigh vacuum, 33, 207, 237
Unit cell, 44

V

Vaccinia virus, 313, 320
Van der Waals
  force, 63–65, 209, 284
  interaction, 288, 297
  surface, 209
Vibration isolation, 20–22
Vicilin, 193–196
Virus propagation, 313–314

Viton, 21
Viton-metal stack, 22
Volatile salt, 261
Voltammogram, 241, 243

W

Water cell, 286–288
Wavelength dispersive X-ray analysis, 141–142
WDX, *see* Wavelength dispersive X-ray analysis
Wedge dislocation, 314
Wentzel, Kramers, Brillouin approximation, 6, 8
Wheat gluten, 179–183
Wiener filter, 42–43
WKB approximation, *see* Wentzel, Kramers, Brillouin approximation

Work function, 7, 9, 151, 154, 217
Working electrode, 93, 231

X

X-Ray diffraction, 171, 260, 277

Y

Young's modulus, 18

Z

Z-DNA, 266–267